Vision or Mirage

Whenever I landed in Jeddah or Riyadh and wanted to discover what was *really* going on, the 'man-in-the-sand' whose expertise I always sought out first was David Rundell, the Quiet American who had the 'inside scoop' on politics, business and, above all, the people of the Kingdom. Dave was always just back from some oasis or tribe or border territory where secret things were happening, or heading for the desert to pow-wow with the king. His brilliant book generously discloses a lifetime of wisdom and insights that take the reader inside one of the world's most enigmatic and crucially important of lands. Saudi Arabia? It's all in here . . .

Robert Lacey, author of *The Kingdom* and *Inside the Kingdom.*

The author of this book is 'pro-Saudi', and at the same time he is *entirely* objective. He reconciles direct opposites not by fudging the differences, but by offering us his uniquely deep knowledge of a country and a state that remain poorly documented. This is a very valuable book.

Dr. Edward N.Luttwak, Senior Associate at the Center for Strategic and International Studies, Washington D.C.

A most impressive account of the cunning manner in which King Salman is attempting to secure his family's place in the 21st century by establishing the fourth kingdom through his son Muhammad bin Salman. David Rundell's insights into historical precedents and personal knowledge of the personalities of the individuals involved is compelling and provides a far more credible narrative of Saudi developments since the death of the late Abdullah bin Abd al-Aziz than other current analysis.

Ambassador Mark G. Hambley, Former American Consul General in Jeddah and Ambassador to Qatar and Lebanon

Saudi Arabia has always been difficult for outsiders to understand, but it will be less so now thanks to David Rundell. With insightful analysis of the roles of the ruling family, the tribal structure, the merchant class and the religious leadership, he forges all the pieces into a coherent whole that will enlighten specialists and novices alike.

Thomas W. Lippman, Former Washington Post Middle East Bureau Chief, Adjunct Scholar Middle East Institute Washington D.C.; author of *Saudi Arabia on the Edge* and *Inside the Mirage, Saudi Arabia's Fragile Relationship with the United States*

David Rundell's eye for detail and meticulous research provide the reader with a compelling story of initial conquest, years of stability, and then a tectonic rupture in the social contract between the ruler, his family, and the population.

Ambassador Robert W. Jordan, Former American Ambassador to Saudi Arabia, Diplomat in Residence, John G Tower Center at Southern Methodist University

Vision or Mirage

Saudi Arabia at the Crossroads

David H. Rundell

I.B.TAURIS

LONDON • NEW YORK • OXFORD • NEW DELHI • SYDNEY

I.B. TAURIS
Bloomsbury Publishing Plc
50 Bedford Square, London, WC1B 3DP, UK
1385 Broadway, New York, NY 10018, USA

BLOOMSBURY, I.B. TAURIS, and the I.B. Tauris logo are trademarks of Bloomsbury
Publishing Plc

First published in Great Britain 2021

Cover design: Adriana Brioso
Cover image @ Pacific Press/Getty Images
Maps and Charts by Daniel Feher @ freeworldmaps.net

A catalog record for this book is available from the British Library.

A catalog record for this book is available from the Library of Congress.

ISBN: HB: 978-1-8386-0593-3
       ePDF: 978-1-8386-0595-7
       eBook: 978-1-8386-0594-0

Typeset by RefineCatch Limited, Bungay, Suffolk
Printed and bound in Great Britain

To find out more about our authors and books visit www.bloomsbury.com
and sign up for our newsletters.

This book is dedicated to the ladies in my life—Virginia, Yin, and Caroline Rundell—who have each contributed to it more than they may realize.

Contents

Part V   Meeting New Challenges

List of Illustrations

Maps and Charts

Maps

Charts

Acknowledgments

A great many people helped to launch this book. My tutors at Oxford, Philip Stewart, Albert Hourani, Roger Owen, and Derek Hopwood started me on my way. Several American Ambassadors for whom I worked, including Richard Murphy, Walt Cutler, Ford Fraker, and Michael Gfoeller encouraged me to record what I had learned about the kingdom. Jim Purvis of Midland, Texas taught me most of what I know about the oil business. Robert Lacey, whose own book *The Kingdom* is still the best history of Saudi Arabia, helped me get started as an author. Karen Elliot House and Dr. John Sfakianakis, both Saudi experts in their own right, were kind enough to read early drafts. My first editor Iradj Bagherzade provided many comments that made this a better book, as did my second editor Sophie Rudland.

I owe my understanding of Saudi Arabia to the patience and kindness of countless Saudis, from kings, crown princes, ministers, and provincial governors; to shopkeepers, shepherds, tribal chiefs; and religious scholars. Over the past forty years, they have often listened to my questions and provided thoughtful, candid answers. Above all, I am grateful to my wife Yin who was a tireless support in writing this book, and my parents who did not protest when I told them I wanted to study Arabic or move to Riyadh.

Arabic Words

This book contains many words and names that will sound strange to those not familiar with Arabia. Any transliteration of these words will inevitably lead to dispute, either confusing the general reader, or annoying the specialist. I have tried to combine accuracy in the Arabic pronunciation with accessibility to the non-Arabic speaker and consistency with accepted English spellings.

For most place names I have relied on the National Geographic Society in Washington DC. For proper names I have used those found in the mainstream media such as *The New York Times*, *The Washington Post*, and *The Economist*. For tribal names I have turned to those who lived among them such as De Gaury, Raswan, Philby, and Thesiger. When there was no consistency among these sources, I followed my own criteria.

Preface

In February 1979, the last Shah of Iran was overthrown by a violent Islamic revolution. Two years later when I arrived in the Middle East, many commentators expected Saudi Arabia's King Fahd would soon be the next monarch to fall. The Washington consensus held that we had paid inadequate attention to events outside of Iran's capital, Tehran. We did not want to repeat that mistake. So, as the American Embassy's most junior political officer, I was assigned to spend ten days a month for nearly two years traveling the byroads of rural Saudi Arabia to see what I could learn. I learned a great deal, and when I was done I argued against considerable skepticism that there would be no Saudi Revolution. Why was that my assessment then—and, more importantly, is it still correct today?

Over the following decades, I continued to live and work in Saudi Arabia. I spent sixteen of my thirty years as a foreign service officer in Riyadh, Jeddah, and Dhahran. At the American Embassy in Riyadh, I was at various times chief of mission, deputy chief of mission, political counselor, economic counselor, and commercial counselor. I was in Saudi Arabia for Operation Desert Storm in 1991. On September 11, 2001, I watched the Twin Towers collapsing on the television in my Jeddah office. I remained in the kingdom throughout the al-Qaeda terrorist campaign of 2003–7. On dozens of field trips, I continued to travel regularly throughout all of Saudi Arabia's thirteen provinces. I helped to negotiate the kingdom's entry into the World Trade Organization. I was fortunate to have had such a rich and varied exposure to such a closed country. I hope to share some of what I learned.

Since my retirement from the Foreign Service I have lived in Dubai and spent many months working in Saudi Arabia as a consultant. During these years of involvement with Saudi Arabia, I developed a model to help me understand the kingdom, assess its stability, and evaluate future trends. That model, which is the result of hundreds of conversations over nearly forty years, forms the framework of this book.

Saudi Arabia remains, without any doubt, an unusual place. Riyadh is the only major world capital where a true king reigns from a palace accompanied by courtiers, falcons, poets, and jesters. Vast oil reserves have created an unconventional economy that is more distributive than productive, and provides the population with a living standard totally divorced from their actual productivity. Unlike many Muslim countries, there is no talk in Saudi Arabia of reintroducing Islamic law because it never disappeared. All of this makes Saudi Arabia different, not only from the West but also from other Arab countries. Anyone seeking to deal successfully with Saudi Arabia would benefit from understanding these differences.

Saudi Arabia also remains an important country. Much of the global economy still relies on stable, predictable Saudi oil production. Whether the pulpit in Mecca is used to promote religious tolerance or intolerance matters to Muslims and non-Muslims alike. Without a stable and cooperative government in Riyadh, confronting terrorism, managing the global economy, containing revolutionary Iran, or resolving the Arab–Israeli conflict would all be more difficult tasks.

Saudi Arabia is in the midst of a remarkable, difficult, and unsettling period of change. Vision 2030, the Crown Prince's blueprint for the country's future, is the most radical set of economic and social reforms seen there for half a century. A newfound sense of urgency and optimism has replaced the glacial pace of change that long characterized the kingdom. How Vision 2030 will affect the long-established balance between tradition and change that has kept Saudi Arabia stable is the principal theme of this book. Will the current reforms prove to be a successful vision or a deceptive mirage? This is the broad question I seek to answer.

Introduction

"The noblest pleasure is the joy of understanding."
LEONARDO DA VINCI

The first question to ask about Saudi Arabia is not when will its government collapse, but why is it still here? Only if we understand how an absolute monarchy survived into the twenty-first century can we reasonably assess how long it will continue. A second question to ask is: Does Saudi Arabia still matter to a world awash in shale oil—and, if it does, how might the West best encourage positive change without compromising Saudi stability? To answer these two questions, the framework of *Vision or Mirage* is divided into five parts each dealing with one of the pillars of Saudi stability. They are: "Creating a New Nation," "Managing Succession," "Balancing Stakeholders," "Delivering Competent Government," and "Meeting New Challenges."

Part 1: Creating a New Nation

This opening section describes how King Abdulaziz al-Saud unified most of the Arabian Peninsula for the first time in a thousand years. Within two generations, his kingdom was transformed from an impoverished, turbulent, politically fragmented backwater into a wealthy, stable unified nation of international consequence. This was accomplished without the benefit of a colonial administration to provide even the most rudimentary physical infrastructure, educational system, or civil service. It was by no means inevitable. King Abdulaziz's achievement continues to supply a sense of historic legitimacy that underpins the House of Saud. What follows is not a detailed or academic account but rather a review of specific, strategic choices that King Abdulaziz made in creating a new nation, and an examination of how those choices still resonate today.

Chapter 1 explains how, in 1744 King Salman's forefathers first used religion to unite the wandering tribes and quarreling towns of Central Arabia. Yet by 1891, all that they had created during the First and Second Saudi States was gone. Chapter 2 explains how, between 1902 and 1926, King Abdulaziz utilized the collapse of the Ottoman Empire and the support of Great Britain to re-establish his family dynasty and create the current Third Saudi State. Chapter 3 looks at how he dealt with armed opposition to his rule from both former allies and new subjects. It illustrates many of the policies that his sons have subsequently adopted to maintain the kingdom's internal security.

Part 2: Managing Succession

Like so many other Arab dynasties, the Second Saudi State collapsed in 1891 because of intra-family succession conflicts. In more recent times, many Arab countries have suffered from coups, revolutions, or civil wars. In contrast, the Third Saudi State has been able to peacefully transfer political power on no fewer than six occasions.

Chapter 4 describes the discord between King Saud and Crown Prince Faisal, which first threatened the dynasty and then led to a system that preserved it for over fifty years. Chapter 5 shows how the Saudi monarchy functioned and evolved during the reigns of Kings Khalid, Fahd, and Abdullah. Chapter 6 discusses the dramatic succession changes that King Salman has implemented as he prepares to move the crown on to the next generation of princes.

Part 3: Balancing Stakeholders

The Al Saud have very consciously based their rule on a coalition of powerful stakeholders. Each group has distinct and often competing interests, which the monarchy seeks to balance. Each group has its own leadership elite, which the Al Saud help to remain prominent. This section examines how balancing the interests of stakeholder groups and maintaining elite cohesion have been the keys to short-term stability in Saudi Arabia.

The stakeholders in the Al Saud's coalition are the tribes, clerics, merchants, technocrats, and members of the royal family itself. Chapter 7 deals with tribes: how they function, how King Abdulaziz dealt with them, and why they remain important. Chapter 8 examines the crucial role of Saudi Arabia's religious establishment. It looks in some detail at the Wahhabi movement and its influence on law, education, and social norms in Saudi Arabia. Chapter 9 describes how the Al Saud promoted national economic integration and helped to transform a small merchant class into a broad-based business community that has provided social mobility in the past and will be essential to job creation and economic diversification in the future.

Technocrats are the newest, most loosely led and often most liberal of the kingdom's stakeholder groups. Their rank and file can be found in the bureaucracy, military, and media, although their most influential members sit in the Council of Ministers and the Consultative Council. Chapter 10 discusses the technocrats, who they are and what they want from their government. Finally, the Al Saud themselves are both a national institution and a family business. Chapter 11 describes the family's unifying vision, governance system, and compensation plan. It also reviews how the Al Saud have used public access, consultation, and consensus building as cornerstones of their rule.

Part 4: Delivering Competent Government

In the short term, balancing stakeholder interests and maintaining elite cohesion, particularly between the royal family and the tribal and religious leadership, has been enough to keep the Al Saud's "show on the road." In the medium term, the majority of the Saudi people also expect their leaders to provide a safe, prosperous, and religiously conservative society. Providing competent government has been another essential element in the monarchy's success. Part 4 examines what that has involved.

Before King Abdulaziz rose to power, Arabia was a violent and dangerous place. For most Saudis, physical security remains the most valuable service that King Abdulaziz and his sons have provided. Between 2003 and 2007, that security was threatened all across the country by a serious and prolonged al-Qaeda terrorist campaign. Chapter 12 looks at how the Al Saud first unwittingly encouraged and then subsequently defeated al-Qaeda with support from the clerics, tribes, and general public. Examining why al-Qaeda failed in Saudi Arabia reveals a great deal about how the Al Saud have succeeded in holding on to power.

King Abdulaziz unified Saudi Arabia before Chevron ever drilled the kingdom's first oil well. His sons' economic policies, on the other hand, have revolved largely around managing the hydrocarbon bonanza that followed. The Saudis have not spent all of their oil wealth wisely, they have at times been handicapped by profligacy, corruption, and mismanagement. Yet by most measures of economic development, they have done considerably better than fellow oil producers in Iran, Iraq, Nigeria and Venezuela. Chapter 13 examines specific choices that Saudi leaders made in order to preserve the independence and efficiency of Saudi Aramco, stop the wasteful flaring of natural gas, create a world-class petrochemical industry, and join the World Trade Organization—all of which contributed to the kingdom's economic prosperity, but at the same time laid the foundation for some of its current economic problems.

For more than eighty years, Saudi Arabia largely avoided foreign wars. Again, this is a notably better record than Egypt, Iraq, Jordan, and Syria. Saudi success resulted not from military strength, but largely from skillful diplomacy. Chapters 14 and 15 describe how the Al Saud have used alliances with powerful friends, a pivotal role in global energy markets, and a central role in the Islamic world to maintain the kingdom's external security.

The Saudi people are conservative, but most are not reactionary. Few would now choose to close automobile showrooms, girls' schools, or television stations—though all three of these developments caused protests, and even violence, when they were first introduced. To an extent seldom appreciated outside of the kingdom, the Al Saud leadership has frequently promoted modernization against significant local opposition. Chapter 16 discusses the Al Saud's role in promoting educational, legal, and social change, including greater religious tolerance and gender equality at a pace that the majority of the population found acceptable.

Part 5: Meeting New Challenges

Saudi Arabia currently faces formidable political, social, economic, and security challenges. Business as usual will no longer guarantee stability. Although no one can assuredly predict how the future will unfold, this section identifies the most significant long-term risks to Saudi stability, examines how the Al Saud are responding to them, and provides a set of specific indicators to suggest the most likely direction of change.

Limits to both the price and volume of oil exports have made the kingdom's oil-funded, distributive economy unsustainable. At the same time, a badly broken labor market has led to productivity and labor force participation rates that will not support the standard of living that Saudis have come to expect. Vision 2030 is an ambitious plan to reduce Saudi Arabia's dependence on oil, diversify its economy, create private sector jobs, and reduce the government's budget deficit. Chapter 17 looks closely at this plan, what it has achieved thus far, and why it may or may not succeed.

Saudi Arabia sees itself surrounded by a hostile, revolutionary Iran determined to fundamentally change the political order of the Middle East. Riyadh's costly military involvement in Yemen grew not from a fear of Houthi tribesmen, but from concern that Iran could establish a beachhead on the Arabian Peninsula. Iranian hostility and now a war in Yemen are King Salman's most challenging foreign policy problems. Chapter 18 examines how he has combined Saudi Arabia's traditional foreign policy tools with new approaches to address these issues, but with mixed results.

As the Arab Spring moved so disruptively across the Middle East, it largely bypassed Saudi Arabia. The Saudi people were not interested in a dramatic, wholesale change of government. Nevertheless, as Chapter 19 discusses, the Saudi monarchy does face political challenges. The Saudi Shia have suffered from a century of systematic discrimination that has left 15 percent of the population outside of the kingdom's political mainstream. Corruption remains an issue that angers Saudis because it adversely affects the income distribution network. Three generations of Saudis have become increasingly more prosperous and well educated; they are looking for more direct participation in their government, as well as greater levels of transparency and accountability. All of these issues carry the seeds of instability if they are not addressed.

Saudi Arabia has long been more stable than many expected. Now because of an exceptionally rapid rate change, it is less stable than many presume. In this evolving society, the old roots are dying while new ones have yet to take hold. Chapter 20 concludes that over the next decade Saudi Arabia will evolve into either a more repressive and unstable monarchy or possibly into a more stable and accountable one. The West has an interest in that outcome—and the means to influence it.

Map 1. Cities and Towns of Saudi Arabia

Part I

Creating a New Nation

In the early hours of November 4, 2017, Saudi Arabia changed forever. Across the country, princes, bureaucrats, and businessmen who had long believed themselves above the law discovered that their immunity had evaporated. Charged with corruption, some were pulled from their beds, while others were summoned to nonexistent meetings or detained as soon as their international flights touched down in the kingdom. Among them were air force generals, navy admirals, and three sons of former King Abdullah, including the commander of Saudi Arabia's National Guard. A former minister of finance and a governor of the Central Bank found themselves confined together with former ministers of commerce, planning, and communications. Along with three brothers of Osama bin Laden, some 300 prominent individuals were taken to Riyadh's Ritz Carlton Hotel, which was converted for the occasion into the world's first five-star prison.

These events marked the end of an order that had governed Saudi Arabia for sixty-five years. Gone was a political structure in which numerous princes made collective decisions. Gone, too, was a military structure in which senior princes controlled independent armies. Some called it the Saudi Spring. Others saw the start of the Fourth Saudi State, or "Salman Arabia." It left some celebrating and others distraught. All agreed that it was part of a revolution that King Salman had initiated when he ascended the throne in January 2015. There was, however, much less agreement about what it would mean for Western diplomatic and commercial engagement with Saudi Arabia.

King Salman bin Abdulaziz al-Saud had been dealt a poor hand. Unlike his half-brother Abdullah, Salman had not inherited an oil boom that would fund enough royal generosity to win him the title "King of Hearts." Instead, in 2015 oil prices were collapsing—and with them, Saudi Arabia's economic growth and government revenues. The Saudi welfare state was fast becoming unsustainable. As one cabinet official told the author, "We were going to hit the wall – and in our own lifetime, not our children's." Unfortunately, in 2015, King Salman lacked the means to right his listing ship.

What he had was a well-established but shambolic and somewhat medieval system of government. Thousands of princes believed that they were entitled to privileged social status, legal immunity, and generous state-funded incomes. The conservative Wahhabi religious establishment, upon whose support the monarchy's legitimacy partially rested, controlled most of the country's legal and educational institutions. Its religious scholars were determined to maintain their historical importance and to block social change, especially when it related to the role of women. The civil service, which employed 70 percent of working Saudis, was filled with sinecures, nepotism, and corruption. This bureaucracy lacked both the motivation and administrative capacity to implement serious reform. Under the Obama Administration, Saudi Arabia's strongest ally, the United States of America, had grown distant, while its greatest adversary, the Islamic Republic of Iran, had grown more menacing. The Saudi military remained loyal and firmly under civilian control, but after years of training and billions of dollars invested in advanced weaponry, it was still unable to defeat a militia of Yemeni tribesmen or protect vital oil installations from inexpensive drone attacks.

Salman bin Abdulaziz had not expected to inherit these problems. He was only a few years younger than his two full brothers, Sultan and Naif. Both of them had been named crown prince and both had died younger than Salman would be when he ascended the throne. Although fate made Salman an unexpected king, he was not unprepared. He had been governor of Riyadh Province for forty-eight years. Intelligent, pragmatic, hardworking, well organized, and disciplined, he was also strict, demanding, and humorless. He made firm decisions and would become known locally as the "King of Decisiveness."

Salman spoke little English and had no university degree, but as he sometimes told visitors, "I did not need to go to the Sorbonne to learn how to be a Prince." He read widely in Arabic and owned one of the largest private libraries of any Saudi prince. Salman was firmly dedicated to preserving the monarchy, but recognized that doing so would require political, economic, and social change. At a minimum, he would need to manage the transition of power from the sons of King Abdulaziz to his grandsons, reduce the country's dependence on oil, and bring its social practices more into line with those prevailing in the rest of the world.

The realm that King Salman inherited was politically stable. Unlike Saddam Hussein's Iraq, it was not a police state ruled solely by fear. Nor was it an inaccessible "hermit kingdom" like North Korea. It was, however, unquestionably authoritarian and culturally isolated. There was no legal mechanism for its people to change their government. Ancient tribal and religious practices, once common in much of the world, remained a part of daily life in Saudi Arabia. Freedoms of speech, press, and assembly were very limited. Although there were far more political prisoners in Egypt or Turkey than in Saudi Arabia, dissent was not tolerated.

In 2020, Saudi Arabia is a very different place than it was in 2015. The complexities and contradictions that long made the kingdom difficult to understand have been compounded by rapid, unexpected, and disruptive change. Politically, the king has sought to prepare the government to pass smoothly on to the next generation of princes. Economically, he has sought to reduce the scope of the Saudi welfare state. Socially, he has sought to transform a closed and conservative society into a more tolerant one. In all of these efforts the king and his son, Mohammed bin Salman, have been partially successful.

However, these unsettling changes have made Saudi Arabia less stable than it was in 2015. Although a majority of the population continues to support the monarchy for a wide range of reasons, voices demanding either much more or much less change have become louder. Although there is more openness to the outside world, at home fear has become more prevalent.

The king and his son are not trying to make Saudi Arabia more democratic, but they are trying to make it more stable, prosperous, and religiously tolerant. They have a vision, but will it prove to be a mirage? Should the West shun them or seek to help them—and, if so, how? To answer these questions, one needs first to understand the legacy of dynastic power, religious reform, and national unification that the king and his son are trying to preserve. The Al Saud have two-and-a-half centuries of local history behind them. That history provides the foundation of the dynasty's legitimacy and a starting point for understanding Saudi Arabia.

The Rise and Fall of the House of Saud

For many centuries, Central Arabia suffered from lawless violence, political instability, and economic backwardness. Known locally as the Nejd, this was a harsh, haggard, barren, unforgiving land that no great power had wished to occupy for long. Since the Bronze Age, it had been a collection of wandering nomads and warring city states. Its local chieftains rose and fell. Their "states" were merely loose alliances of townspeople and nomads; the *Hadhar* and the *Bedu*, respectively. The nomads lived by animal husbandry and raiding; the townspeople engaged in farming, trade, and simple manufacture. These "states" provided only minimal administration and had no well-defined borders. Every town and tribe had its own independent, hereditary leadership. The urban notables were known as *aiyan* and the tribal chiefs as *sheikhs*.

When the founder of modern Saudi Arabia, Abdulaziz bin Abd al-Rahman al-Saud (1876–1953) was born, the Arabian Peninsula was as fragmented as Germany had been before Bismarck, or Italy before Garibaldi.[1] Saudi Arabia did not exist. East of Riyadh in Al Hasa, the Ottoman Turks controlled some of the world's largest date-palm groves in what would become the world's largest oil fields. Along the rest of the Gulf coast from Kuwait to Oman, Britain had established effective control over the local rulers, whom it administered as part of its Indian Empire. In Mecca, the Hashemite Sharif Hussein ibn Ali governed and administered the annual Hajj pilgrimage on behalf of the Ottoman Sultan. To the north lay Ha'il, seat of the Al Rasheed dynasty, which based its power on the large Shammar tribal confederation. In the southwest, Imam Yahiya Mohammed Hamid al-Din ruled Yemen from his mountain capital at Sanaa. By 1930, all but Imam Yahiya and the British would be gone. By the 1960s even they would disappear from Arabia's political landscape and a new nation state, the Kingdom of Saudi Arabia, would have effectively united most of the Arabian Peninsula for the first time in 1,000 years.

The religious practices of the Arabian Peninsula were as diverse as its political loyalties. Most inhabitants were Sunni Muslims, but there are four branches or schools of Sunni law and practice. The inhabitants of the Nejd were largely followers of the conservative Hanbali interpretation of Islamic law. The Ottoman Turks in Al Hasa were Hanafi and much of the Hejaz (Western Arabia) was Shafi. All four schools, including the Maliki, were present in Mecca.

There were Shia Muslims on the peninsula as well, and they too were divided according to whether they followed the twelfth, seventh, or fifth *imam*. In the date groves of Al Qatif, Hofuf, and Medina were large populations of "Twelver," or Ithnashari

Shia. In the south, near Najran, there were pockets of "Sevener," or Ismaili Shia. Further south in Yemen were the Zaydis, or "Fivers," some of whom have become today's Houthis. In the Hejaz, Islam's mystic Sufis were common. By the end of King Abdulaziz's life, there would be only one centralized Hanbali religious and legal establishment throughout his realm.

Across the empty deserts of Arabia roamed the *Bedu*—nomads with no fixed location or political loyalty beyond those of their tribe, clan, and family. They lived off their camel herds. In summer, when temperatures reached 120 degrees (50 degrees celsius) and grass withered, they could be found camped by deep, permanent wells or at trading centers where they exchanged livestock, wool, butter, and cheese for dates, grain, bullets, tobacco, and coffee.[2] In winter, when the grass grew lush from the autumn rains, they could be anywhere. Their *sheikhs* came from well-recognized noble families within each tribe and maintained their own legal systems, tax codes, and military forces. Within two generations, this political and economic independence of the Saudi tribes would be broken. Unlike in Yemen, Afghanistan, or parts of Iraq today, tribes would no longer challenge the political authority of the central government or act as an alternative source of social services.

All of these changes were brought about by the Al Saud family—and, most notably, by King Abdulaziz, or Ibn Saud as he is often known in the West. Their story began in 1744, when the dynasty's founder, Imam Mohammed al-Saud, ruled only the small Nejdi village of Dir'iyyah. As in all the surrounding villages, its inhabitants eked out a subsistence living. They tended date palms, grew grain, dug wells, and prayed for rain. They engaged in simple handicraft manufacture and traded with the surrounding Bedouin tribes for livestock or wool. Their trade with other oasis settlements was limited. Their contact with any central government in Istanbul or Mecca was nearly non-existent. Over the centuries their practice of Islam had incorporated many folk customs, such as worshiping saints and praying at graves. Imam Mohammed al-Saud died in 1765, and in such a backwater would have remained as obscure as the village that he ruled had he not met Mohammed Abd al-Wahhab (1703–1792).

Mohammed Abd al-Wahhab was something of an Arabian Martin Luther with a touch of John Knox thrown in. He was charismatic, possessed significant political skills, and saw himself as a religious reformer. Some would say he was a fanatic; he was certainly fervent in his beliefs and today would be called a fundamentalist. Like all fundamentalists, Abd al-Wahhab accepted a literal interpretation of his holy book and, like Luther, wanted to rid his religion of practices for which he could find no basis in scripture. Among these practices were: sorcery, idol worship, sun worship, fortune-telling, animism, the cult of ancestors, seeking intercession from saints, and even worshiping stones, tombs, and trees.[3] Above all, he emphasized the unity of God (*tawheed*) and the avoidance of innovation (*bid'a*), by which he meant anything not found in the Quran or known to the *Salaf*, the pious ancestors of Islam's first three generations from whom the term *Salafi* is derived.[4]

The descendants of Mohammed Abd al-Wahhab are known as the family of the Sheikh or the Al al-Sheikh. Their alliance with the Al Saud is deeply rooted in the teachings of a fourteenth-century scholar Mohammed ibn Taymiyyah (1263–1328), who is the intellectual godfather of what is today known as the Wahhabi movement. Ibn Taymiyyah saw the lay rulers (*ummara*) and the religious scholars (*ulama*) as the two branches of an ideal Islamic government. The ruler was charged with providing security and enforcing Islamic, or *sharia* law, while the scholars were responsible for interpreting that law.[5] There was no need for a legislative branch of government, since God's eternal laws had already been revealed in the Quran.

In a Wahhabi state, religious scholars do not govern or wield executive authority, but they do control the legal system and furnish the principal check on absolute executive authority. They provide advice, consent, and legitimacy to a Muslim ruler who must govern according to God's law as expressed through the principles of the *sharia*. Commanding what the law requires and banning what it prohibits make a ruler legitimate. Any ruler who flagrantly violates Islamic law or orders his people to do so is, by definition, illegitimate and subject to removal.[6] Although the Al Saud's interpretation of this ideal Islamic polity has always been shaped by self-interest, local custom, and tribal law, Ibn Taymiyyah's political framework governs Saudi Arabia in a recognizable form to this day.

This framework creates a very different state from the European models envisioned by Niccolò Machiavelli, who urged Renaissance-era secular princes to rid themselves of interfering religious authorities; or John Locke, who viewed legitimate government as a contract in which an inherently free people conditionally transferred some of their natural rights to a government, so long as that government protected those rights and provided for the public welfare. Ibn Taymiyyah did not look to secular concepts such as consent of the governed or majority rule. For him, Islamic government was based on immutable, indisputable, divine revelation. The emphasis was always on how a ruler governed rather than how he came to power. His legitimacy was derived from upholding an Islamic social and political order, not from elections.

Islam cannot be practiced properly in isolation by a hermit for there can be no Islam without the *ummah* or community. Moreover, there can be no community without the *imam* or leader. Because political stability is required for a functioning Islamic community, Abd al-Wahhab called for obedience to any established ruler who acted justly and defended Islam. This doctrine explicitly calls for obedience to a ruler unless he commands the violation of Islamic law. It puts Wahhabi religious scholars in the position of either defending a ruler or offering quiet, behind-the-scenes criticism. It contrasts sharply with the view of Islamic revivalists such as the modern-day Muslim Brotherhood, who have no compunction about openly denouncing rulers or even striving to depose them.[7]

Mohammed Abd al-Wahhab started preaching in the Nejdi villages of Huraimila—where he wrote his most important work, *The Oneness of God*—and then in Uyainah—where, again like Martin Luther, his preaching made him more than a few enemies. Forced to flee, he sought the protection of a neighboring prince and, although Frederick the Elector of Saxony sheltered Luther, it was Mohammed al-Saud who gave refuge to Abd al-Wahhab. In the summer of 1744, Mohammed Abd al-Wahhab took his message to Dir'iyyah, where he entered into a compact with Mohammed al-Saud, sealed with the same oath that the Prophet Mohammed and the men of Medina had sworn in order to cement their alliance some 1,400 years ago. The reformer promised the ruler that if he held fast to the doctrine of God's unity his domain would expand, and the ruler declared his readiness to undertake *jihad* in defense of Islam.[8]

Finance was as much a part of this new alliance as religion. Abd al-Wahhab declared payment of the Islamic alms tax (*zakat*) to be obligatory rather than voluntary, and thus transformed Al Saud revenue collection into an unchallengeable religious doctrine. Intolerance was also there from the outset. Any Muslim who heard the Wahhabis' call or *dawah* and did not join them was an infidel whose property was forfeit. This conviction that any opponent was an infidel or *kaffir* united, inspired, and disciplined the Wahhabis. From the outset, military conquest was part of their ideology, as commonplace raiding for loot was transformed into a holy war for religious reform.[9]

In 1744, Mohammed al-Saud married a daughter of Mohammed Abd al-Wahhab, adding family ties to the alliance in which for more than 200 years each family has lent legitimacy and support to the other. As the supreme leaders of the Wahhabi movement, the Al Saud became known as *imams*.[10] They govern in political and economic matters, but seek advice from the Al al-Sheikh scholars on religious and social issues.[11] Their position differs from that of the Shia *imams* and is more analogous to the British monarch who, as "Defender of the Faith" (*fidei defensor*), is technically the supreme governor of the Church of England but leaves ecclesiastical matters to the bishops.

Since 1744, the protection and propagation of Islam has remained the Al Saud dynasty's motivating ideology. It contributed significantly to the success of King Abdulaziz's nation-building program and was a strategic choice at odds with that of more secular nationalist Muslim leaders such as Mustapha Kemal Ataturk or the Shah of Iran. Article 23 of Saudi Arabia's Basic Law of Governance remains clear: "The State shall protect Islam, apply Islamic Law, promote Good, prohibit Evil and carry out the duty of spreading Islam." How King Salman will adjust the relationship between ecclesiastic and temporal authority in the increasingly secular twenty-first century remains to be seen, but that change is both inevitable and potentially destabilizing to his kingdom.

Aggressive Wahhabi evangelism created the First Saudi State (1744–1818) when, as in Islam's first three centuries, religion fueled military conquest. By the time he died in 1765, Mohammed al-Saud had, through war and diplomacy, gained control over most of the Nejd. In 1802, his grandson, Saud al-Saud, led the Wahhabi army into the Hejaz. He captured Mecca in 1803 and Medina in 1805. In 1806 the Wahhabis sacked Karbala in Iraq, the resting place of the Prophet's martyred grandson, Hussein, where they destroyed important Shia shrines. They forced the rulers of Oman to pay tribute and even raided as far as Aleppo in northern Syria.

During these early advances, two prominent ideological characteristics became apparent. First, the Wahhabi movement contained a strain of Arabian nationalism. In Mecca and Medina the Wahhabi puritans destroyed not only shrines where pilgrims prayed to saints rather than God, but also all the hookahs and stringed instruments they could lay their hands on, as they made a conscious effort to obliterate all traces of Ottoman Turkish influence.[12] Their rejection of tobacco, silk clothes, and music was a rejection of Ottoman culture and rule as much as a protest against innovation (*bid'a*). Second, the Wahhabis' militant anti-Shiism contained the seeds of their current struggle with the Persian carriers of what they deemed the Shia heresy.[13]

Meanwhile in Istanbul, the Ottoman Sultan began to take notice of the Wahhabis' growing influence. The Ottomans had controlled Mecca and Medina since the early sixteenth century. They ruled through the Sharifs of Mecca, whom they appointed, but who always came from the Hashemite family of the Quraish tribe that claimed direct descent from the Prophet Mohammed. Control of the Islamic holy places was important to the Ottomans. It strengthened the Sultan's legitimacy as Caliph, or supreme leader of Islam, and provided a steady stream of income from pilgrim taxes. The loss of Mecca threatened Sultan Mahmud II's authority and finances, and in 1811 he ordered the Ottoman Governor of Egypt, Mohammed Ali Pasha, to expel the Wahhabis.

Mohammed Ali Pasha and his son, Ibrahim Pasha, campaigned for seven years to destroy the First Saudi State. When Imam Saud al-Saud died in 1814, his son, Imam Abdullah, withdrew to the Nejd pursued by Ibrahim Pasha and an army of 5,000 infantry, 2,000 cavalry, and 150 artillerymen with howitzers, mortars, and cannon.[14] This largely Egyptian army besieged the then Saudi capital of Dir'iyyah, where the vastly outnumbered and outgunned Wahhabis, who had no artillery, held out for six months before surrendering on September 11, 1818.

Many members of the Al Saud and Al al-Sheikh families died in the fighting. Many more were tortured and executed in its aftermath, and still others were taken to Egypt as prisoners.[15] The Shah of Iran and the Ottoman Sultan sent Mohammed Ali Pasha letters of congratulation. Festivals were held in Cairo and Istanbul to celebrate the end of the Wahhabi "heresy" and its Al Saud leaders. Abdullah al-Saud was taken in chains to Istanbul, where the Ottoman religious scholars tried to convince him to renounce Wahhabi doctrines—particularly, its extreme monotheism. When Imam Abdullah refused to recant, he was paraded through the city streets for three days and then beheaded at the gate of the Topkapi Palace. His decapitated head was crushed to rid the world of his bad ideas. His headless body was hung at the palace gate before being thrown into the Bosphorus.[16] However, today in Riyadh, Imam Abdullah is remembered not as a defeated political leader but as a martyr who died for an Islamic reform movement.

The Al Saud capital of Dir'iyyah was pillaged and then razed to the ground. Its walls were mined, its wells filled in, houses burned, date palms and fruit trees cut down. When a British army officer, Captain George Sadler, passed through the following year, he found many Nejdi villages laid waste and not a single person living in Dir'iyyah.[17] The First Saudi State ceased to exist, but the surviving Al Saud learned two important strategic lessons: first, you must obtain modern military equipment and second, you can lose everything if you quarrel with the superpower of the day. King Salman, the current ruler, has not forgotten either of those lessons.

In 1818 the Al Saud were down but by no means out. Within six years a new Saudi government emerged. The Second Saudi State (1824–1891) never reached Mecca, though it did at times stretch from the Persian Gulf to the edge of the Hejaz, and from Ha'il in the north to Yemen and Oman in the south.[18] It, too, faced Ottoman invasion, assassinations, and rebellion—but it survived and even flourished; first under Imam Turki al-Saud (1755–1834) and later under his son, Imam Faisal bin Turki (1785–1865). Unfortunately, after Faisal's death his eldest sons, Abdullah and Saud, fought for control of the new Saudi capital at Riyadh.

The prolonged and complex struggle between Abdullah, Saud, and Saud's sons greatly weakened the dynasty. In 1871, the Turks took control of Al Hasa, today's Eastern Province, claiming to have come to restore order for the Al Saud. In 1887. the Al Rasheed clan, who had governed Ha'il on behalf of the Al Saud, set themselves up as rulers there. When Imam Abdullah's own nephews imprisoned him, the Al Rasheed restored him to power effectively making the Al Saud their vassals.[19]

In 1891, Imam Faisal bin Turki's last surviving son, Imam Abd al-Rahman, supported a failed effort to drive the Al Rasheed from the southern Nejd. The Al Rasheed's victory at the Battle of Mulayda forced him to flee from Riyadh, and the second Saudi State came to an end.[20] From that bitter defeat, the Al Saud learned another strategic lesson: above all else, do not use force against each other; keep family disputes peaceful and private; and unite quickly and firmly against anyone who violates this rule. Modern-era Kings Faisal, Khalid, Fahd, and Abdullah all respected this stabilizing principle. King Salman and his ambitious Crown Prince, Mohammed bin Salman, have not.

The future King Abdulaziz was an eight-year-old boy the night his soon-to-be-homeless family fled Riyadh. For two years they lived with the nomadic Murra tribe in southern Arabia's vast, trackless desert, the *Rub al-Khali* or Empty Quarter. During those years, Abdulaziz acquired a valuable understanding of the *Bedu* and desert warfare. The family eventually settled in Kuwait as guests of Sheikh Mubarak al-Sabah, who had the young Abdulaziz taught history, geography, and mathematics.[21] The exiled Al Saud remained guests in Kuwait for a decade, a debt that they would repay when the Al Sabah fled to Saudi Arabia and established their own government in exile following Iraq's 1990 invasion of Kuwait.

In the 1890s, Kuwait offered a dramatically more cosmopolitan and commercially vibrant environment than Riyadh. As a result, by his middle teens the future King of Saudi Arabia had acquired firsthand experience of dynastic politics, humiliating exile, and desert warfare. He spoke some English and had watched Sheikh Mubarak conduct commercial and diplomatic relations with Europeans. He was a very unusual young man for his time and place, and he stood six foot, four.

Mubarak of Kuwait sheltered the Al Saud because, like them, he had no time for the Al Rasheed of Ha'il, and in 1900 he began several unsuccessful campaigns against them. The young Abdulaziz participated in these efforts, trying and failing to capture Riyadh in 1901. It was only the following year that he became a local legend by successfully recapturing his ancestral home. Against huge odds, Abdulaziz restored his family's battered fortunes with little more than audacity, careful planning, stealth, and luck. Remember that formula; his grandson does. More than a century later, Mohammed bin Salman bin Abdulaziz is once more counting on audacity, planning, stealth, and luck to preserve his family's fortunes.

That capture of Riyadh by the young Abdulaziz is the foundation myth of modern Saudi Arabia. Like all myths, it has numerous versions, but its essence is clear and inspiring. In the spring of 1902, Abdulaziz secured permission from his father and Sheikh Mubarak of Kuwait to try once again to capture Riyadh. He left Kuwait with only sixty men on second-rate camels. Most of his companions were close relatives, fellow exiles, and slaves—for in old Arabia slaves were often raised to a position of trust, and those who fought valiantly could win their freedom.[22]

This band proceeded to Al Hasa, gathered more *Bedu* followers and raided tribal allies of the Al Rasheed. When the Turks threatened retaliation, the always fickle *Bedu* abandoned Abdulaziz, and his father sent a letter ordering him to return home. Instead, Abdulaziz read the letter to his remaining followers emphasizing the dangers they all now faced. He said he would not return to Kuwait and invited those who wished to go with him to stand on his right and those who wished to return to Kuwait to step to his left. Some accounts maintain that not a single man chose to go home, others say a few chose the safer option. But it is generally agreed that at the Haradh Oasis, which would one day become a major Aramco gas-processing facility, a striking force of some forty men pledged to follow Abdulaziz to Riyadh. This was not the last time that the young prince would use a dramatic gesture to galvanize support at a difficult moment, nor would it be the last time that he relied on patience, secrecy, and surprise.[23]

Abdulaziz disappeared into the desert with his crew and waited quietly for an entire month until the Al Rasheed concluded that he had gone home. Having been to Riyadh the previous year, he knew he could not storm the city with forty men, but, as would often be the case in his career, he had good intelligence. He knew that the population was dissatisfied with Al Rasheed rule and he knew that the wary Governor Ajlan slept in the Mismak Fort each night before returning to his home in the morning. With a dozen followers, Abdulaziz entered the city at night over a broken part of the wall, occupied the governor's house, and waited for him to come home. In the morning, he ambushed the governor. Ajlan struck repeatedly with his sword, which Abdulaziz parried with his rifle before his cousin, Abdullah bin Jaluwi, killed the governor.[24]

An hour after sunrise on that cold, January morning in 1902, the young Abdulaziz appeared on the battlements of the Mismak Fort and threw Ajlan's severed head to the gathering crowd.[25] Thus began the Third Saudi State (1902 to the present day). With their governor dead, the Al Rasheed garrison quickly surrendered and the city fathers welcomed the return of their traditional rulers. Today, the Mismak Fort is a national museum and Saudi schoolchildren come to see the spot where King Salman's father grappled with Ajlan.

The future King Abdulaziz did not capture Riyadh in the name of Islam. In fact, the Al Rasheed were devout Wahhabis just like the Al Saud. The young prince risked his life to restore his family's honor and reclaim its leadership position in Central Arabia. As his Lebanese companion and biographer Ameen Rihani wrote, "it seems clear that the dominant motive in Ibn Saud's recapture of Riyadh was the need to restore the patrimony of the Al Saud," for he said "Know by God, that what was the right of our forefathers is ours and if we cannot get it by friendly means we will take it with the sword."[26] In 2020, maintaining their historic patrimony still remains the first order of business for Abdulaziz's heirs.

The new Amir of Riyadh immediately sought support from the Al Saud's traditional allies, the Al al-Sheikh family, and the senior religious scholars promptly declared the Al Saud to be Riyadh's legitimate rulers. Abdulaziz and two of his brothers then began a pattern of political marriages by each marrying a daughter of the Grand Mufti Abdullah bin Abd al-Latif Al al-Sheikh. One product of this marriage into the Al al-Sheikh family was the son who would one day become King Faisal (1906–1975).[27]

The young victor then demonstrated another fundamental, stability-generating, Al Saud principle—a respect for age. He asked the people of Riyadh to swear allegiance not to himself, but to his father, Imam Abd al-Rahman, who was still in Kuwait. Only after his father declined did Abdulaziz assume effective control. For the rest of his life, Imam Abd al-Rahman led the prayers in Riyadh with his son respectfully standing a few paces behind him.[28] Again, this is a principle that King Salman has largely ignored.

The capture of Riyadh did not re-establish Al Saud control over Arabia, or even the Nejd; far from it. Abdulaziz had, in fact, barely begun the long struggle for unification that would test his skills as a tribal politician and desert warrior for the next thirty-two years. However, having won one a huge gamble when he had very little to lose, Abdulaziz became more conservative. Caution, careful calculation, and reconciliation with defeated opponents now became his signature tactics. After taking Riyadh, he always wanted the odds to be on his side—and, during the decades of near-constant campaigning and over fifty battles that it took him to create modern Saudi Arabia, they usually were.

Abdulaziz was no Napoleon or Alexander. Most of his battles were relatively small engagements fought by irregular troops. Still, he had the sense, for instance, not to allow himself to be caught in a walled city by a superior force. He did not wait for the powerful Al Rasheed Army to arrive from Ha'il to besiege Riyadh. Instead, he slipped away into the desert to recruit new tribes to his cause. Many of his early followers came from Mohammed Abd al-Wahhab's tribe, the Bani Tammim, or from his mother's tribe, the Dawaisir.[29] With them, he first consolidated control over the regions south of Riyadh that were traditionally most loyal to the Al Saud and farthest from the Al Rasheed capital at Ha'il.

Over the next five years, Abdulaziz personally fought against the Al Rasheed and their Turkish allies in many small engagements as well as several significant battles. The Turks sent professional troops from Iraq and Medina to support the Al Rasheed, but the Amir of Riyadh, as he was then known, harassed their supply lines and defeated them—first at Al Bukayriyah, where he was severely wounded, and later at Shanuna and Tarafiya.[30] In these early battles, Abdulaziz relied almost entirely on support from the settled population. Those *Bedu* who did fight with him proved extremely unreliable.[31]

In 1904, Abdulaziz pressed northward into Al Qasim Province, where Unayzah, Buraydah, and Ar Rass are the principal towns. There he developed a pacification technique he would use repeatedly. He agreed with the local population that they would always be ruled by one of their own nobles. In Unayzah, he appointed Sulaiman Sulaim as governor, a position his family holds to this day. A similar arrangement prevailed in Buraydah until it became a provincial capital with a royal governor. In Ar Rass, the Al Assaf family surrendered the city and still serve as its governors. Local *amirs* and leading merchants from towns like Uyainah, Radwa, and Tharmada all made peace with Abdulaziz, and their leading families—such as the Al Mu'ammar, Al Madi, Hejailan, and Al Angari—all remain prominent in Saudi government and business circles today.[32] Gradually, Abdulaziz united a collection of local Nejdi dynasties into a regional state, one in which the traditional elites remained recognized and privileged participants in the new political order.

In 1906, Saudi forces surprised a much larger, but sleeping, Al Rasheed army at Rowdhat Muhanna, where the competent and respected Abdulaziz Al Rasheed died trying to rally his surprised forces. His death created a leadership vacuum in Ha'il that set in motion a violent intra-family struggle, which ultimately led to the collapse of the Al Rasheed dynasty. Abdulaziz al-Saud took the signet ring from the slain Abdulaziz al-Rasheed's finger and used it to convince the townspeople and tribes of Al Qasim that the ruler of Ha'il was dead, and that their *bay'ah* or sacred oath of allegiance given to him was no longer binding.[33]

After Rowdhat Muhanna, Abdulaziz's survival in the central and southern Nejd was never seriously in doubt, but the Amir of Riyadh was still only one of many players on the Arabian Peninsula.[34] He was surrounded by the Ottomans, Hashemites, Al Rasheeds, and British. He may have been heir to a distinguished family history and something of a local hero, but he was in fact nothing more than a traditional local chieftain whose opportunities for expansion did not appear promising. He had no permanent military force and depended solely on the temporary service of settled villagers or unreliable nomadic tribesmen. During his first decade in Riyadh, Abdulaziz had not used religion as the ideological motivation for his campaigns of military conquest.[35] But that was about to change.

The Wars of National Unification

Abdulaziz al-Saud started from a weak position. Control of Mecca brought his Hashemite rivals international recognition and a steady stream of pilgrim taxes. The Al Rasheed of Ha'il led the large Shammar tribe, dominated important caravan trade routes, and were close allies of the Ottoman Sultan. The Al Saud had none of these advantages. Abdulaziz needed a competitive edge, and he found it in the Ikhwan Movement which, during the 1920s, became the dominant military force on the Arabian Peninsula. The Ikhwan armies he created provided the ideological energy and physical muscle for Saudi expansion. They were a unique, strategic innovation that permanently set Abdulaziz apart from his rivals.[1]

The *Bedu* of Central Arabia herded camels in empty deserts where there were no mosques or schools. They produced warriors, not scholars, and their understanding of Islam was limited. The original idea behind the Ikhwan, or Brotherhood Movement, was to settle the *Bedu* in order to improve their religious practices. Abdulaziz may not have conceived of the idea himself—that credit often goes to his father-in-law Abdullah bin Abd al-Latif Al al-Sheikh—but from the outset, Abdulaziz strongly supported the movement for his own reasons with money, seeds, and farming equipment.[2]

Abdulaziz wanted settled farmers whom he could tax rather than wandering nomads whom he could often not even locate. He wanted to replace the divisive tribalism that motivated the *Bedu* with a unifying Islamic identity, and, above of all, he wanted a dependable military force. He was partially successful in meeting all three objectives. Many *Bedu* did settle down and, at least temporarily, tried farming. They stopped smoking tobacco, renounced praying at tombs, and put on the dreaded white turban that signified Ikhwan membership. Eventually there were more than 100 Ikhwan settlements scattered across the Nejd. Never more than a day's march apart, they created an extraordinary military network that provided Abdulaziz with some 60,000 men of fighting age.[3]

These Ikhwan settlements created religious, agricultural, and military communities that were something of a cross between an American Indian reservation, religious revival meeting, and organized military camp. Despite their original goal of replacing tribalism with Islam, most remained tribally based. Founded around 1913, the first and most famous settlement was Artawiyah, base of the Mutair tribe led by the shrewd and ambitious Faisal Daweesh. Others included Ghatghat, home of the Utaibah, led by the devout—but less politically astute—Sultan ibn Bijad, and Duhna where the Harb were led by Ibn Nuhait.[4]

Abdulaziz paid the Ikhwan leaders and armed their followers. He dispatched Wahhabi missionaries (*mutawa'a*) to recruit new followers and preach in established settlements. His missionaries taught the importance of obeying the *imam*—for Islam could not be dominant without unity, and unity required a leader. In an effort to replace tribal justice with *sharia* law, Abdulaziz also sent trained judges to the Ikhwan settlements. For as Mohammed Abd al-Wahhab had written, it was absolutely necessary to have an *imam* who would enforce *sharia* law and lead the Muslim community against its enemies.[5]

These missionaries and judges explicitly used religious education and *sharia* law as a means to extend Al Saud authority.[6] For example, the *mutawwa'a* were authorized to flog those who neglected their religious duties or failed to pay Abdulaziz's tax collectors. Nevertheless, many Ikhwan retained their tribal values, prejudices, and loyalties such that, even today, the Al al-Sheikh and Al Saud strive to replace tribalism with Islam and, more recently, with Saudi nationalism.

Most *Bedu* joined the Ikhwan Movement as part of a tribal unit rather than as individual recruits. When they joined their lives changed dramatically. They were strictly forbidden from raiding tribes under Abdulaziz's protection, they were banned from collecting protection money from villages and weaker tribes, and they were prohibited from charging fees to travelers and merchants crossing their territory. Their once independent leaders now paid the religious tax (*zakat*) to Abdulaziz. Their only remaining opportunity for capital accumulation was attacking "polytheists," by which they meant any Muslim who had not yet embraced the Wahhabi movement. Before the First World War, that included nearly everyone in the Arabian Peninsula outside of the central provinces of Riyadh and Al Qasim. Attacking polytheists became the Ikhwan's last chance for loot, and something they pursued with vigor.

In 1913, Turkish control of Al Hasa, today's Eastern Province, was weak. The Ottoman Sultan maintained a few hundred troops in major towns but exercised little authority over the countryside. Abdulaziz captured the principal town of Hofuf in a surprise night attack, in which he made his first limited use of the Ikhwan as a fighting force. This was a giant leap forward for the fledgling Third Saudi State. The defeated Turks retreated to Iraq and the new province brought large, taxable date groves as well as a seaport through which Abdulaziz could conduct his own trade and gain a steady source of income from customs duties.[7] Though he did not realize it at the time, this conquest had also brought his family control of the oil fields that would sustain their rule well into the next century.

Abdulaziz quickly made his presence felt in Al Hasa. He appointed his trusted cousin, Abdullah bin Jaluwi, governor. He stopped the *Bedu* from raiding caravans, expelled foreign merchants, and imposed an 8 percent customs duty on imports.[8] He again relied on support from local Sunni merchant families such as the Al Gosaibi, and Sunni sheikhs such as the Al Shuraim of the Murra tribe.[9]

Inevitably, tensions arose between the local Shia population and the Sunni Wahhabis from the Nejd. Differences in theology and rituals between Sunni and Shia Muslims exist across the Islamic world. The dispute originates in a disagreement over Mohammed's rightful immediate successor as political and spiritual leader of the Muslim community; Ali—the Prophet's cousin and son-in-law, whom the Shia believe was chosen by Mohammed as his successor—or Abu Bakr, whom Sunnis chose by consensus believing Mohammed had not designated a spiritual heir. Mohammed's only male descendants were Ali's sons, Hasan and Hussein. All Shia regard Ali and his descendants as their rightful *imams*, or spiritual leaders, though as has been mentioned, they differ among themselves as to whether there were five, seven, or twelve historical *imams*. Most Shia accept *ismah*, the belief that all historical *imams* within their different traditions were infallible in their religious judgments. Most Shia who follow the Twelver tradition also believe in occultation, or *ghaybah*, the view that the last *imam* remains alive though hidden, and that as the Mahdi he will return to restore Islamic order and justice to the world.[10] Unlike the Sunnis, the Shia depend primarily on financial support from their congregation rather than the state, and have a well-organized hierarchy led by the ayatollahs and grand ayatollahs.

Despite these differences, most Sunnis and all Shia still regard each other as fellow Muslims. The Wahhabis do not. They refer to the Shia as rejectionists (*raffidah*). In his detailed "Treatise on the Denial of the Raffidah," Mohammed Abd al-Wahhab made it very clear that attributing infallibility to *imams* is a form of polytheism.[11] Even today, many devout Wahhabis continue to regard the Shia as dangerous heretics whose reverence for their *imams* violates the Wahhabi's most fundamental doctrine of the unity of God (*tawheed*).

The Ikhwan demanded that Abdulaziz convert, expel, or execute the Shia of Al Hasa. Abdulaziz did not agree, nor would he ever ban Shia from making the Hajj as the Ikhwan subsequently demanded. Some sources believe that Abdulaziz negotiated a pact with the Shia *ulama* that would guarantee their religious freedom in return for their loyalty.[12] If this was the case, his promise of religious freedom was never fulfilled. Instead, Abdulaziz and his successors have sought to position themselves as buffers between the Saudi Shia and the Sunni *ulama*. They have imposed numerous restrictions on Shia religious practices such as processions during the month of Muharram, gathering in mourning houses (*hussaniyyat*), and even the building of new Shia mosques. At the same time, they have protected the Shia from more extreme Wahhabi demands for their expulsion or forced conversion.[13]

For decades, Britain propped up the faltering Ottoman Empire as a source of regional stability and a bulwark against Russian expansion towards India. In 1914, when the First World War erupted and the Ottomans sided with Germany, that policy had to change. Now London policymakers worried that the Turks themselves would advance towards the Suez Canal or the British base at Aden in present-day Yemen. The Royal Navy had become increasingly dependent on oil from Turkish-controlled Iraq, and British leaders hoped to deny their enemies access to the recently discovered oil fields of Iran. These officials also feared that the Ottoman Sultan, in his capacity as Caliph, could incite millions of Indian and African Muslims to wage a holy war against their Christian, colonial masters. All of this lent new importance to events in Arabia.

The opening years of the First World War did not go well for Britain in the Middle East. Plans to take Istanbul ended in failure at Gallipoli. An attempt to take Baghdad resulted in disaster at Kut Al Amara, 100 miles to the south of the capital, where more than 10,000 British and Indian troops surrendered, many of whom died in Turkish prison camps. Meanwhile, although Britain sought to exploit a growing sense of Arab

Nationalism in the Arab provinces of the Ottoman Empire, prolonged negotiations failed to persuade Sharif Hussein of Mecca to declare an Arab Revolt.

Throughout their long and difficult campaign in what would become Iraq, British generals saw the Ottoman-allied Al Rasheed in Ha'il as a threat to their left flank. They repeatedly urged Abdulaziz to attack them—and in January 1915, with a force of Ikhwan and Ajman tribesmen, he confronted the Al Rasheed Army at Jarrab near Zilfi, northwest of Riyadh. The battle proved inconclusive.

After Jarrab, and despite numerous British appeals, the Amir of Riyadh seldom took an active part in the conflict.[14] He occasionally harassed pro-Turkish forces but did not stop Nejdi merchants from trading with them. At times he sold camels to the Turks; at other times he confiscated Turkish-bound camels and gave them to the British. Militarily, he expended more effort dealing with revolts among the Ajman and Murra tribes than confronting the Turks. Essentially, he sat on the sidelines waiting to see who would win the Great War.[15] The celebrated Arab Revolt, in which T. E. Lawrence played such a prominent role, took place in the Hejaz where, unlike the Nejd, there were still Turkish troops to rebel against.

Politically, Abdulaziz sought to counter the Al Rasheed's firm alliance with Turkey by drawing closer to Britain. In December of 1915, with a bullet still in his arm from his last encounter with the Ajman, the Amir of Riyadh met the British Political Agent for the Gulf, Sir Percy Cox, at Al Qatif, on the Arabian Peninsula's eastern coast, and signed the Treaty of Darin.[16] Britain formally recognized Abdulaziz al-Saud and his heirs as the rulers of Nejd, Al Hasa, and Al Qatif. It also agreed to protect the Al Saud from foreign threats, while Abdulaziz pledged to neither conduct an independent foreign policy nor grant concessions to non-British firms. He further agreed not to attack Kuwait, Bahrain, or the Trucial States (today's United Arab Emirates and Qatar), as these were already under British protection. He made no such commitment regarding Sharif Hussein in Mecca, who was still an Ottoman subject.

This was Abdulaziz al-Saud's first formal appearance on the international stage, and it came with the first Saudi arms package. Cox agreed to provide him with 300 rifles, ample ammunition, and £5,000 in cash every month thereafter.[17] Now Britain was effectively arming and funding the Ikhwan, much as the United States would do for the Afghan *mujahideen* half a century later. These British payments continued even after the First World War ended, and were essential to the Al Saud's success. With them, Abdulaziz built Ikhwan settlements and paid for wells, seeds, arms, and loyalty. His start-up firm had obtained crucial financial backing from Britain, but the world's leading colonial venture capitalist got a seat on his board, and more than once vetoed Al Saud expansion plans.

The collapse of the Ottoman Empire at the end of the First World War fundamentally changed the political landscape of Arabia. In Ha'il, the Al Rasheed lost their powerful protector and, without Ottoman subsidies to distribute, the Shammar tribe soon began to desert them. Sharif Hussein of Mecca, who had eventually proclaimed the Arab Revolt against the Turks, now became an independent ruler in the Hejaz. In recognition of his wartime service, Britain installed his sons as the kings of its two newly created mandates in Iraq and Jordan. Yemen and Asir, a region lying south of the Hejaz between Jeddah and Yemen, had also revolted against the Ottomans, but had made no contribution to the war effort. They were now independent, but weak.[18] All across the

Arabian Peninsula, local chieftains tried to use the new order to strengthen their positions. Most failed. It was Abdulaziz al-Saud who most skillfully managed his military affairs and diplomatic relations in order to pick up the pieces.

At the San Remo Conference in April 1920, the victorious allies carved up the carcass of the Ottoman Empire. Britain would take control of mandates in Iraq, Jordan, and Palestine, while France received Lebanon and Syria. Neither party wanted the trouble or expense of administering the worthless deserts of Central Arabia, leaving that to trustworthy local chieftains would be more than adequate. Among those chieftains, Sharif Hussein ibn Ali al-Hashem in the Hejaz and Amir Abdulaziz ibn Abd al-Rahman al-Saud in the Nejd, were the two most powerful.

Unfortunately, these two local leaders were also longstanding political rivals. After launching the Arab Revolt against the Ottomans in 1916, Sharif Hussein had begun to style himself King of the Arabs. When the war ended, he called upon all Arab leaders to respect this claim. The Amir of Riyadh made it very clear that with the Ottomans gone, he had no intention of even nominally accepting any new overlord.

The Amir and the Sharif also followed very different interpretations of Islam. Abdulaziz once told the British explorer H. St. John Philby that he had more respect for Christians than for the Muslims of the Hejaz because the Christians, though misguided, at least followed their book correctly.[19] Wahhabi religious scholars regarded Hejazis as polytheists or *mushriqeen* because they invoked the intercession of saints and prayed at the shrines of holy men. In Ikhwan settlements across the Nejd, Wahhabi missionaries constantly preached the duty to fight such infidels. Thus, Abdulaziz had a military force looking for work, but except for the occasional Jewish merchant, there were no non-Muslim infidels in the Nejd or the Hejaz. The non-Wahhabi Muslims of the Hejaz soon became the Ikhwan's next target.[20]

The looming struggle between the House of Hashem and the House of Saud initially focused on who would collect taxes from tribes that grazed their livestock between the Hejaz and Nejd—particularly, the Harb and Utaibah. Tensions came to a head at Al Khurmah, a collection of oases in the deserts between Mecca and Riyadh. The population had long been loyal to the Sharif, but after 1912 under the influence of Wahhabi missionaries, they sought to embrace the Islamic revival movement and change allegiance. As a result, in 1919 Sharif Hussein sent his second son, Abdullah, with more than 1,000 well-armed, professional soldiers to occupy the oasis. The inhabitants of Al Khurmah appealed to Abdulaziz for help, and he duly dispatched Sultan Ibn Bijad with the Ghatghat Ikhwan to protect them.[21]

For the Ikhwan, the matter was straightforward. Their brothers were being persecuted and needed assistance. In yet another of their trademark surprise night attacks, they massacred the Hashemite army at Turabah, east of Mecca. With his army destroyed, the Sharif's son fled in his nightshirt. Very few escaped, and as late as the 1970s the bones of the Hashemite dead could still be found on the field.[22]

Until Turabah, Abdulaziz had been very secretive about the Ikhwan's strength and true purpose. Although Sharif Hussein had sent numerous warnings about this emerging military force, the British had continued to regard the Ikhwan as purely a Nejdi religious movement.[23] Now it became clear that the Al Saud had created an army that the Hashemites could not contain. The road to Mecca was open, but the founder of the Third Saudi State was more circumspect than his ancestors had been a century earlier. He did not proceed before gauging British reaction, and at that moment His Majesty's Government still saw a place in Arabia for both the Hashemites and the Al Saud. It advised the Amir of Riyadh not to advance into the Hejaz and sent aircraft to Jeddah to reinforce the point.[24]

Abdulaziz was determined and consistent in pursuing his overall objective of expansion, but flexible and pragmatic in how he proceeded. In 1920, he turned away from the Hejaz, which Britain protected, and began his conquest of the Asir. The following year he led a largely Ikhwan army north to Ha'il, where the Al Rasheed's rule had become completely dysfunctional.[25] Three of the last four Al Rasheed amirs had been murdered by members of their own family. Their internally divided regime based on one powerful tribe was no match for Abdulaziz, who by now had united many tribes under the Wahhabi banner of religious reform. With his usual concern for weaponry, Abdulaziz also dragged Turkish cannon captured at Al Bukayriyah in 1904 some 400 miles from Riyadh to Ha'il. After a two-month siege, the leading citizens of Ha'il sent a member of the al-Subhan family to negotiate and open the city gates to the Al Saud.[26] The last Al Rasheed Amir, Mohammed bin Talal, retreated into his citadel, but he too eventually surrendered peacefully and was taken to live out his life as a "guest" in Riyadh, where his descendants still receive a small monthly pension.[27]

Abdulaziz prevented the Ikhwan from looting Ha'il, and appointed Ibrahim al-Subhan governor.[28] He allowed many of the Al Rasheed and their Shammar tribal supporters to leave for Iraq—where they remain today, creating a deep source of blood relations and Saudi sympathy for Iraq's Sunni population. Following his policy of political marriages, Abdulaziz married several Al Rasheed and Shammar women including King Abdullah's mother, Fahda bint Asi al-Shuraim. Several of Abdulaziz's sons, including Kings Saud and Fahd, also married into the Al Rasheed family. The deposed Amir of Ha'il then married one of Abdulaziz's own daughters. At the time, the British diplomat and traveler Gertrude Bell predicted that "the conquest of Ha'il will have far reaching consequences for it will bring Ibn Saud into Trans-Jordanian politics." And so it did for, after the fall of Ha'il, Abdulaziz changed his title from Amir of Riyadh to Sultan of the Nejd.[29]

Britain did not object to the Al Saud conquest of Ha'il. Nor did it protest the following year when Ikhwan armies took the northern oasis of Al Jawf from the paramount sheikh of the Ruala tribe, Nuri Shaalan. However, as Abdulaziz expanded northward, Britain did seek to formalize the border between the Nejd and its new mandate in Iraq. In November 1922, Abdulaziz and Sir Percy Cox met again—this time at Al Uqair, a small port north of Dhahran on the Persian Gulf. The treaty signed at Al Uqair was a significant event in Saudi history. For the first time, it gave formal, internationally recognized borders to the Al Saud's realm.

"Cockus," as the Arabs called him, wanted lines on a map denoting international boundaries. Abdulaziz wanted the allegiance of as many tribes as possible so that he could tax them and divert their trade to ports where he collected customs duties. For him what mattered was a tribe's allegiance, not which side of an artificial line it happened to be grazing on. The *Bedu* saw it the same way. For them, the desert was like an ocean where the nationality of a ship depended on the flag it flew, not which part of the sea it was sailing upon at any particular moment.

Abdulaziz and the *Bedu* lost the argument. Cox insisted on a line and drew it much further south than the new Sultan of the Nejd had hoped. However, as he was still receiving a substantial British subsidy of £60,000 a year, the Sultan was in no position to argue.[30] He was obliged to give up any claim to southern Iraq together with the tribes that lived there, such as the Muntafiq, Amarat, and Dhafir. Abdulaziz also accepted restrictions on annual migrations north by the Nejdi tribes, which was particularly disruptive to the grazing patterns of the Mutair. Cox agreed to send the Dahamsha south with no right of return. His decisions disturbed many border tribes, much as Saudi efforts to demarcate their own border with Yemen would eighty years later.

Sir Percy reportedly consoled the unhappy Sultan of the Nejd by saying, "My friend I know exactly how you feel and for this reason I gave you two thirds of Kuwait's territory."[31] Whether he actually said this or not, it is essentially what Cox did. Abdulaziz negotiated, complained, and compromised. Tribes were divided and ancient grazing patterns permanently disrupted, but in the end the Al Saud got nearly as much territory from Kuwait as they lost to Iraq—again, a bit of history that the Amirs of Kuwait have not completely forgotten.

Meanwhile, back along the Red Sea coast, the political leadership of the Hejaz had been in the hands of the Bani Hashem clan of the Quraish tribe since 961 AD. Known as Sharifs or *Ashraf*, they traced their lineage back to the Prophet Mohammed's daughter, Fatima, and her son, Hasan. The Sharif of Mecca was subordinate to the Ottoman Sultan, who appointed him and paid him a monthly stipend from the Ottoman treasury as well as separate funds to maintain the holy places, but he and his relatives were regarded as special.[32] They were addressed as sir or *sidi* and wore unique attire, including distinctive sandals and daggers. They were accorded both social and legal privileges, with anyone striking a Sharif losing the offending hand.[33]

After the Turks abandoned Mecca in July 1916, Sharif Hussein ibn Ali became an increasingly oppressive and unpopular ruler. When the cost-conscious British government reduced his wartime subsidy, Hussein tried to make up the loss by imposing higher taxes both on local residents and on visiting pilgrims. In protest over the transfer of Syria to French control, he refused to sign the Treaty of Versailles—a decision that cost him membership in the League of Nations.[34] He refused to accept an Anglo–Hejaz Treaty or the British and French mandates in Iraq, Palestine, and Syria. He complained endlessly about Jewish immigration to Palestine and opened diplomatic relations with the Soviet Union. In short, he remained a staunch Arab Nationalist when that cause was no longer useful to Britain. Hussein had become an irritant; "his pretensions bordered on the tragicomic"; and, when the crisis finally came, a British decision to abandon him was not difficult.[35]

For centuries, the Ottoman Sultans had been recognized as Caliphs—for Sunnis the spiritual successors of the Prophet Mohammed. In 1924, the leader of post-Ottoman Turkey, Mustapha Kemal Ataturk, constitutionally abolished the Caliphate. Sharif Hussein immediately sought to fill the vacancy, and with great ceremony had himself proclaimed Caliph. His presumption outraged many Muslims including the Ikhwan, who already regarded the Hejaz as a sinkhole of iniquity and heresy. Combined with restrictions that Hussein had placed on Wahhabis making the Hajj, this was all the pretext Abdulaziz needed to send a largely Ikhwan force into the Hejaz.[36]

The Ikhwan soon stormed Taif, a mountain town near Mecca, and what happened there is disputed to this day. It is clear that many unarmed local inhabitants were killed when the Ikhwan looted the town. Exactly how many died is less clear, with estimates ranging from fifty to more than 500.[37] Whatever the number, it was more than enough to terrify the citizens of Mecca, many of whom were already disillusioned with Sharif Hussein and hoping for better government under the Al Saud. In October 1924, the Ikhwan entered Mecca without opposition, and this time were much more restrained. Abdulaziz remained in Riyadh long enough to gauge British reaction, and when there was none, proceeded to walk into Mecca in the garb of a humble pilgrim: bareheaded, barefooted, and wrapped only in the traditional seamless white cloth. After this public show of piety, he sent his army on to besiege the Sharif who had fled to Jeddah.

In the vain hope that Abdulaziz would accept the existing political order with a different ruler, the Hejaz's religious scholars and urban notables, the *ulama* and *aiyan*, persuaded Sharif Hussein to abdicate in favor of his son, Ali.[38] Abdulaziz was not prepared to accept cosmetic changes. Declaring that his sole purpose was to guarantee access to the holy places for all pilgrims, he demanded that the new Sharif Ali leave the Hejaz as his father had already done.[39]

The siege of Jeddah began in January 1925. The Ikhwan commanders urged Abdulaziz to simply storm the city, but he was cautious and concerned about international public opinion. The Sultan of the Nejd did not want another Taif-style, Ikhwan rampage in the Hejaz. Reader Bullard—then a young British Consul in Jeddah, who would later become British minister in Tehran during the Second World War— wrote in his weekly report to London.

> There is little doubt that Ibn Saud could take Jeddah if he made a serious attack. His failure to do so is attributed here mainly to the presence of foreigners and to Ibn Saud's fear that his followers might get out of hand and indulge in indiscriminate killing and looting ... The last thing he wants is to begin his rule in Jeddah falling foul of the foreign powers.

A week later Bullard added, "No communication whatever from Ibn Saud who sits quietly in Mecca doing nothing much, but getting married to a lady who is said to be his 107th wife."[40]

The siege of Jeddah lasted nearly a year and reduced the city to starvation. Shut within its walls, people died by the hundreds and bodies rotted in the streets. "Men and women mad with thirst and skeletons from hunger killed the scavenger dogs for food and fought for offal to eat."[41] In response to an inquiry from London, Reader Bullard replied, "There are two foreign institutions left in Jidda: a bank which is closed and a cemetery which is open."[42] While these painful events are not entirely forgotten locally, Abdulaziz managed his international public relations well—unlike the current Saudi war in Yemen, the siege of Jeddah did not expose the Al Saud to harsh foreign criticism.

The new Sharif, Ali bin Hussein, repeatedly asked Britain for support—but this time it was not forthcoming. His Majesty's Government did, however, take the opportunity to clarify some additional border issues—and in September 1925 sent Sir Gilbert Clayton to negotiate. Abdulaziz was in Mecca and, since the Christian Clayton could not enter the Muslim holy city, they met at two locations between Jeddah and Mecca. The resulting Treaty of Hadda dealt with the Hejazi–Jordanian border. What the British wanted, and got at Hadda, was the long, narrow panhandle of northeastern Jordan, which connected their mandates in Palestine and Iraq. Abdulaziz effectively gave up his claims to Al Aqabah and Ma'an, which had traditionally been considered part of the Hejaz but are now in Jordan. He lost sections of the Bani Sakr and Huetat tribes but managed to keep much of the Ruala and Shararat. The Treaty of Barha dealt with border issues between the Nejd and Iraq. It further regulated tribal migrations and, for the first time, criminalized the age-old sport of cross-border raiding.[43] Although nothing was ever made explicit, it seems that the quid pro quo for these concessions was de facto British acquiescence to Al Saud conquest of the Hejaz.[44]

Medina surrendered on December 5, 1925. In order to prevent Faisal Daweesh and his Ikhwan from sacking the city, Abdulaziz sent his third son, Mohammed, to accept the surrender. On December 20, 1925 the last Hashemite Sharif of Mecca gave up the fight. He boarded the HMS Cornflower in Jeddah harbor and sailed off into history—or at least as far as Aden, where he took another ship to join his brother, Faisal, in Iraq. Three days later the Mayor or *Qaimaqam* of Jeddah, Hajji Abdullah Alireza, formally handed Abdulaziz the keys to the city gate and his own resignation. The Sultan of the Nejd accepted the keys but not the resignation, and subsequently demolished the city wall.[45] Alireza remained at his post, and his family remains one of the city's most prominent to this day.

This was not Abdulaziz's only gesture of reconciliation. Amnesty was granted to those who had supported the Hashemite government, and most civil servants maintained their positions. Funds were provided to ship the Sharif's remaining Egyptian, Syrian, and Yemeni mercenaries home, and the Hashemites were allowed to maintain ownership of their private land-holdings—very valuable real estate, much of which still belongs to Sharif Hussein's great-great-grandson, King Abdullah of Jordan.

Jeddah was, by Arabian standards, a large and very modern city. It had electricity, telephones, automobiles, and a professional civil service—none of which existed in Riyadh at that time. It had been a functioning part of the Ottoman Empire and sent delegates to the Ottoman Congress in Istanbul. Its cosmopolitan merchant class was familiar with the ideas of both Arab Nationalism and self-determination. Some of them had hoped to replace the Ottomans with an elected government. Very few welcomed the Ikhwan, whom they regarded as overly zealous bumpkins.

Abdulaziz recognized that the Hejaz was very different from the Nejd, and understood that the Hejazi notables did not wish to be incorporated into his Wahhabi domain. He agreed to administer the Hijaz and Nejd as two separate countries. He established a Consultative Council or *Majlis al-Shura* in several Hejaz cities in order to represent the views of leading families, local *ulama*, and most importantly, the merchants.[46] Because the title *imam* was closely associated with the Wahhabi movement, Abdulaziz did not use it in the Hejaz. He remained there for more than a year but, unlike in the Nejd, never married into the local elite.

Abdulaziz gradually expanded Nejdi political influence in the Hejaz. Although he continued to employ much of the existing Hashemite civil service, he appointed Nejdi police chiefs in Jeddah and Mecca. He allowed the local *ulama* to administer schools, mosques, and religious endowments, but placed members of the Al al-Sheikh family at the top of all these institutions. For example, the Nejdi scholar, Abdullah Bulayhid, became the chief judge of Mecca, while in Medina, the *imam* at the Prophet's Mosque was "replaced by two Wahhabis and a negro from Timbuktu."[47] The traditional Hejazi overcoat and small turban, known respectively as the *jubba* and *imama*, were gradually displaced by the robe and head scarf, or *thobe* and *guttra*, of the Nejd. Today, while there is no explicit prohibition against using the term Hejaz, it is seldom heard in Saudi Arabia, and is seen to contravene the official emphasis on national unity.

While Abdulaziz was asserting political control, the Ikhwan and Wahhabi *ulama* attempted to replace the Hejaz's liberal, urban values and Shafi legal code with the socially conservative Hanbali code and tribal customs of the Nejd. They sought to stamp out tobacco, alcohol, and music. They destroyed ancient shrines, which they regarded as idolatrous. They enforced prayer, fasting, *zakat*, and modest dress without gold or silk.[48] Like his grandson, Mohammed bin Salman, Abdulaziz did not tolerate dissent as he strove to consolidate political power and change social practices. Spreading "harmful ideas" or participating in anti-government meetings resulted in prison terms of two to five years.[49]

Abdulaziz was willing to use force but preferred to craft compromises between his different sets of subjects. In Mecca he placed the Ikhwan leader, Khalid ibn Lu'ay, in charge of military matters but made the former Egyptian schoolteacher, Hafiz Wahba, the civil governor. He allowed tobacco importation to please the Jeddah merchants, but banned smoking in public to placate the Ikhwan.[50] Today, this contradictory balancing act remains very much a part of the Al Saud's system which seeks compromise between liberal and conservative as well as urban and tribal values. In 2018, this unusual arrangement led the same government that issued 120,000 driving licenses to women for the first time to also arrest seven women who demanded an immediate end to guardianship policies that limited a woman's independence, and then the following year to abolish many of these gender-based restrictions.

The economy of the Hejaz was based on trade and religious tourism controlled by large merchant families and hereditary guilds. There were guilds for water carriers, butchers, jewelers, tour guides, and street sweepers. Like the Bedouin tribes, these guilds were organized, powerful, and politically independent. They had their own guild masters and ran their own courts. They legally exercised monopoly power to set prices, collect fees, share profits, and create barriers to entry. For centuries they had run the Hajj purely for their own profit. Those pilgrims who were not robbed by *Bedu* on their way to Mecca were often fleeced by the guilds once they arrived.

Unlike Sharif Hussein, Abdulaziz was determined to run a good Hajj. He brought the private guilds firmly under government control, breaking their independence just as he would eventually do with the nomadic tribes.[51] He significantly improved the health conditions and transportation services offered to pilgrims. He ended the right of Hejazi tribes, most notably the Harb, to tax passing pilgrims, and severely punished those who did not obey. Reader Bullard, wrote:

> Material conditions in Jeddah were greatly improved ... But the greatest change was in the establishment of security ... Security meant much to the pilgrims. The pilgrimage dues were still high, forming as they did one of the chief mainstays of the Saudi revenues, but at least the pilgrims now got something in return.[52]

The Dutch Consul in Jeddah, David Van der Mulen, held a slightly different view:

The first benefit Ibn Saud brought to the pilgrims was complete security. Pilgrims were no longer killed or robbed. From the third year of his accession, no major epidemic marred the pilgrimage. As Jeddah depended on condensed sea water, a second condenser was ordered from Scotland and a capable British mechanic put in charge. Ibn Saud made the Hajj a safe, healthy and easy undertaking. At the end of his life when he was no longer dependent on the income of the Hajj, he made it cheap. In doing all this he made the Hajj spiritually cheap as well.[53]

In 1926, Abdulaziz appointed his second son, Faisal, Viceroy of the Hejaz, and returned to Riyadh. The geographic unification of what is today the Kingdom of Saudi Arabia was largely complete. The political unification of the kingdom was, however, far from secure, for Abdulaziz's most perilous challenge was still to come. As would happen in Afghanistan three quarters of a century later, a tempest of religious passion and intolerance nurtured to achieve short-term goals would re-emerge to haunt the Al Saud.

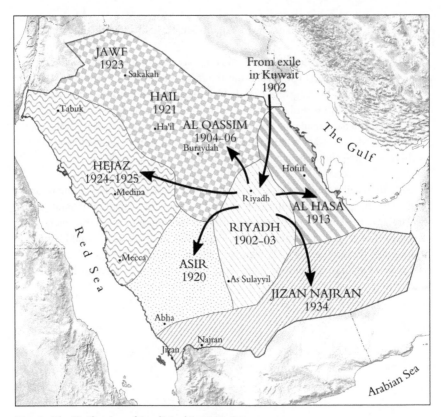

Map 2. The Unification of Saudi Arabia, 1902–34.

The Ikhwan Revolt: Reaping What You Sow

The most valued service that Abdulaziz brought to his new subjects was physical security—and that remains true of his son's government today. To regions once characterized by frequent acts of random violence, the Al Saud have brought law and order. Captain John Bagot Glubb—a British officer serving in Iraq after the First World War who later, as Glubb Pasha, commanded Jordan's Arab Legion—wrote at the time, "When the central government is strong, like the present Saudi regime, then the tribes are made to keep the peace and behave themselves, but when the strong hand is weakened or withdrawn, then merry hell breaks loose with every tribe at its neighbor's throat."[1]

The great irony of Abdulaziz's achievement was his use of the lawless Ikhwan to establish law and order. The Ikhwan enforced *sharia* law with extreme rigor, beheading for murder, cutting off a hand for theft, and flogging for immorality.[2] Their fanatical hatred of settled citizens and nomadic tribes that would not join them became a political problem for Abdulaziz, while their independent identity posed a genuine security threat.

The British observer Captain Norman Bray wrote:

Ibn Saud is uncertain of his own position and fearful of what may result ... he recreated the Ikhwan to provide himself with a means of self-defence, but the horse of religious ferment has exceeded his expectations in its power and the rider finds it harder to control than he had anticipated.[3]

Moreover, by 1926 the Ikhwan had ample reason to resent Abdulaziz and, like some of the Saudi *mujahideen* who went to Afghanistan half a century later, these original Ikhwan turned on their creator.

Having contributed significantly to Al Saud victories, the Ikhwan expected material and political rewards. Instead, they were prevented from looting and chastising heretics in the Hejaz and rode home with little sense of triumph. It became clear to their leaders—Faisal Daweesh of the Mutair, Sultan ibn Bijad of the Utaibah, and Dhaidhan ibn Hethlayn of the Ajman—that they would never be made governors of Ha'il, Mecca, or Medina. These prominent holy warriors returned to the Nejd angry and resentful. They had financial grievances as well. In the early 1920s, Abdulaziz began taxing market transactions between nomads and townspeople and—after improving the port facilities at Jubail and Al Qatif, where he collected customs duties—he prohibited Nejdi tribes from trading through Kuwait.[4]

Above all, the Ikhwan resented Abdulaziz's efforts to establish a strong central government and limit the independence of their tribes. They sought to preserve what Daweesh called the "old system": *al nizam al qadim*.[5] The Ikhwan opposed Abdulaziz's use of radio and telegraph systems, ostensibly because they were un-Islamic innovations, but in fact because they dramatically improved his communications advantage over them. They opposed the automobile as another innovation, but in reality, understood that motor transport meant the end of their livelihood as breeders of transport camels. The Ikhwan protested Abdulaziz's close relations with the British, ostensibly because he was dealing with infidels, but in fact because he sometimes allowed British-controlled Iraqi tribes to graze their herds in Saudi territory. And they repeatedly criticized Abdulaziz for the tolerance he displayed towards the Shia of the Eastern Province, whom they wanted forcibly converted to Sunni Islam or expelled from the kingdom. Abdulaziz tried to compromise. He lowered taxes and publicly destroyed his new, "un-Islamic" phonograph. He would not, however, abandon his radios or relationship with Great Britain. Nor would he expel the Shia.

The Ikhwan possessed certain tactical advantages over other tribes and military forces in Arabia. They had a strong esprit de corps based on religious zeal, as well as relatively sophisticated systems of leadership, intelligence, and logistics. They relied on simplicity of plan, speed of movement, tactical surprise, and terror to demoralize their enemies. Their innovations from traditional Bedouin raids included long marches, a desire for martyrdom, and the slaying of all males in the enemy camp—sometimes including children. Like Abdulaziz, they were fond of the surprise night attacks that the Bedouin call *al-hijad*.[6]

No longer allowed to raid in the Al Saud's realm, the Ikhwan turned their predatory attentions to Iraq, where pedestrian shepherd tribes provided easy targets for their camel-mounted forces and many of the shepherds were Shia, whom the Ikhwan despised. These were not minor incursions, they were large-scale raids involving hundreds of camelmen stealing thousands of sheep. They were also exceptionally violent. Captain Glubb was one of the few Westerners to witness these raids. He wrote:

> The advancing Ikhwan were now about one thousand yards away from us. If they had ridden straight at us they would have been on us in a few minutes, but they lost time rounding up the flocks and killing the shepherds ... It was tragic to pass through the innumerable little companies of donkeys laden with the humble belongings of the shepherds, a couple of cooking pots and a few little bags of rice and dates, a goat skin filled with water and a rolled up tent. The convoy was urged on by the women running barefoot behind the donkeys, frantic with fear carrying babies at their breast and beating the tiny donkeys, constantly glancing over their shoulders ... To be overtaken by those camelmen meant not only the loss of everything they owned in the world, but death to their fathers, husbands, brothers and sons.[7]

As such accounts make clear, similarities between the Ikhwan's economically motivated, religiously justified terrorism and that of ISIS in Iraq ninety years later are more than superficial.

Abdulaziz was not himself a zealot. H. R. P. Dickson, the British Political Agent in Bahrain in 1920 wrote, "No one who knew him intimately could accuse Ibn Saud of being a Wahhabi fanatic or ever having possessed the religious frenzy he managed to instill in his Ikhwan followers."[8] Captain Glubb wrote of him that, "If he encouraged fanaticism, it was to use it as an instrument to achieve his objective: he was never himself a fanatic."[9] Perhaps the clearest indication that Abdulaziz was no uncompromising zealot, came from the Lebanese author Ameen Rihani, who was present at the 1922 Uqair Conference and wrote that Abdulaziz kept Sir Percy Cox's tent well supplied with Johnnie Walker whisky and Havana cigars.[10]

Cigars or not, the British authorities in Iraq were becoming increasingly alarmed by these raids and Abdulaziz's refusal, or inability, to control them. They built fortified border posts and sent out armed patrols. In 1927, the Royal Air Force began bombing the Ikhwan in Saudi territory. From the air it was often difficult to distinguish the Ikhwan from peaceful tribes, and many innocent people were killed.[11] These operations were in many ways the 1920s equivalent of twenty-first century drone strikes, and unsurprisingly proved an embarrassment to Abdulaziz. Not only were the British putting his house in order for him but also, like Afghan leaders a century later, he faced angry protests when his people were killed by foreign bombs in their own homes.

Abdulaziz repeatedly condemned Ikhwan raids into Iraq but claimed to lack the public support needed to stop them. As he often did when in a weak position, he played for time. H. R. P. Dickson, by then the British political agent in Kuwait, wrote:

> It certainly seems as if old Ibn Saud has been playing his cards well ... the task of bringing Faisal Daweesh to heel has been left to the Royal Air Force while Ibn Saud has been able to make political capital out of our military incursions into his territory ... and by remaining in inaccessible places or feigning illness, he evades personal interview ... until his position becomes more favourable.[12]

To strengthen his position, Abdulaziz convened a General Assembly (*Al Jam'iayah al-Umumiyah*) in November 1928. Often referred to as the Riyadh Conference, it was something of a Saudi Estates General with 800 invited religious scholars, urban notables, tribal sheikhs, and Ikhwan leaders all seated in their respective groupings. The two most prominent Ikhwan leaders, Faisal Daweesh and Sultan ibn Bijad, did not attend—though Daweesh sent his son, Azaiyiz.[13]

The *ulama* were concerned about the telegraph, which some considered a type of sorcery. Bedouin demanded the abolition of taxes on sheep brought to town for sale. The Ikhwan wanted to know by what right Abdulaziz sat with Christians to draw lines on a map that kept them from grazing or raiding where their forefathers had for generations. Specifically, they wanted the British police post at Busaiya in Iraq dismantled. The urban notables, many of whom were merchants, were most concerned about security for trade.[14]

Abdulaziz began by reminding the assembly, "When I came to you I was weak and had only forty men. As you all know, I had no strength but God. Yet, I have made you into one people and a great people."[15] Then, in a dramatic gesture, he offered to abdicate. If people felt his governance was not in accordance with Islamic law, then they should select another member of the Al Saud family to lead them. He had no desire to rule a people who did not wish to follow him, and he would not punish any who spoke against his leadership. This was more than political theater. Each member of the assembly had once voluntarily sworn their *bay'ah*, or oath of allegiance, to Abdulaziz—an act with both legal and religious significance. He was now offering to release them from it. Abdulaziz's offer was a reminder to his audience that although he had seized power by force of arms, the foundations of his legitimacy remained his enforcement of *sharia* law and the dynastic role of his family.[16]

Abdulaziz understood that he was speaking to a group of people who distrusted each other much more than they disliked him. As he had no doubt anticipated, they demanded that he remain their *imam*.[17] The Wahhabi *ulama* feared anarchy. These scholars understood that the Wahhabi movement and their own careers depended on stable government, much as they still do today. The *ulama* spoke first and emphasized the fact that not only was Abdulaziz fit to rule but that his subjects had a religious obligation to obey him. They declared Faisal Daweesh a usurper (*baghi*), whose elimination was religiously sanctioned.[18] The merchants understood that the real choice before them was between settled, urban government or a nomadic, tribally based leadership. The Ikhwan were prepared to abandon Faisal Daweesh, provided Abdulaziz convinced the British to remove the border forts. The conference ended with Abdulaziz's eldest son, Saud, introducing each guest as they once again pledged allegiance to his father.[19]

Then Abdulaziz did something revolutionary. He dismissed the leaders of two major Ikhwan tribes. He announced that Abdulaziz Daweesh would replace Faisal Daweesh as chief of the Mutair, and Ibn Ruba'yan would replace Sultan ibn Bijad as paramount *sheikh* of the Utaibah.[20] This was unprecedented. Abdulaziz was the *imam* of the Wahhabis, just as King Salman is today. As such he was the community's supreme political and spiritual leader, but no existing tradition allowed him to depose tribal chiefs. It was not at all clear that Abdulaziz could enforce such changes, but it was very clear that he intended to limit tribal independence and create a strong central government.

In hindsight, many of Abdulaziz's guiding principles were on display at the 1928 Riyadh Conference. He assembled and consulted all the stakeholders in his realm—the tribal chiefs, religious scholars, urban notables, and the Al Saud family. He reasserted the authority of the Al Saud family above any individual family member. He reconfirmed the alliance between the Al Saud and the Wahhabi *ulama*. He explained the importance of good relations with the superpower of the day and made clear his intention to reduce tribal independence. In most instances his sons have continued to follow these policies, and in most instances the groups that he assembled still trust their king more than they trust each other.

In February 1929, the Ikhwan committed a fatal blunder. So long as they had only robbed Iraqi shepherds, it had been difficult to organize Saudi public opinion against them. Faisal Daweesh understood this. Ibn Bijad was less astute. His followers began to rob and murder local merchants traveling in the Nejd. Mohammed Asad was a Polish, Jewish journalist who had converted to Islam and become an outstanding spy for Abdulaziz. He also wrote one of the most elegant English translations of the Quran and ended his career as Pakistan's ambassador to the United Nations. He left this description of events:

> Mysterious emissaries rode on fast dromedaries from tribe to tribe. Clandestine meetings of chieftains took place at remote wells. Central and Northern Arabia became the scene of widespread guerrilla warfare as the proverbial public security of the country vanished and chaos reigned in the Nejd; bands of rebel Ikhwan swept across it in all directions attacking villages, caravans and tribes that had remained loyal to the king.[21]

This situation led to outrage among merchants and panic among farmers, who no longer felt safe from the Ikhwan—even in Al Saud territory. It was a direct challenge to Abdulaziz who claimed to have brought security to the Nejd, but it also created the groundswell of public support that he had been waiting for. As Glubb wrote, "whether or not Ibn Saud would have taken action against Ibn Bijad, the massacre of the Qasimi merchants created such an outcry amongst the townspeople of the Nejd, as almost to compel him to take action."[22] When Abdulaziz raised his standard against his long-time allies, Nejdi townsmen, the Qahtan, Shammar, and Harb tribesmen, as well as some sections of the Utaibah led by Ibn Ruba'yan and the Mutair led by Ibn Busayis, joined him. All who came were paid in gold, and many came.

Abdulaziz's war against the Ikhwan was long and difficult. The principal battle occurred on March 30, 1929 at Sabillah near Al Artawiyah, in the center of the peninsula, where the determined, battle-hardened, but badly outnumbered Ikhwan faced the Al Saud's hastily gathered army of largely inexperienced Nejdi townsmen and loyal *Bedu*. The king, dressed for battle with bandoleers strapped across his chest, commanded the Saudi center. His brother, Mohammed, and eldest son, Saud, their flanks. Senior *ulama*, Abdullah al-Anqari and Abu Habib al-Shithri, led prayers before the battle. The king then dismounted from his war horse and, in accordance with the tradition of the Prophet Mohammed, took a handful of sand and threw it in the direction of his enemy.[23]

In some ways it was a thoroughly medieval scene and in others it reflected the Al Saud's ability to move with the times. The battle lasted barely an hour for, unlike the ill-fated defenders of Dir'iyyah in 1818, Abdulaziz had made certain that he commanded superior firepower. The climax came when an Ikhwan charge ran straight into well-concealed British machine guns that Saudi merchants had obtained in Kuwait. The devastating result left scores of Ikhwan dead or wounded. Faisal Daweesh was badly wounded and allowed to depart in the expectation that he would not live long. Expecting similarly lenient treatment, Ibn Bijad accepted Abdulaziz's invitation for a meeting at nearby Shagra, where, with strains of the more recent Ritz Carlton episode, he was unceremoniously arrested and imprisoned for the rest of his life. The king's brother, Abdullah, marched to the Utaibah Ikhwan's base located at Ghahtghat between Riyadh and Mecca and razed it to the ground. Well satisfied with his efforts, Abdulaziz took a break, traveled to Unayzah and, as a mark of respect and gratitude for the town's strong support, married a particularly beautiful girl from the prominent Al Shubaily family.[24]

Four months after the battle, Abdulaziz addressed 2,000 Utaibah tribesmen at Ad Dawadimi, west of Riyadh, where he emphasized his vision for bringing stability and cohesion to the Nejd.[25] The freedom of nomadic tribes represented by the Ikhwan would be restricted and replaced by a settled, organized government. The people of the Nejd would have to accept a pacified, centralized state under Al Saud authority. The Wahhabi *ulama* would be the guardians of the state's religion and ideology. There would be respect for the *ulama* and their rulings as well as respect for the lives and property of all Muslims. Partisan, independent religious interpretations were forbidden. Those who attended seemed to agree with the program. Unfortunately, Faisal Daweesh did not attend. Instead, he recovered from his wounds and was soon back on the warpath.

Abdulaziz responded by calling his third national conference in less than a year. The meeting was held in September 1929 at Ash Sha'ra between Riyadh and Mecca, and this time it was decisive. The king sought and received firm public support for harsh retribution against Daweesh and his followers. The resolution stated that those who revolted would now be tried and punished as common criminals according to Islamic law for the deaths they had caused. Anyone giving them aid would have their property confiscated.[26] Those living in the "corrupted" Ikhwan camps would be evicted. This was the turning point in Saudi Arabian history that established the supremacy of the central government over the tribes. Those who supported the old ways were now outlaws, not holy warriors. Nothing like this had ever happened in Arabia before—and still has not happened in either Yemen or Afghanistan.

Faisal Daweesh's son, Azaiyiz, chose to disregard the resolutions from Ash Sha'ra. Instead, he led the elite of the Mutair Ikhwan on a large raid into Transjordan and Iraq. It was summer, not the best time for raiding, and Azaiyiz was inexperienced—or perhaps, just over-confident. While the more prudent Ikhwan headed back to Saudi territory through Iraq, Azaiyiz took the direct route by way of the water wells at Umm Urdhumah north of Ha'il. That was a fatal mistake. Ibn Musa'id bin Jaluwi—the Al Saud governor of Ha'il, as well as Abdulaziz's cousin and father-in-law—sent a strong force of Shammar to occupy the wells before Azaiyiz arrived. The result was a desperate struggle in which the Mutair had to take the wells or die of thirst. The famous Shammar *sheikh*, Ibn Naheer, and many others were killed in the fighting but the Ikhwan could not take the wells. Some 500 Mutair Ikhwan, including Azaiyiz, died fighting or of thirst. The remaining 250 surrendered and were promptly beheaded as murderers by Ibn Musa'id.[27] In terms of total fatalities, Umm Urdhumah was a greater defeat for the Ikhwan than Sabillah had been.

In preparation for a final showdown with the Ikhwan, Abdulaziz made some highly innovative plans. He ordered four DH9 British, biplane bombers and recruited a number of British pilots to fly them. He initiated plans for a Marconi radio network to connect all his provincial capitals.[28] He had noted how effective British-armored cars had been against the Ikhwan in Iraq, but he had no armored cars of his own. In fact, the only cars of any kind in his realm were in the Hejaz, nearly 1,000 miles from the Kuwaiti–Iraqi border region where the last Ikhwan bands had gathered. So Abdulaziz organized a collection of nearly 200 automobiles and had them driven across Arabia, through deserts without roads and through villages that had never seen a wheeled vehicle of any type. Those that arrived were stripped down and mounted with machine guns.[29] These became the decisive weapon in his last confrontation with the Ikhwan, for they far outpaced and outgunned even the fastest, best armed camelmen.

In late 1929, Abdulaziz's increasingly modern army took to the field against the remaining Ikhwan. For the first time, the Saudi Army deployed Bedouin fighting in multi-tribe units rather than single-tribe formations. These units were the origin of today's Saudi Arabian National Guard. As Abdulaziz advanced towards the Mutair heartland near Kuwait, the Awazim, Bani Khalid, Qahtan, Sahul, and Dawaisir tribes joined his forces. Ibn Musa'id's force of loyal Shammar and Harb again moved north from his headquarters at Ha'il towards the Transjordan border. The redoubtable Khalid ibn Lu'ay, who had helped to destroy the Hashemite Army at Turabah in 1919, left his base at Al Khurmah with the loyal Utaibah to seal the Western Nejd. It was a well-coordinated campaign.

The demise of the Arabian Ikhwan resembled that of the American Plains Indians, for these two nomadic, tribal, warrior cultures had much in common. Muhsin al-Firm and his Harb tribesmen defeated the increasingly desperate Faisal Daweesh at Hafr Al Batin in northeastern Saudi Arabia. Like the Sioux fleeing the United States Cavalry into Canada, small bands of Ikhwan sought to escape into British-controlled Iraq. Unlike the more welcoming Canadians, however, the British mandatory authorities in Iraq and Jordan prevented the Hashemite kings from assisting the Ikhwan, while British aircraft chased them out of Kuwait.[30] Trapped between British and Saudi forces, the Ikhwan were defeated in numerous small engagements. Wherever he went, Abdulaziz confiscated their camel herds, horses, and arms, which he divided among his loyal troops. The king's sixteen-year-old son, Khalid (1913–1982), participated in this campaign—and was the last future king of Saudi Arabia to have actually fought in the Wars of Unification.

Faisal Daweesh with the last bedraggled, starving remnants of the once-feared Mutair Ikhwan, surrendered to the British north of Kuwait on January 10, 1930. The British refused to grant the last Ikhwan leader political asylum and instead flew him back to Saudi Arabia, where he was immediately arrested. Some say Faisal Daweesh died in prison the following year. Others say he was executed before he ever got there. Whatever the truth, with him died the last significant tribal threat to the Al Saud. The aggressive independence of the Nejdi tribes was irreversibly broken and an important step in nation-building had been taken. In reaching his goal, Abdulaziz had consistently observed the lessons learned from the First Saudi State's demise: cooperate with the superpower of the day and obtain modern military technology.

Al Saud rule had been imposed on the Hejaz by force and, not surprisingly, led to resentment.[31] Some local notables had hoped to create an independent Hejazi state when the last Hashemites fled Jeddah. Those who championed European political ideas had been disappointed when the elected consultative councils that Abdulaziz created turned out to lack meaningful authority and lasted for only two years. Others resented Abdulaziz's management of the Hajj—particularly, the limits that he placed on service charges paid by pilgrims. Many simply disliked the Al Saud, their Bedouin army, and Wahhabi clerics. The most discontented formed the Hejaz Liberation Movement.

Led by several members of the Dabbagh family, one of whom had been the Hashemite finance minister, this new political party wanted to free the Muslim holy places from Wahhabi "infidels" and establish representative government in the Hejaz. The Dabbaghs sought funding from King Fuad in Egypt and Amir Abdullah in Transjordan. In the northern Hejaz along the Jordanian border they encouraged armed rebellion by the Billi tribe, who had never been Wahhabis or Ikhwan and who resented the centralizing efforts of Abdulaziz.[32]

By this time, the Third Saudi State had become very much part of British strategic thinking for the Middle East. So, in another significant sign of support for Abdulaziz, the British government instructed Amir Abdullah of Transjordan not to assist the Billi rebellion. British units sealed the Transjordanian border and British warships arrived at Al Aqabah to prevent the rebels' resupply. When they received no local support, the Billi, led by Ibn Rifada, hid in the hills behind Dhuba, a small Red Sea port north of Yanbu. Abdulaziz lured them down with a letter from local leaders promising their support. When Ibn Rifada descended, he fell into a trap, much as Ibn Bijad had done after Sabillah, and was soundly defeated.[33]

In the southern Asir, along the Yemeni border, the Dabbagh brothers encouraged a second rebellion by Hasan al-Idris, who had governed a semi-independent protectorate there for nearly a decade and now sought to regain full independence. In response, Abdulaziz sent Khalid ibn Lu'ay on his last major campaign to subdue the southern Asir. Ibn Lu'ay brought Hasan al-Idris back to Mecca where he lived out his life as a "guest" with a pension, while Abdullah Dabbagh fled to Hashemite-ruled Iraq.[34]

In 1935, with the Ikhwan and Hejaz rebellions over, Abdulaziz offered a general amnesty to those who had opposed him. Majid bin Khathila had been one of Ibn Bijad's deputy commanders and now helped to found the Saudi National Guard. The Shaalan family maintained its prominence in Al Jawf despite the fact that some members had joined the Ikhwan rebellion.[35] Abdullah Dabbagh eventually returned from Iraq. The Mutair were famous for their herd of specially bred black camels, known as *As-Shurf* or honored ones, which they drove before them in battle. Abdulaziz had confiscated the herd and given it to his third son, Mohammed. Now he returned some of the camels to their former owners and Faisal al-Shiblan, a Mutair Ikhwan commander, rose to become the keeper of all the king's camels. The widows of some Ikhwan began to receive monthly pensions.

Members of the Daweesh family continued to govern Al Artawiyah. Faisal Daweesh's surviving sons were treated as respected tribal leaders, and some became National Guard officers. Faisal Daweesh's son Azaiyiz left two widows; Abdulaziz married one and his brother, Abdullah, married the other.[36] Abdulaziz's sons, including the future King Salman, also married into the families of former Ikhwan leaders. Thus, Abdulaziz established reconciliation with those willing to submit as an important principle of Al Saud rule—a principle that his sons have put into practice over time with their Nasserite, Shia, and al-Qaeda opponents.

Deciding what not to do is a key element of any successful strategy. In business, many new firms fail when they allow ambitious expansion plans to outstrip their resources. The same can be true in politics. The First Saudi State (1744–1818) overreached and collapsed. Abdulaziz did not repeat that mistake, he knew when to stop. He eventually prevented Ikhwan raids into British-controlled Jordan and Iraq. He never seriously threatened the small Gulf sheikhdoms that Britain protected, and, perhaps most importantly, he retreated from Yemen.

Abdulaziz incorporated Jizan Province, just north of Yemen, into his realm in 1930. This led to border tensions with Imam Yahiya in Sanaa, which were exacerbated by differing religious doctrines. Yahiya was a Zaydi, and while the differences in dogma between this moderate Shia sect and the Wahhabis were relatively minor, each group considered the other to be heretics.[37] Although armed conflict over charges of heresy may seem extraordinary to the modern Westerner, it was common enough in Europe before the Peace of Westphalia in 1648. In 1932, these existing religious tensions and territorial ambitions turned into armed conflict when Imam Yahiya occupied the great oasis of Najran, which lay across important trade routes from Yemen into the southern Nejd.

After several failed attempts to negotiate a settlement, Abdulaziz invaded Yemen. One Saudi column led by his eldest son, Prince Saud, captured Najran and advanced to Sa'dah, the center of today's Houthi movement. Facing tremendous difficulties with mountainous terrain and tribesmen, he subsequently had no more success than the Roman General Gallus had had 2,000 years earlier or the Royal Saudi Air Force would have eighty years later. A second column led by the second son, Prince Faisal, was more successful. Using motor transport and modern weapons paid for with a loan from the newly arrived Standard Oil of California (today's Chevron), Faisal advanced rapidly down the flat Red Sea coast.[38] The Yemeni coastal tribes—notably, the Zaraniq—are Shafi Sunnis and were happy to join the war against the Zaydi Shia. They facilitated the surrender of the coastal city of Hodeidah without a fight.[39] Meanwhile, Abdulaziz remained in Mecca coordinating the campaign through his newly installed radio network.[40]

Abdulaziz now had to make a choice. Yemen was poor, mountainous, and far away. It had a large population, many of whom were Shia. Its people were tribal but mostly settled farmers, not wandering herdsmen. A worried Abdulaziz asked H. St. John Philby, "Where will I get the manpower to govern Yemen?"[41] Moreover, neither Britain nor Italy was enthusiastic about Saudi expansion into Yemen. Britain had interests in Aden and Italy in the Horn of Africa. Both nations had treaties of friendship with Yemen. The British government advised Abdulaziz that staying in Yemen could risk war with Italy and that London would not support him if it did.[42]

For the first and only time in his career, Abdulaziz ordered a strategic retreat, pulling out of Hodeidah and Sa'dah. Although he withdrew from large parts of what is today northern Yemen, the 1934 Treaty of Taif confirmed Saudi control of Najran and Jizan provinces. Across the border, many Yemenis continue to harbor irredentist claims on this Saudi territory, which they sometimes refer to as Historic Yemen.

In 1932, Abdulaziz united the Kingdom of Hejaz and the Sultanate of Nejd, to create the Kingdom of Saudi Arabia. It is this date that the Saudi government uses to mark the founding of the nation.[43] Ultimately, political skill, military success, and British support had enabled Abdulaziz to eliminate the independence of numerous tribes, local dynasties, and commercial guilds. He offered his subjects physical security and, by the standards of the time in Arabia, competent administration. Each success was further proof that his regime was durable and that the Turks, Al Rasheed, Hashemites, and

Ikhwan were not coming back. Acquiescence to the House of Saud became the wisest choice, and for nearly a century Abdulaziz's legacy contributed to the monarchy's stability.

In 2020, that legacy has lost some of its luster. Like the poverty and insecurity that their grandparents once knew, Abdulaziz's nation-building exploits are a fading memory for the youthful 50 percent of Saudis who are under twenty-five. Likewise, some of Abdulaziz's guiding principles have been ignored. King Salman has clearly used mild force against some members of the royal family and dispensed with the seniority system that long guided family politics. Foreign public opinion, which Abdulaziz carefully monitored and courted, has in more recent times turned against the House of Saud.

Although the kingdom's diverse tribes and regions have learned to coexist, they have not yet fully assimilated into one Saudi people. Resentment lingers among urban Hejazis, Eastern Province Shia, and some tribal people over their involuntary incorporation into the Third Saudi State. The unelected Saudi government must constantly balance and accommodate the competing interests of different stakeholders in the coalition Abdulaziz built, but some of these interests are now being undermined by necessary social and economic reforms. Ultimately, his father's historic role in creating Saudi Arabia contributes to King Salman's legitimacy—but Abdulaziz's achievements can no longer guarantee popular acquiescence to the Saudi monarchy.

Figure 1. Al Saud forces on the march in 1911. Like the Saudi flag today, their banner bears the words "There is only one God and Mohammed is his prophet".

Figure 2. The town of Hofuf in Al Hasa or the Eastern Province which Abdulaziz took from the Ottoman Turks in 1913.

Figure 3. Downtown Riyadh in 1914. A walled city of 30,000, photographed by the ill-fated British political agent to Kuwait Captain William Shakespear.

Figure 4. Abdulaziz's 1916 meeting with Sir Percy Cox, the British Political Resident in the Persian Gulf, and Gertrude Bell at Basra, Iraq.

Figure 5. An Ikhwan warrior wearing the group's distinctive white turban that came to signify terror across the Arabian Peninsula.

Figure 6. Sharif Hussein ibn Ali al-Hashimi (1853-1931) who lost control of Mecca to Abdulaziz in 1924.

Figure 7. The fort at Ad Dawadimi between Riyadh and Mecca in 1935. Like other forts Abdulaziz built, it housed troops to patrol the roads as well as water, food, and fuel for travelers and a radio station for his communication network.

Figure 8. A bedouin encampment in 1937, when most of Saudi Arabia's population was still nomadic.

Figure 9. An early Aramco drilling operation. The 1938 initial discovery of 20 percent of the planet's oil reserves greatly accelerated the end of a way of life that had changed little for centuries.

Part II

Managing Succession

King Abdulaziz ruled for fifty-one years and saw his realm transformed from a remote, desert chieftaincy into a founding member of the United Nations. Yet he created no governing institutions beyond himself. His rule was personal, built purely on relationships with tribal, religious, and business leaders. Although he selected his immediate successor, he also left behind a family rife with rivalries.

In 1936, the British political agent in Kuwait, H. R. P. Dickson, wrote a paper entitled, "The Future of Saudi Arabia," in which he predicted that "when Ibn Saud disappears, Arabia will probably collapse into chaos."[1] He was not alone in anticipating that succession would prove to be the Al Saud's Achilles' heel. Of the three principal dangers facing the Saudi monarchy—foreign invasion, domestic unrest, and intra-family conflict—the third has often proved the most difficult to avoid.

The Al Saud are a large family with many thousands of members, essentially a hereditary political party in a one-party state. Different family sections compete vigorously for a limited number of senior positions and finite financial resources, as the cone of privilege narrows with each passing generation. Like all dynasties, the Al Saud are most vulnerable to internal conflict at times of succession. As was mentioned in Chapter 1, succession disputes between the sons of Imam Faisal bin Turki led to the collapse of the Second Saudi State in 1891.

The Third Saudi State has been more successful at managing succession. Over the past sixty years, the Al Saud have removed an incompetent king; dealt with the assassination of another; and weathered the long, debilitating illness of a third. Since the abdication of King Saud in 1964, the sons of King Abdulaziz have transferred political power four times without violence or public protest: to Khalid in 1975, Fahd in 1982, Abdullah in 2005, and Salman in 2015. This is a much better record than many of the Arab World's so-called republics. In a region where violent coups and revolutions have been more common than orderly political transitions, the Al Saud's consistent ability to transfer power swiftly and peacefully has contributed to their legitimacy.

King Salman now faces the tremendous challenge of assuring that this pattern of orderly succession continues when a grandson of Abdulaziz becomes king. He has a plan that seems to be working, but his replacement of two respected and experienced crown princes with his much younger and less experienced son was unsettling to many in the royal family. King Salman has not broken any princely heads in his succession planning, but he has changed the rules and bruised many royal egos. This section of the book examines how the Al Saud have managed succession and, more importantly, assesses whether King Salman is likely to maintain their record of success.

Kings Saud and Faisal: A House Divided

According to tribal tradition, on the Arabian Peninsula, power passes from one ruler to another within a particular family, but not necessarily from father to eldest son. Primogeniture, while common, is not automatic. Uncles, nephews, brothers, and younger sons are all legitimate contenders. In the First Saudi State (1744–1818) the title *"Imam"* as the Al Saud rulers were originally known, was transferred from father to son four times until Abdullah bin Saud was executed by the Ottomans.[1] The Second Saudi State (1824–1891) had a much more confused pattern of succession, with most of its leadership changes involving assassinations or coups and eventually, civil war. At the time, such chaos and bloodshed was common among the settled communities of the Nejd, where many ruling families failed to develop reliable mechanisms for orderly succession.

King Abdulaziz was well aware of this problem. He knew that family conflict had destroyed the Second Saudi State and contributed to the collapse of his great rivals, the Al Rasheed of Ha'il. He was determined to avoid dynastic suicide in the Third Saudi State (1902 to the present). Although Abdulaziz did very little to institutionalize government administration, the system that he established for an orderly transfer of royal power was nearly as important to the monarchy's future as his unification of the kingdom or the discovery of oil.

Abdulaziz's only full brother, Saad, died fighting for him in the Wars of Unification, but he had six half-brothers as well as many uncles, cousins, and nephews who were all legitimate contenders for his throne. In addition, there were numerous cadet branches of the Al Saud family, all with various claims to power, including the Saud Al Kabeer, Bin Jaluwi, Al Thunayyan, and Al Farhan.[2] King Abdulaziz respected, consulted, and depended on all of the Al Saud, but his first strategic choice regarding succession was that power would transfer to his sons rather than his brothers or cousins. From the outset, he intended to marginalize all but his direct descendants, much as King Salman appears to be doing today.[3]

In the Treaty of Darin, where the British government first recognized Abdulaziz and his sons as rulers of the Nejd, "sons" was very clearly used in the plural. The treaty, which was signed in 1915, also stated that the living ruler would select his successor from Abdulaziz's "descendants."[4] Abdulaziz carefully studied and revised several versions of the document. Even at this early date, he clearly intended power to be transferred first between his sons and then among his grandsons. Had he envisioned a system of primogeniture, one that included all members of the House of Saud, or a collective selection process, the treaty would have been drafted very differently.

Abdulaziz confirmed this interpretation of the treaty in 1933 when, shortly after unifying the Nejd and Hejaz into one kingdom, he designated his eldest son, Saud, the crown prince. Al Saud princes, religious scholars, tribal leaders, and urban notables—the *ummara, ulama, sheikhs,* and *aiyan*—all swore allegiance to the new crown prince. This was a significant event that established the monarchical principle for a unified kingdom. Only Abdulaziz's half-brother, Mohammed, would not swear allegiance to Crown Prince Saud. Mohammed did not dispute his older brother's right to rule, but he had been part of the band that took Riyadh in 1902 and had fought in many campaigns during the Wars of Unification. Mohammed felt that he or his eldest son, Khalid, had a legitimate right to be considered Abdulaziz's successor. It was the future King Faisal who finally persuaded his uncle to pledge allegiance to Crown Prince Saud—thus playing a role very similar to the one his own son, Khalid al-Faisal, would play sixty years later in resolving another succession dispute.[5]

Abdulaziz's second strategic choice regarding succession was that, unlike Alexander or Charlemagne, his realm would not be divided between his generals or sons. Determined to preserve the unity of the kingdom he had created, Abdulaziz rejected any thought that his three eldest sons would partition Saudi Arabia, with Faisal ruling an independent Hejaz, Saud ruling the Nejd, and Mohammed becoming king of the Eastern Province. Only preserving the monarchy itself remains more important to the ruling family than maintaining the unity of their kingdom. The Al Saud firmly believe, as do many of their subjects, that should they ever lose power, the dismemberment of the nation they created would not be far behind.

Abdulaziz had long harbored reservations about his eldest son's ability but would not pass him over. Tall and handsome, Crown Prince Saud looked like his father. He had been born on the very day his father captured Riyadh to a mother from the Arayar clan, hereditary chiefs of the Bani Khalid tribe. Saud had lived most of his life in the Nejd and, like his father, married into many prominent urban and noble tribal families. The crown prince frequently traveled and hunted with the tribes. He generously supported them because—again, like his father—he believed that tribal loyalty would be essential to maintaining stability during any future succession.[6]

Saud was also devoted to his father. In 1936, when a knife-wielding Yemeni assassin attacked King Abdulaziz in Mecca, the crown prince spontaneously stepped in front of his father and took the blade in his own shoulder. Many believe it was this incident that compelled Abdulaziz not to sideline Saud for his more able half-brother, Faisal.[7]

After the Second World War, Abdulaziz's health began to fail. His old war wounds, some of which had been very serious, sapped his strength. He could not walk easily and his eyes gave him trouble. Yet he saw clearly enough that his two eldest sons, Saud (1902–1969) and Faisal (1906–1975), did not get along. The tension between them was long-standing and their father tried to ease it by insisting that Saud's eldest son, Fahd, marry Faisal's daughter, Anud. More importantly, he made the brothers swear upon the Quran seven times that they would not quarrel and that Saud's sons would not inherit the throne. Faisal would become king and eventually select his own crown prince from among his brothers, not his own sons. Then Abdulaziz reportedly told Saud, "Faisal is smarter than you are, you should take his advice."[8]

The aged king could have handed power to his eldest son and left it at that. Instead, he clarified his long-term intentions. His decision to transfer power among his sons was a strategic choice and something of an innovation. This inclusive approach contributed significantly to family unity and regime stability for the next sixty-five years. It gave more people a stake in the system—far more, in fact, than the very limited number of royal players in the failed monarchies of Iraq, Iran, Egypt, or Libya. It would, for instance, have been very difficult to murder all of Abdulaziz's sons in one night, as happened in 1958 to the Hashemite ruling family in Baghdad, or to ship them all to Egypt, as happened to the Shah of Iran in 1979.

By the summer of 1952, Abdulaziz was too ill to make the Hajj, and it was only then that he began to hand significant responsibility to Crown Prince Saud. He sent Saud to manage the Hajj in Mecca, where he stayed after the pilgrimage, reorganizing the provincial government that Faisal, as Viceroy, had managed for years. In August 1953, Abdulaziz finally transferred real authority by placing Saud in charge of the armed forces and police.[9] Then in October 1953—in a belated effort to establish some sort of institutional structure that could replace his fading personal rule, and in what was to be his last decree—Abdulaziz created the Council of Ministers to serve as a cabinet. He appointed Crown Prince Saud to be the first prime minister or, as the position is known in Saudi Arabia, President of the Council of Ministers.

Some argue that King Abdulaziz never gave Crown Prince Saud enough authority to prepare him to lead. Others believe that the steps he finally took to reshape his government to function without him were simply too little, too late.[10] As King Salman makes preparations for the next generational transition within the House of Saud, he has not repeated those mistakes. King Salman made his son, Mohammed, minister of defense and then appointed him chairman of a newly created Council of Economic and Development Affairs, which controlled all the civilian elements of government. He placed him in charge of Saudi Arabia's sovereign wealth fund and eventually appointed Mohammed bin Salman deputy crown prince, and then crown prince. Unlike his own father, King Salman has given his son ample opportunity to succeed or fail and learn before he finally ascends the throne.

King Abdulaziz passed away at Taif, near Mecca, on November 9, 1953. It is said that he listened to a blind Quran reciter until just before his heart stopped beating.[11] Of his thirty-four living sons, only Faisal was with him at the end. When Saud arrived from Jeddah, Faisal rose to greet him and immediately swore allegiance to his older brother as the new king. Faisal then took a ring off their dead father's finger and presented it to his brother in a gesture of loyalty.[12] As his father had instructed him to do, King Saud publicly proclaimed Faisal his crown prince and confirmed his appointment as Viceroy to the Hejaz. King Saud remained prime minister; Crown Prince Faisal became deputy prime minister and established the link between those two positions that continues today.

More than 100 princes soon filed past the two brothers to give the traditional oath of allegiance, "We pledge our allegiance to you on the Book of God and the traditions of God's Messenger."[13] King Abdulaziz's body was flown to Riyadh and, in accordance with Wahhabi tradition and Mohammed Abd al-Wahhab's strict interpretation of monotheism, was buried in an unmarked grave so there would be no shrines, pilgrimages, or praying there. "Today one must ask of the desert winds and the cold Arabian stars where to find his resting place."[14]

King Abdulaziz had ruled a feudal society with a pre-industrial economy. Although he was a master of domestic politics, neither he nor his finance minister, Abdullah Suleiman, had the financial background needed to deal with the nation's growing oil wealth. In just one year, from 1950–1, Saudi oil revenue doubled from $56 million to $110 million. The national budget, which had been only $7 million in 1947, rose to $205 million in 1952.[15] Having been hard pressed for cash most of his life, Abdulaziz now spent generously. Treating the state treasury as his private purse, he often approved development projects without understanding their full cost—the railroad from Dhahran to Riyadh being a prime example. Saudi Arabia had been facing financial difficulties well before Saud came to the throne, but the new king quickly made matters much worse.[16]

King Saud was very fond of women and luxury. He had 108 children and a particular weakness for Cointreau, which he consumed in large quantities. Enclosed by a seven-mile-long pink wall, his Nasariya Palace outside Riyadh was staffed by more than 1,000 servants. Lit with massive crystal chandeliers and neon lights flashing verses from the Quran, it had a geyser-like fountain illuminated by shifting colored lights, a swimming pool scented with Chanel No. 5, and was said to consume half the city's electricity.[17] King Saud not only lived lavishly himself, but tolerated widespread graft and corruption among his retainers. His chief of staff was his former chauffeur who allowed the value of construction contracts to be greatly inflated. Bills often went unpaid, as those who should have handled them pocketed the money instead. Palace fittings were "lost" and then sold back to the court chamberlain. Even food disappeared from palace kitchens, only to be resold to the cooks.[18] The British explorer, H. St. John Philby, wrote to his wife, "It is sad to see such widespread decadence in a country that had a greatness of its own when it was poor."

Five years into King Saud's reign, the country was facing bankruptcy. Civil servants, soldiers, and contractors were not being paid. The American-owned oil company Aramco refused to make additional loans against future Saudi production.[19] The Saudi Riyal was devalued by 50 percent, inflation soared, and there was labor unrest in the Eastern Province. By 1957, the kingdom was forced to seek a loan from the International Monetary Fund, which insisted on seeing the country's first detailed budget. That financial plan sharply reduced royal family living expenses. Over the next six years privy-purse expenditures fell by two thirds. In Riyadh, many half-finished palaces were abandoned, and infuriated princes correctly blamed their distress on King Saud's financial mismanagement.[20]

Meanwhile, after nationalizing the Suez Canal in 1956, Egypt's President Gamal Abdul Nasser became the first Pan-Arab hero since Saladin (1138–1193). Many Arab military officers idolized the former Colonel as the man who had finally stood up to Britain. Having already overthrown Egypt's King Farouk, Nasser now encouraged other military officers to do likewise. Eventually those in Iraq, Syria, Libya, Yemen, and the Sudan followed his advice and overthrew their monarchs.

Nasser's secular, Arab Nationalist, Arab Socialist message was beamed to a vast audience across the Middle East by Cairo's Voice of the Arabs. The station gleefully broadcast stories of King Saud's extravagance and womanizing. It called all kings obstacles to Arab progress. Like Qatar's Al Jazeera news network four decades later, the radio station became a thorn in the side of the House of Saud. The Al Saud felt justifiably threatened by Nasser and the small class of Western-educated Saudi intellectuals who began calling for elections, a written constitution, labor unions, a free press, and closing the American airbase at Dhahran.

King Saud did not respond effectively to the very real threat that Arab Nationalism and its anti-Western sentiments posed for the Saudi monarchy. Instead, in 1957 he became the first Saudi king to visit the United States. When he landed in New York, Mayor Robert Wagner refused to greet him saying, "This fellow comes from a country where slavery is legal, in the Air Force you cannot have any Jewish boys and a Catholic priest can't say mass"[21]—all of which was of course true. The Saudis had an image problem long before oil embargoes, 9/11, or the Khashoggi affair, but they also had a pragmatic friend in the White House.

In Washington, President Dwight Eisenhower was more concerned with America's strategic interests in the Cold War than with political point scoring. Dhahran was the only US Air Force base in the region capable of supporting strategic B-29 bombers, and had thus become an important Cold War asset on the southern flank of the Soviet Union. In 1957, Dhahran was every bit as important to American security as the sprawling Al Udeid Air Base in Qatar is today. So, instead of shunning King Saud, President Eisenhower met him on the tarmac at National Airport, something he had never before done for any foreign leader. Eisenhower then arranged for the king's route from the airport to be lined with military troops and bands. In return for a large American loan and additional military training, King Saud renewed the Dhahran basing rights free of charge.

President Nasser loudly denounced King Saud's visit to Washington as more evidence of the Al Saud's reliance on Western imperialists.[22] Then in March 1958, the Arab world was shocked when Egypt's official newspaper, *Al Ahram*, reported a failed Saudi plot to assassinate the Arab world's most popular leader. The paper claimed that King Saud had paid a Syrian intelligence officer £1.9 million as the down payment on another £20 million if he could blow up Nasser's aircraft. Although these charges were never proven conclusively, the Syrians held a press conference at which they produced canceled checks and an incriminating letter from King Saud's private office to support the charge.

Other Arab papers began to denounce King Saud's repression of Aramco workers and his support for the reactionary Hashemite kings of Jordan and Iraq. In Damascus, the daily *Al Nasr* somehow got hold of the confidential Dhahran basing agreement, which it published as proof that the Saudi monarchy "opposed the interests of Arab Nationalism."[23] Nasser began to openly call for the overthrow of the Al Saud. In Riyadh, King Saud's brothers became convinced that his foreign policy bungling, combined with his economic mismanagement, was putting their family at risk.

The elder brothers agreed that Saud should keep his throne but relinquish all executive authority to Faisal. King Saud accepted this arrangement in March 1958. Crown Prince Faisal became prime minister, appointed himself finance minister, and began to balance the kingdom's budget. He cut spending across the board, suspended development projects, canceled agriculture subsidies, delayed payments to contractors and tribal *sheikhs*, imposed import controls on luxury goods, and devalued the riyal. He reduced stipends for royal family members and obtained new loans from Aramco as well as leading merchants, including Osama bin Laden's father Mohammed.[24] At the same time, oil production increased by more than 50 percent from 1 million barrels a day in 1957 to 1.6 million barrels a day in 1962.[25] The kingdom's budget was balanced and its currency stabilized. The inflation rate fell sharply. By 1960, Faisal's austerity had reduced not only the national debt, but also his own popularity with the tribes, merchants, and princes.

King Saud never intended to quietly fade away. Now he saw a chance to mobilize discontented clerics, merchants, tribal leaders, and some mid-ranking princes. He began blocking Faisal's appointments for judges and governors. In November 1960, King Saud refused to approve Prime Minister Faisal's proposed budget. A frustrated Faisal submitted a letter stating that "As I am unable to continue, I shall cease to use the powers vested in me as from tonight."[26] Faisal did not use the word resign, but that was how Saud chose to read it. King Saud used Faisal's letter as the pretext to reclaim his role as prime minister, and then created a new cabinet in an alliance with the so-called free princes.

The free princes were a group of King Abdulaziz's younger sons led by Prince Talal, who felt politically marginalized and claimed to support a constitutional monarchy with an elected parliament and supreme court. In King Saud's new cabinet, Prince Talal became finance minister while another free prince, Abd al-Mushin, assumed control of the critical Ministry of the Interior. Saud's own son, Mohammed, became defense minister. There was a distinctly liberal flavor to this cabinet, particularly among the technocrats, which did not reflect the conservative king's views so much as those of the new allies whom he thought he needed.

Instead of helping King Saud regain authority, his left-leaning cabinet further alienated many traditional stakeholders in the Al Saud's coalition. The *ulama* feared that a more liberal, secular, nationalist agenda would undermine their role in society. Having seen what had happened to property owners in Nasser's Egypt, the Nejdi *aiyan* and Hejazi merchant class had no use for Arab socialism. In Washington, the Kennedy administration became alarmed when King Saud announced that he would not renew the lease for the Dhahran airbase.

King Saud eventually recognized that this liberal cabinet was costing him more support than it produced, and that the free princes had political aspirations of their own. In September 1961, he fired them and appointed more of his own sons to senior positions.[27] The hapless king might have recovered his balance with this new cabinet had his own health not deteriorated as events in Yemen spun out of control.

When Yemen's Imam Ahmad bin Yahiya (1881–1962) died in his sleep, Republican military officers quickly sought to overthrow the ancient, religiously-based Hamid al-Din dynasty. The Republicans claimed that Saudi Arabia had unjustly seized the Jizan and Najran provinces from Yemen in 1934 and demanded their return. King Saud's government rejected that claim and supported the Yemeni royalists with arms, money, and subsidies to cooperative tribes. Egypt's President Nasser—who supported the socialist, Arab Nationalist Republicans—hoped to add Yemen to the United Arab

Republic that he had created with Syria, and to use the country to overthrow the House of Saud.[28] Today, King Salman fears that Iran has similar intentions in Yemen.

Saudi Arabia soon found itself engaged in a proxy war with Egypt, which eventually sent 70,000 troops to Yemen, including one Colonel Anwar Sadat. Some 1,000 Soviet technicians maintained Egyptian aircraft, which dropped napalm and, on at least two occasions, poisonous gas on Yemeni civilians.[29] Egyptian aircraft began bombing the Saudi cities of Jizan, Khamis Mushayt, and Abha. The damage was significant, and in one Egyptian air raid thirty-six patients in an Abha hospital were killed.[30] The situation called for strong leadership, but that was something that King Saud could not provide.

Having abandoned the free princes, dismissed the liberal technocrats, frightened the *ulama*, alienated merchants, and cooled relations with the United States, King Saud had few friends left. In October 1962, the Al Saud family and the *ulama* again pressured him into accepting the return of Crown Prince Faisal as prime minister. Faisal immediately removed Saud's sons from the cabinet and installed the team of brothers and half-brothers that would govern Saudi Arabia for the next fifty years. Faisal's half-brothers—Khalid, Fahd, Sultan, Abdullah, and Salman—became, respectively, deputy prime minister, minister of the interior, minister of defense, commander of the National Guard, and governor of Riyadh. All but Sultan would eventually become kings. All but Salman have now passed away.

In March 1964, King Saud's health had improved, and he wrote to Prime Minister Faisal demanding the full restoration of his authority. Neither Faisal nor the Al Saud family would agree. Instead, the third eldest brother, Mohammed bin Abdulaziz, marshaled family support for a religious ruling or *fatwa* that would permanently reduce King Saud to a respected head of state. Faisal would remain prime minister and no longer need to consult King Saud on any internal or external matters. King Abdulaziz's two surviving brothers, Abdullah and Musa'id, took the idea to the *ulama*, who issued their *fatwa* on March 29, 1964. The following day seventy-two princes met at the home of Prince Mohammed bin Abdulaziz and endorsed the *fatwa*. The Cabinet issued a supporting resolution, transferred control of the Royal Guard to the Ministry of Defense and cut King Saud's income in half.[31]

In October 1964, King Saud tried one last time to regain power by formally asking the *ulama* to reconsider their March *fatwa*. The king argued that it was unnatural, against tradition, and wrong to have a king with no authority. By making that claim, King Saud effectively brought about his own deposition. Instead of returning Saud to power, his senior brothers agreed with him that the current situation was unnatural. Princes Fahd and Mohammed led the effort within the family to force Saud's abdication. Future kings Khalid and Abdullah, with their own strong tribal connections, sought agreement from the tribal *sheikhs*.[32]

Eventually, the *ulama* met at the home of the Grand Mufti Mohammed Ibrahim Al al-Sheikh, while the most senior princes met at the home of Prince Khalid bin Abdulaziz to consider the matter. More than 100 princes including sons, grandsons, brothers, and nephews of King Abdulaziz met with several dozen religious leaders at Riyadh's Sahara Palace Hotel. A consensus was reached that King Saud should be removed and Faisal declared king. The *ulama* drew up an appropriate *fatwa*.

All of the Saudi elite and much of the general public had sworn an oath of personal allegiance or *bay'ah* to King Saud. This was a serious matter that, like swearing an oath in court, had both legal and religious significance. It could be annulled only by a senior religious figure such as the grand mufti, and this is not something that he would undertake lightly. Mohammed Ibrahim was very hesitant to annul the *bay'ah* given to King Saud. He did so only when the family's senior member, Abdullah bin Abd al-Rahman, told him clearly that King Saud was going to be removed. Only then did the grand mufti become convinced that the public order that he so highly valued, not to mention his own career, would best be served by offering religious justification for a settled political decision. The blind grand mufti prepared a tape recording welcoming the new king, but not all of the senior Wahhabi clerics, scholars, and judges agreed with him. Some who did not were detained in Riyadh's Yamamah Hotel until the *fatwa* annulling the *bay'ah* given to King Saud had been broadcast on the radio and the abdication was complete.[33]

Long in coming, the succession went smoothly. King Saud abdicated formally and peacefully in November 1964. With the Minister of Defense Prince Sultan and the Commander of the National Guard Prince Abdullah aligned with Faisal, Saud had little choice. The *ulama* issued another religious ruling declaring the abdication process *sharia*-compliant. Formally led by King Abdulaziz's eldest surviving brother, Abdullah; his third eldest son, Mohammed; and family elders who had fought in the Wars of Unification, the Al Saud family closed ranks and swore allegiance to the new king. The Council of Ministers approved the decision in a document signed by Prince Khalid as the deputy prime minister. The Consultative Council or *Majlis al-Shura*—which at the time was a largely symbolic assembly of Meccan notables—approved the decision. From Washington, President Lyndon Johnson sent Faisal his congratulations.[34] The fifty-two sons of King Saud waited a month, but then they too swore allegiance to Faisal.[35]

What had finally united his family against King Saud was his violation of two governing Al Saud principles. First, Saud had appointed his sons to be minister of defense, commander of the National Guard, commander of the Royal Guard, governor of Riyadh, and governor of Mecca. He also appointed the son of his deceased full brother Turki to be minister of the interior. All of these young men were less experienced than their uncles. Some were inept, and the quality of government provided by the Al Saud began to decline. More importantly, by appointing so many sons to senior positions, King Saud made clear his intention to pass power on to them rather than to his brothers. This was neither King Abdulaziz's intention nor something King Saud's brothers would abide.

Second, King Saud appeared ready to use force against family members. He sent a very threatening letter to his half-brother, Faisal; sought to recruit tribal allies against Faisal; and, at one point, retreated into his palace with his own guard force. This, neither the Al Saud nor the *ulama* would tolerate. Mohammed bin Abdulaziz told King Saud directly that his letter was unacceptable.[36] Dreading civil unrest or *fitna*, the grand mufti informed King Saud that if his quarrel with Faisal resulted in bloodshed, the *ulama* would immediately withdraw their allegiance to him.[37]

It is also worth noting that the struggle between Saud and Faisal occurred on several levels. First, it reflected two distinct views of the monarchy's future. Saud wished to rule as his father had done—as a supreme, autocratic chieftain allied with traditional tribal, religious, and merchant leaders. Faisal sought a more collective family rule, employing a newly institutionalized bureaucracy. Second, it was personal. Although Faisal was willing to share power and wait his turn, Saud seemed determined to sideline his brother and put his own sons in power. Third, it reflected divisions among the stakeholders of the Al Saud's coalition. King Saud was supported by his sons, the younger sons of King Abdulaziz, liberal technocrats, and many of the tribes. Faisal, on the other hand, was supported by Abdulaziz's older sons, the senior *ulama*, the merchants, and, tacitly, by the United States. Of these parties, it was the support of the senior brothers and the *ulama* that proved decisive, though when the exiled King Saud assembled an armada of aircraft with which he hoped to retake his throne, it was the CIA that broke up the plot.[38] Finally, King Saud was removed because he failed to deliver the competent government upon which Al Saud legitimacy partially rests. Saud's bungled economic and foreign policies united the family against him, because his blunders put the monarchy at risk.

Throughout their struggle, Crown Prince Faisal sought to preserve family unity. He could have seized the throne on numerous occasions but waited until there was an overwhelming consensus within the Al Saud family to remove his brother. As the American ambassador to Saudi Arabia at the time, Parker Hart, wrote:

> As the relationship between the two men unfolded before my own eyes over a number of years, I must credit both men, who were reputed never to have been on the best of terms, with a real measure of loyalty to their very exacting oath. In 1964 Saud precipitated a head on confrontation that ended badly for him. Faisal's position was simply put: I will do nothing against my brother, but I cannot govern if authority is removed from me.[39]

Faisal removed his brother with dignity. He made a point of publicly seeing Saud off at the airport when he flew into exile. When Saud died in 1969, his body was flown back from Athens to Riyadh and buried in the al-Ud cemetery near his father. His sons have since held provincial governorships and important military commands.

The abdication of King Saud made it clear that the sons of King Abdulaziz would follow their father's intentions regarding succession. Saudi Arabia would not become a monarchy based on primogeniture. It would be a family enterprise in which both age and ability shaped a consensus as to who should lead the family and the kingdom. The manner in which the struggle was resolved demonstrated why Saudi princes have generally spent more energy preserving the family franchise than fighting each other for absolute power. Senior Saudi princes do not need to fight to the death. Unlike Ottoman princes, the sons of Abdulaziz who lost out did not end up blinded or at the bottom of the Bosphorus, instead, they remained wealthy and respected—benefiting from generous stipends, business opportunities, and a very elevated social status.

Moreover, who became king did not matter greatly to the majority of junior princes who were never competing for the position in the first place. What mattered to them was that whoever ascended to the throne was competent, preserved their family's privileged position, and continued to dedicate a portion of the kingdom's growing oil fortune to their royal stipends. Like the religious scholars, junior princes accepted the outcome of a process in which they played only a limited role provided their collective interests were protected.

When he came to the throne in January 2015, King Salman abandoned many of these principles. Having been on the throne for barely ninety days he removed Crown Prince Muqrin, the youngest surviving son of King Abdulaziz, and replaced him with a grandson of the kingdom's founder, Prince Mohammed bin Naif bin Abdulaziz. Two years later he removed Mohammed bin Naif to make way for his own son, Mohammed bin Salman. King Salman has used limited force against his opponents, has not treated losers with the respect they might have expected, and has curtailed the privileges of many junior princes.

In doing all this, King Salman overturned a system that had produced political equilibrium for fifty years. However, it is equally true that the partnership of brothers he overturned, regardless of how successful it may have been in the past, could not have continued much longer. Soon there will be no sons of Abdulaziz left alive. King Salman had to create a new system, and this was never going to be easy or painless. It now remains to be seen how his efforts will affect the family unity upon which Saudi Arabia's political stability has depended for so long.

Kings Khalid, Fahd, and Abdullah:
A Team of Rivals

Since 1964, the number of players directly involved in the Saudi succession process has steadily declined. In March of that year a delegation of senior *ulama*—including the Grand Mufti Mohammed Ibrahim Al al-Sheikh, his brother, Abd al-Malik, and the future Grand Mufti Abdulaziz bin Baz—went to King Saud and demanded that he surrender control of the Royal Guard to the minister of defense.[1] Seven months later, more than sixty *ulama* from all parts of the country attended the meeting at the Sahara Palace Hotel in Riyadh to discuss deposing the king, and it was a delegation of *ulama* and princes that went to ask Faisal to replace his brother.[2] Never again would Saudi Arabia's religious scholars play such a direct role in selecting, as opposed to confirming, a Saudi ruler.

Nor would the extended Al Saud family ever again play such a direct role in selecting the king. There were more than 100 princes at the Sahara Palace Hotel meeting to depose King Saud, representing a very broad cross section of the kingdom's royal family. They included King Abdulaziz's brothers, nephews, cousins, and distant relatives. Within a few years, all of Abdulaziz's brothers would be dead. The last to hold office was Minister of Finance Musa'id bin Abd al-Rahman, who retired in 1975. Many of Abdulaziz's cousins had been prominent provincial governors. By the mid-1970s the last of them, Fahd bin Saad bin Abd al-Rahman in Ha'il and Fahd bin Abdullah bin Abd al-Rahman in Al Qasim, had been replaced by Abdulaziz's younger sons and grandsons. The Bin Jaluwi branch of the Al Saud had ruled the Eastern Province as a semi-autonomous fiefdom for three generations. They were replaced by one grandson of Abdulaziz in 1984 and another in 2013. In 2015, only direct descendants of King Abdulaziz played a role in selecting King Salman.

The succession struggle between Saud and Faisal resulted in four changes of government in five years. This was a disconcerting episode for the family and the public, which the Al Saud have sought to avoid repeating. For example, during King Saud's chaotic reign the prime ministership shifted back and forth three times. This division of power between the king as "Head of State" and the prime minister as "Head of Government" became a source of stalemate and instability. It was ultimately the political platform that Crown Prince Faisal used to confront his brother. When he became king, Faisal eliminated this potential problem by serving as both king and prime minister.[3] All of his successors have done likewise. In a further effort to avoid destabilizing disputes, King Faisal established a Council of Senior Princes to advise him on succession issues and to supervise succession in the event of his death. This Council initially included two of Abdulaziz's brothers, Abdullah and Musa'id, as well as five of his sons: Khalid, Fahd, Abdullah, Sultan, and Nawwaf.[4]

Faisal carefully structured this council to balance competing factions within the family. Its aforementioned two eldest members were not direct descendants of Abdulaziz. Princes Fahd and Sultan came from the powerful block of full brothers known as the Sudairi Seven.[5] Princes Khalid and Abdullah represented the large, but fragmented, group of non-Sudairi brothers. They also had strong connections to the kingdom's tribes, which they had used to winnow away support for King Saud. The final member of the council, Prince Nawwaf, was a full brother of the free prince leader, Prince Talal. As the son of an Armenian concubine, he had little prospect of becoming king. His inclusion reflected Faisal's willingness to rehabilitate repentant free princes, even those who had previously endorsed Nasser's calls to abolish the monarchy. Other former free princes whom King Faisal recruited for important positions were Badr (deputy commander of the National Guard), Abd al-Muhsin (governor of Medina) and Fawwaz (governor of Mecca).[6] All of these half-brothers owed their positions to Faisal and were loyal subordinates. The independent, princely fiefdoms that would characterize the cabinets of Kings Fahd and Abdullah were yet to emerge.

When Faisal became king, the crown prince position became vacant. Abdulaziz's third son, Mohammed, was experienced. He had taken Medina for his father and drafted the letter telling King Saud that he must abdicate and either go into exile or remain in the country under house arrest.[7] But, Mohammed was known as Abu Sharrain (the "father of two evils") because of his violent temper and heavy drinking. He was considered unlikely to make a good ruler.

As a result, Faisal had appointed Mohammed's younger full brother, Khalid (1913–1982), deputy prime minister in 1962. That implied, but did not confirm, that Khalid would become crown prince. Only in 1965 did King Faisal officially decree that Khalid would be his successor.[8] In doing so, he followed his father's instructions to select his successor from among his brothers, not his own sons—thus confirming that the succession would be lateral, not linear. In skipping Mohammed, Faisal also confirmed that the crown would pass to the "eldest most able," not simply the most senior brother in age. Needless to say, Prince Mohammed was very well compensated financially for giving up his birthright.

King Abdulaziz had selected his next two successors informally with a pact between his two eldest sons. King Faisal sought to make this process more secure, transparent, and predictable. Prince Fahd (1921–2005) was the seventh of Abdulaziz's surviving sons. He had strongly supported Saud's abdication, and as the oldest of the Sudairi Seven brothers had thrown their considerable political weight behind Faisal.[9] Thus in 1967, Faisal selected Fahd to fill the newly created position of "second deputy prime minister." Again, although this appointment did not confirm that Fahd would become crown prince, it did for the first time give a strong indication of who the heir to the heir would be.[10]

King Faisal is often credited with saving the Saudi monarchy. He certainly centralized and institutionalized it. He created the title "Your Royal Highness" for direct descendants of King Abdulaziz, as opposed to "Your Highness" for other members of the extended

Al Saud family. He reduced the number of princes receiving government stipends and formalized the system of payments. Royal stipends became a regular monthly salary, no longer dependent solely on the king's discretion. For the first time, Faisal drew a clear line between the state's budget and the royal family's spending money.[11]

King Faisal was assassinated in March 1975 by his nephew, Prince Faisal bin Musa'id bin Abdulaziz. Numerous reasons have been given for this murder. The young man's mother was from the deposed Al Rasheed of Ha'il and his ultraconservative brother, Khalid, had been killed by the police during an attack on Riyadh's first television station, which he had regarded as an un-Islamic innovation. His fiancée was a daughter of the deposed King Saud, and, finally, having been connected to drug dealing while a student in the United States, he was mentally unbalanced. Whatever his motivation, Faisal bin Musa'id arrived at the king's palace unannounced, was allowed to enter, and shot his uncle as he was meeting with the Kuwaiti oil minister. The assassin prince was publicly beheaded several weeks later.

The assassination of its leader tests any political system, yet within hours of Faisal's death the Council of Senior Princes declared Crown Prince Khalid the new king. As expected, the Second Deputy Prime Minister Fahd became the new crown prince.[12] In what had become an established pattern, the new king became prime minister and the new crown prince became deputy prime minister. The putative third in line, Abdullah, became the new second deputy prime minister. Fahd's two older half-brothers, Nasser and Saad, formally renounced their claims to the throne. The royal family's support was proclaimed by its senior member, King Abdulaziz's eldest surviving brother, Abdullah bin Abd al-Rahman. The *ulama* gave their formal consent. Following a smooth and undisputed succession, tribal leaders, urban notables, military officers, and ordinary people arrived by the thousands to swear allegiance to the new king and the new crown prince.

King Khalid's mother was from the Bin Jaluwi branch of the Al Saud. In 1934, Khalid had negotiated the Treaty of Taif, which ended the Yemen War. In 1943, he had accompanied Prince Faisal to Washington to initiate the Saudi–American relationship. He had served as Viceroy of the Hejaz and minister of the interior. Though well qualified, Khalid had never sought the throne—and by 1975 he was in poor health, more interested in falconry and horse breeding than in governing.

Because of his declining health and preference for desert camping over cabinet meetings, Khalid allowed Crown Prince Fahd to act as day-to-day ruler. A Royal Decree issued in May 1975 granted Fahd full responsibility for the routine management of the kingdom. However, this document was very different from those issued during the reign of King Saud which had formally stripped the king of executive authority. Khalid was relieved of routine decision-making but remained prime minister and the final authority in all matters. He was never a mere figurehead.[13] King Khalid was the ultimate source of his crown prince's authority and legitimacy, just as King Salman remains today for Crown Prince Mohammed bin Salman.

In Faisal's increasingly institutionalized kingdom, power came to depend more upon which ministry a prince controlled than on which tribe he was close to. King Khalid attempted to distribute senior positions among different family sections, but Crown Prince Fahd and his six full brothers clearly emerged as the dominant bloc.

Their mother, Hussa bint Ahmed al-Sudairi was reputed to have been the most intelligent of King Abdulaziz's many wives. Her seven sons remained close and lunched together with her once a week as long as she lived.[14] Between them, these brothers held the positions of minister of defense, vice-minister of defense, minister of the interior, vice-minister of the interior and governor of Riyadh Province. Hardworking and loyal to each other, the Sudairi Seven looked set to provide the next several kings after Abdullah—and this assumed that Abdullah would become king because his position as second deputy prime minister implied, but did not confirm, that he would become crown prince.

Prince Abdullah led a faction composed of nearly everyone who was not part of the Sudairi bloc. Most of these sons of Abdulaziz had few, if any, full brothers. Their bloc included Abdullah as commander of the National Guard; his half-brother, Badr, as deputy commander of the Guard; Mit'ib at the Ministry of Housing; and Majid at the Ministry of Municipal and Rural Affairs as well as the sons of Abdulaziz who governed Al Qasim, Tabuk, and Mecca provinces.[15]

The sons of King Faisal remained prominent, with Saud as foreign minister, Khalid as governor of Asir, and Turki as head of the External Intelligence Service. In 1975, members of the Al al-Sheikh family also formed something of a bloc, serving as ministers of justice, higher education, and agriculture.[16]

King Khalid served as a mediator between these factions, but in 1977 he was ill in a London hospital with an uncertain prognosis. Rumors, and perhaps more than rumors, began to circulate that Defense Minister Sultan wished to skip ahead of Abdullah to become crown prince when Khalid died. To eliminate that threat, King Khalid took a page from his father's playbook. He summoned Fahd and Abdullah to London, where at the King's bedside they swore an oath to abide by the existing succession plans.[17]

The Al Saud are most firmly united when the monarchy is threatened. Then they close ranks, as they eventually did against King Saud. The most threatening period of King Khalid's reign began on the first day of the Islamic year 1400—November 29, 1979—when several hundred Islamic extremists seized the Great Mosque in Mecca. Led by a former Saudi National Guard corporal, Juhayman Utaibi, who was both the son of an Ikhwan rebel and a former student of the Grand Mufti Abdulaziz bin Baz, these terrorists held a very odd mixture of Wahhabi and millenarian beliefs. They proclaimed Mohammed Qahtani the Mahdi, or Messiah, whose arrival would precede the Day of Judgment. They called for the overthrow of the House of Saud, which they accused of deserting Islam. For good measure they also demanded the expulsion of all non-Muslims from Saudi Arabia, the abolition of television, and a complete embargo on oil supplies to the United States.

With Crown Prince Fahd in Tunisia, responsibility for handling the crisis fell to Defense Minister Sultan and his brother, Interior Minister Naif.[18] They began by closing the Prophet's Mosque in Medina; sealing Saudi Arabia's land borders to non-Saudis; and imposing a total communication blackout on all overseas phone calls, telexes, and telegrams. They gathered various military forces but, as all Muslims know, the Quran prohibits the use of force in Mecca: "Do not fight with them in the Sacred Mosque, until they fight with you in it. But if they do fight you, slay them; such is the recompense of the unbelievers."[19] Because of this Quaranic injunction, the princes were finding many of their troops reluctant to assault the Great Mosque without unambiguous authorization from the kingdom's senior religious scholars.[20]

The Wahhabi scholars had no trouble concluding that Mohammed Qahtani was not the Mahdi, but took their time deciding whether he and his followers were indeed nonbelievers. Eventually, they responded to King Khalid's request for a religious ruling, stating, "The *ulama* unanimously agree that fighting inside the Sacred Mosque has become permissible ... All measures may be used." But the *ulama* also made it very clear that the Al Saud themselves would now need to be more scrupulous about fulfilling their own Islamic duties.

Even with their new religious ruling, the princes found it difficult to dislodge Juhayman. Two major assaults and numerous probing attacks were beaten back with heavy casualties before government forces used anti-tank missiles to knock out snipers in the minarets and armored personnel carriers to break through the mosque gates. Even after they cleared the mosque, the terrorists retreated into the labyrinth of cellars beneath it. Only when they used canisters of paralyzing CS gas supplied by France were Saudi security forces, wearing gas masks, able to clear the cellars.[21]

The cost in lives of the two-week siege was high. 127 soldiers were killed and 461 wounded. More than 100 extremists, including the so-called Mahdi, were killed, along with an unknown number of worshipers, many of whom had been taken hostage and died in the re-taking of the mosque. Sixty captured extremists were publicly beheaded on the same day in eight Saudi cities. Riyadh governor Salman personally oversaw the executions in that city's central square. Other male extremists were executed later. Women who had occupied the mosque and fired upon security forces were not executed, and eventually released. In typical Al Saud fashion, Juhayman's sons were not punished but instead went on to become National Guard officers.[22]

The aftershocks of the Mecca siege within Saudi society were profound. Coinciding as it did with Iran's Islamic Revolution, widespread rioting among the Saudi Shia, and the Soviet invasion of Afghanistan, the seizure of the Great Mosque by Sunni religious extremists severely unsettled the House of Saud. Their response was a dramatic shift towards a more conservative Islamic society as they energetically sought to keep their bargain with the *ulama* and prevent another Juhayman. There would be no more mixed-gender swimming pools, movie theaters, or women reading the television news. Restrictions on women working were intensified, as were the prohibitions against celebrating Christmas, Halloween, and Valentine's Day. Western firms were required to dismiss female employees. Music stores were closed. Importing dolls became illegal. No longer content with simply enforcing prayer on Muslims, the religious police began to harass unveiled women, raid Western compounds, and smash photography shops. Alcohol became harder to find. In Saudi schools, more and more time was devoted to religious studies and Islamic history.[23] To emphasize the Islamic roots of Al Saud rule, King Fahd adopted the old Ottoman title Custodian of the Two Holy Mosques or, more precisely, Servant of the Two Noble Sanctuaries.

Change was not confined to Saudi Arabia. For the first time, religious attaches appeared in Saudi embassies intent on funding new mosques and encouraging existing mosques to spread the Wahhabi Call, or *Dawah Wahhabiya*.[24] Billions of dollars were poured into this international proselytizing effort, which became a significant source of profit and promotion for the Saudi religious establishment. Juhayman had sought to halt and, if possible, roll back King Faisal's social liberalization; to a very considerable extent, he lost a battle and won the war.

In recent years, long after the storm had passed, it became fashionable to condemn Kings Khalid and Fahd for their concessions to the religious fervor sweeping the Islamic world in 1979. What would have happened had they, like the Shah of Iran, "determined to proceed with radical social reforms he knew would anger the clergy"?[25] We will never know, for unlike the Shah—who once spent the holy month of Ramadan skiing in Switzerland—the kings of Saudi Arabia have made a great show of attending Friday prayers, fasting during Ramadan, and building mosques within their palaces. In 1979 the Al Saud bent with the prevailing wind, strengthened their appearance of piety, and survived.

King Khalid was only seventy when he died of a heart attack in the summer of 1982. Led by King Abdulaziz's eldest surviving son, Mohammed, the royal family swore allegiance to the new King Fahd. Within a few hours of Khalid's death, the government announced that senior princes had approved Abdullah as the new crown prince and that King Fahd had appointed Sultan the new second deputy prime minister.[26] Abdullah remained commander of the National Guard, Sultan remained minister of defense, and Fahd's four older brothers each formally renounced their claim to the throne. Once again, after three days of mourning, hundreds of princes, religious scholars, tribal chiefs, military officers, and then thousands of ordinary citizens came to Riyadh's central square to give the new king their formal pledge of allegiance.[27]

King Fahd (1921–2005) came to the throne at the age of sixty-one, with a lifetime of government experience behind him. He had traveled to San Francisco with Faisal in 1945 for the founding of the United Nations. He had served as minister of education (1954–1960), minister of the interior (1962–1975), and crown prince (1975–1982). He had a record of supporting social change, strong relations with the United States, and the advancement of his six full brothers. When he left the ministry of the interior to become crown prince, he made certain that his younger full brother Naif replaced him.

The Sudairi brothers would have preferred that Abdullah relinquish command of the National Guard when he became crown prince, but Abdullah had no full brother to replace him and refused. Several alternative proposals were considered—to combine the National Guard with the Ministry of Defense, for instance, or to reduce the Guard from a light-infantry organization to a police force. Defense Minister Sultan even promoted national conscription, which would have reduced the appeal of the National Guard to the tribal youths from whom it recruited. All of these plans came to nothing. The National Guard remained the most dependable military force in the kingdom, and it remained firmly under the control of Crown Prince Abdullah.

This division of authority, in which the minister of defense and the commander of the National Guard came from competing family factions, had been introduced by King Faisal. It created an important, if unwritten, balance of physical as well as political power within the family. It hindered any attempt to consolidate control in the hands of one prince or family bloc. Under King Faisal, all ministers had clearly served at the king's pleasure, but during his successor Khalid's reign the ministries of defense, the interior, foreign affairs and municipal and rural affairs – as well as the National Guard – all became important, independent power bases and patronage mechanisms for prominent princes. This too, helped maintain a balance of power within the family. As we shall see, King Salman has in recent years systematically dismantled this balancing apparatus.

In March 1992, King Fahd issued the Basic Law of Governance, which constituted a major step in codifying the Saudi succession processes. The law stated that the country was a monarchy with the following attributes:

> Governance is limited to the sons of the Founder King Abdulaziz bin Abd al-Rahman Al Saud and the sons of his sons. Allegiance shall be pledged to the most suitable among them to rule on the basis of the Quran and the Sunnah ... The King shall select and relieve the crown prince by Royal Order ... The crown prince shall assume the powers of the King upon his death ... Citizens shall pledge allegiance to the King on the basis of the Quran and Sunnah in times of hardship and ease, fortune and adversity.[28]

The Basic Law of Governance made it clear that the king could, entirely at his own discretion, select or dismiss the crown prince; that ability as much as age would shape his choice of a successor; and that grandsons of Abdulaziz were eligible to become king even if sons remained alive. The law also made it clear that the *bay'ah* was given only to the king, and that the oath of allegiance given to a crown prince was not an independent act but part of the oath sworn to the king. None of this had ever been so clearly stated before. All of it would become vitally important two decades later, when King Salman dismissed two crown princes and appointed his own son as his successor.

King Fahd suffered a stroke in 1995, which left him increasingly weak and unable to govern during the last decade of his life. When he died in August 2005, his younger half-brother, Abdullah (1924–2015), who had been crown prince for twenty-three years and effective regent for ten, was immediately declared king. Fahd's younger full brother, Prince Sultan (1928–2011), remained minister of defense and became crown prince and deputy prime minister. The number-three post of second deputy prime minister, which King Faisal had created, was left vacant for the first time in thirty-eight years.

Many had expected this third position to go to Sultan's full brother, Interior Minister Naif (1934–2012), but King Abdullah baulked at the prospect of two full Sudairi brothers becoming king one after the other. In fact, from the beginning of his reign, King Abdullah sparred with the six remaining Sudairi brothers, who still firmly controlled the Ministries of Defense and Interior as well as the governorship of Riyadh. Only in 2009, when Crown Prince Sultan's health had deteriorated to the point at which it became clear that he would never be king, did King Abdullah declare Prince Naif second deputy prime minister.

King Abdullah presided over a royal family increasingly concerned that the powerful Sudairi brothers—particularly Sultan, Naif, and Salman—would engineer a takeover of the Saudi government. The king sought to prevent this by recruiting allies from among more marginal princes and creating the Allegiance Council, in which the Sudairis could be outvoted.[29] The Council, created in 2007, represented an interesting but ultimately unsuccessful attempt to institutionalize the Saudi succession process. It consisted of King Abdulaziz's surviving sons and a grandson representing each son who had died. It was intended that a new king would propose up to three candidates for crown prince to the Allegiance Council, which would attempt to form a consensus around one name. If the Council rejected all of the king's nominees, then it would be free to propose its own candidate to the king. Then, "In the event that the King rejects the Council's nominee, the Allegiance Council will vote to choose between the King's candidate and its own. The nominee who secures the majority of votes will be named Crown Prince."[30]

The Allegiance Council never became an independent institution. When Crown Prince Sultan died in 2011 it promptly approved King Abdullah's choice of Naif as the new crown prince. This happened again when Naif died in 2012 and King Abdullah named Salman crown prince. The Council never presented King Abdullah with its own nominee. When the monarch created the entirely new position of deputy crown prince, as opposed to deputy prime minister, he did not assemble the Allegiance Council. He gave this new position to King Abdulaziz's youngest surviving son, the non-Sudairi Prince Muqrin.

Had King Abdullah overseen the effective use of the Allegiance Council, the evolution of succession in Saudi Arabia might have been very different. As it was, when the Council approved Mohammed bin Naif as crown prince in 2015 and Mohammed bin Salman as crown prince in 2017, it was regarded as little more than a rubber stamp for the king's decision. As stated in the Basic Law of Governance, succession remained very much the prerogative of the king.

King Salman and Mohammed bin Salman:
A New Royal Order

King Abdullah died of pneumonia in the early hours of January 23, 2015. By the conservative standards of Saudi Arabia, he had been a reformer who left behind a full treasury and a solid legacy of economic development, social change, and security. He doubled the number of universities in the kingdom, sent thousands of students abroad to study, and founded his namesake university, which he hoped would become a world-class research institution. Saudi Arabia, with a gross domestic product twice that of Egypt, and foreign currency reserves behind only those of China and Japan, had become the leading economic power in the Arab world and joined the G20 international forum for the world's largest economies. Yet the country still remained culturally at odds with the rest of the world and economically dependent on a single, volatile commodity.

The genuine sorrow over King Abdullah's passing was accompanied by relief at the smooth succession that followed. Crown Prince Salman immediately became the new king and promoted his half-brother, Deputy Crown Prince Muqrin, to be the new crown prince. Prince Mishal, now King Abdulaziz's eldest surviving son, led the family and the nation in the *bay'ah* ceremonies as the sixth transition of power in the Third Saudi State took place quickly and peacefully.

Salman was the sixth brother of the Sudairi Seven and the last survivor of the team that King Faisal had installed in the early 1960s. He was generally regarded as one of King Abdulaziz's most intelligent and experienced sons. Salman had become the governor of Riyadh Province in 1960. He held the post for the next forty-eight years, guiding the city's growth from a population of barely 200,000 to more than seven million. Having governed much of Central Arabia for most of his life, Salman was intimately familiar with the tribes and leading families of the Nejd, who remain the backbone of Al Saud rule. He was married into the prominent Ajman and Sbaie tribes as well as the aristocratic Al Sudairi family—all powerful stakeholders in the Al Saud's coalition. Because he was the governor of Riyadh, where most Al Saud princes lived, King Fahd had assigned him the role of "referee" in family disputes and disciplinarian for wayward princes. Sometimes referred to as the "Prince of Princes," Salman maintained a private jail for princes and was well aware of which family members abused their royal status.

The new king possessed a detailed knowledge of his father's career and had made a point of studying King Abdulaziz's methods. Like his father, Salman was a committed autocrat who believed in the Al Saud family's right to rule and feared that without them, Saudi Arabia would quickly and violently fragment into competing tribes and regions. He could be stern with princes, clerics, tribal chiefs, and religious policemen. In 1979, he had firmly supported executing the zealots who had seized Mecca's Great Mosque. Thirty-five years later, he ordered the beheading of forty-five convicted al-Qaeda terrorists whom King Abdullah had kept on death row for nearly a decade.

His stern demeanor notwithstanding, Salman was a popular governor and effective politician who regularly paid condolence calls upon the death of prominent citizens or attended the weddings of their children. He took his job seriously and arrived for work at 8:00 a.m. every morning for half a century. By Al Saud standards, he was neither rich nor corrupt. Salman had memorized the Quran by the age of ten, and firmly supported the Al Saud's alliance with the *ulama*, but he was no reactionary. As a governor, he had both funded and, at times, restrained the religious police. He sought social change, but at a pace that would not provoke a conservative backlash—always a delicate task in the xenophobic and religiously conservative Nejd.

Salman's management style was clearly on display at the Riyadh Development Authority (RDA), which he created to supervise the city's rapid expansion. Essentially a planning task force that reported directly to the governor, rather than a minister, the RDA was widely regarded as more competent and less corrupt than most Saudi government agencies. There were no princes in the RDA bureaucracy. Salman regularly warned employees that anyone caught embezzling would spend the rest of their life in prison, and an Arabic-speaking Western attorney kept a close watch on contracts for thirty years. At the time, no other municipality in Saudi Arabia had such a lean management structure. When he became king, Salman would establish similar bodies in eight other major cities across the kingdom.

Salman's longstanding emphasis on effective administration, opposition to corruption, and unforgiving disciplinary style were all well known when he became king. These traits have characterized his reign ever since. However, coming to the throne at the age of 78, Salman was not the young man who had first taken charge of Riyadh. His general health, stamina, and concentration for extended periods were legitimate concerns. Although the king continued to preside over weekly cabinet meetings and meet foreign leaders, he began to delegate much of the daily management of government to younger princes—especially his son, Mohammed bin Salman or MBS.

The new Crown Prince Muqrin was the youngest living son of King Abdulaziz. Initially a fighter pilot, he had more than forty years of government service, which included postings as governor of Ha'il, governor of Medina and head of Saudi Arabia's external intelligence service, the General Intelligence Presidency or *Al Mukhabarat Al A'ammah*. Unlike King Salman, Muqrin was not part of the family's Al Sudairi branch. In fact, his Yemeni mother was from neither the Al Saud nor an aristocratic tribal family. King Abdullah had intended Muqrin's appointment as deputy crown prince to preserve the decades-old Sudairi/non-Sudairi balance within the family. However, with no full brothers, intra-family connections, or tribal allies, Muqrin lacked a firm base of support.

By 2015, the transition to third-generation princes was imminent and managing that process would be King Salman's most important challenge. When King Abdulaziz died in 1953, it had taken a decade of periodic crises to resolve the manner in which the second generation of princes would govern. The system that they eventually created of thirty-four brothers sharing power, served the kingdom well for many years. The king was always first among equals with final authority and some kings were clearly more dominant than others, but all had sought to maintain family unity. This brotherly, consensus-oriented system had promoted stability, slowed change, and was rapidly running out of princes.

Clearly, most of King Abdulaziz's grandsons could never attain the wealth and power that his sons had enjoyed. Because a power-sharing consortium of 500 cousins would produce a weak and unwieldy government, real political authority needed to remain concentrated in relatively few hands. This winnowing process would inevitably create resentment, anger, and potential instability. Hundreds of grandsons were theoretically eligible to become the next deputy crown prince. Some of the senior third-generation princes, such as King Faisal's sons, were very well qualified but they were now well into their seventies.

On the day of King Abdullah's death, King Salman appeared to resolve this difficult issue by appointing his nephew, the 55-year-old Prince Mohammed bin Naif or MBN to be the deputy crown prince and third in line of succession. This move to a third-generation prince was an historic event, which effectively ended the political hopes of Abdulaziz's few remaining sons. Many in Riyadh believed that by making the transition to third-generation leadership while a second-generation king was still on the throne to supervise the process, King Salman had taken the most important step of his reign on his first day in office.

Mohammed bin Naif was minister of the interior and had served as the deputy minister for thirteen years. Trained by both Scotland Yard and the FBI—the foremost law-enforcement agencies in Britain and the United States, respectively—he is widely credited with defeating the al-Qaeda insurrection that raged across Saudi Arabia between 2003 and 2007. Experienced, popular at home, and well-regarded in foreign capitals, he was a son of Salman's full brother Naif and thus from the Sudairi branch of the family. His appointment seemed to again reinforce the Sudairi/non-Sudairi leadership pattern that had existed since King Faisal's death, but it would soon become clear that neither Muqrin nor Mohammed bin Naif were really part of King Salman's plan.

The new king quickly moved to secure, some might say buy, both popular support and cooperation from the kingdom's tribal, religious, and business leaders. The treasury was full and Salman authorized $32 billion-worth of new spending. Civil servants, military personnel, students, retirees, and handicapped persons were all given two months' extra pay. Licensed charities, sports clubs, and literary clubs all received new grants. No fewer than 2,500 prisoners were pardoned and $5 billion was committed to improve water and electrical services. State-controlled firms such as Saudi Aramco and the petrochemical giant SABIC followed the government's lead on bonuses. Private firms were encouraged to do the same.

The former governor of Riyadh understood that to ignore the kingdom's tribes was to court disaster. Accordingly, none of the agricultural subsidies that tribesmen enjoyed were reduced. Officially recognized tribal *sheikhs* received generous bonuses, and the number of official tribal *sheikhs* receiving regular payments was increased to include a number of minor chiefs. In a completely new outreach to the tribes, King Salman formalized payment to the *muariffeen*, the individuals who determine legal membership of a tribe. This can be important as membership in a Saudi tribe confers valuable state-funded benefits, but in border areas near Jordan, Iraq, or Yemen it may not be clear exactly who belongs to the Saudi branch of a tribe. For many years, the *muariffeen* were paid only when their services were needed. King Salman regularized and increased their salaries. This was a significant benefit increase for the *muariffeen*, which raised their status within their tribes and strengthened their allegiance to the Al Saud.

Religious conservatives had long boasted that when King Abdullah died, they would remove the liberal, religious-police chief that he had installed, restore the conservative *ulama* that he had fired, and shut down the co-educational university that he had founded. King Salman initially gave them much of what they wanted. In the streets of Riyadh, religious policemen celebrated the sacking of their chief, Abd al-Latif Al al-Sheikh, who had sent them to English classes and sensitivity training. They joyfully slaughtered camels and goats in public while coffee houses they favored offered free drinks. The festivities grew to such an extent that the new religious-police chief publicly demanded greater respect for his predecessor. The progressive Justice Minister Mohammed al-Issa was removed from both his ministry and the Supreme Council of Scholars. The conservative Saleh Al al-Sheikh was reinstated to his post as minister of Islamic affairs. The conservative Saad al-Shithri whom King Abdullah had removed from the Scholars Council, was brought back as a Royal Court advisor.

However, religious conservatives did not get everything they wanted. Salman stopped short of closing his predecessor's co-educational university. Instead, he released women activists who had been jailed for trying to drive across the Saudi-UAE border. King Abdullah had appointed the first women to the *Majlis al-Shura*, the government's consultative council. King Salman welcomed them as the first women ever to attend the important *bay'ah* ceremonies. Salman confirmed that women would be allowed to vote and run for office in the upcoming municipal elections. He signed a new labor law that banned women from working in "dangerous occupations" such as mining, but also introduced maternity leave for working mothers. During his first weeks in power, King Salman performed the traditional Al Saud balancing act of give-and-take with the country's religious conservatives.

The new king met with more than 150 senior business leaders. He assured them that he would support pro-business policies and made some labor law adjustments that they had requested. He did not lay out any plans for dramatic economic reform but noted that sharply falling oil revenue would require budget adjustments. The business community accepted that the king could not control commodity price cycles and, as investors, they shared the *ulama's* overriding concern with political stability.

Many Al Saud family members continued to hold senior positions at the Ministries of the Interior, Defense, and Foreign Affairs. Sons of Kings Abdullah, Faisal, Khalid, and Saud, as well as former Crown Princes Sultan and Naif all continued to hold provincial governorships. King Abdullah's son Mit'ib remained commander of the National Guard. Despite growing budget deficits, the privy purse continued to pay thousands of Al Saud family members a monthly stipend that ranged from token sums to millions of dollars a year. Princes remained exempt from paying electricity bills for their palaces. All cabinet members but one were formally reconfirmed in their posts, including eight new ministers whom King Abdullah had appointed less than two months before his death. In short, during the first few weeks of his reign, King Salman made very few changes as he sought to embrace and reassure all stakeholders in the Al Saud's coalition.

Saudis expected this customary, gradual transition process from one king's team to the next, but in hindsight, one cabinet change marked the beginning of the end for "business as usual." Salman had appointed his son Mohammed head of the crown prince's Diwan in 2013 and minister of state without portfolio in 2014. Now, at age twenty-nine, MBS became head of the Royal Diwan—essentially, chief of staff for the king's executive office—and minister of defense. In the latter capacity, MBS controlled the Saudi Army, Navy, and Air Force, along with a quarter of the government budget. Although no one had ever been both minister of defense and head of the Royal Diwan, few at the time anticipated that within three years Mohammed bin Salman would displace both Crown Prince Muqrin and Deputy Crown Prince Mohammed bin Naif.

King Salman soon quickened the pace of change and has not slowed down since. Over the following three months, he issued more than 100 royal orders, which substantially reorganized the government of Saudi Arabia. He fired thirty-three senior officials and appointed thirty-five new ones. He abolished a dozen long-standing cabinet subcommittees, which, though they had maintained bloated staffs, had been moribund for years. For example, the Supreme Petroleum Council convened less than once a year and the Supreme University Council hardly met at all. All the existing councils were replaced by two new ones: the Council for Political and Security Affairs (CPSA), chaired by Mohammed bin Naif; and the Council for Economic and Development Affairs (CEDA), chaired by Mohammed bin Salman in his capacity as a minister of state. These new councils met weekly, and quickly sped up the decision-making process. Only Mohammed bin Salman and the minister of information sat on both new councils. The rapid pace in which these changes were made, demonstrated clearly that considerable forward planning had long been underway.

As the Chairman of CEDA, Mohammed bin Salman assumed control not only of most civilian ministries but also of state-owned companies. Within a few months, the king appointed new chief executive officers for most of these, including Saudi Aramco, the Saudi Arabian Basic Industries Company (SABIC), the Saudi Telecom Company, the Saudi Railroad Organization, the Civil Aviation Authority, the Saline Water Conversion Authority, and the Military Industries Company. MBS took effective control of Saudi Aramco away from the Ministry of Petroleum. He moved the Public Investment Fund, which holds hundreds of billions of dollars of government assets, from the Ministry of Finance into his direct control.

In his first 100 days, Salman replaced half the Council of Ministers, but not with the usual bureaucrats and academics; now, there was a clear emphasis on practical experience. The new minister of agriculture was a former executive at Almarai Dairy, and the new minister of information a former journalist; the new minister for the civil service had a background in executive recruitment; the new minister of justice was a practicing attorney, not a religious scholar; and the new head of the General Authority for Civil Aviation came from the airline industry.

Many Al Saud princes suddenly found their services no longer required. King Abdullah's two sons, Turki and Mishal, lost their posts as governors of Riyadh and Mecca. Prince Bandar bin Sultan, the long-serving former ambassador to Washington, was relieved of his role as secretary general of the National Security Council. Prince Khalid bin Bandar was removed from his post as head of the external intelligence agency. There were numerous other examples of princes being replaced by technocrats, just as had long been the case at the Riyadh Development Authority.

On April 29, 2015 it was announced that the 70-year-old Crown Prince Muqrin had asked to be relieved of his duties. Few in Riyadh believed that he had decided to retire just three months after having taken up his new job. Most accepted that after attending a diplomatic dinner, Muqrin had been summoned by the king and asked to resign on the understanding that he could keep his new palace and a good deal more. Muqrin quickly agreed and left—relieved of a burden that many thought he had never really wanted, and financially much better off than he had ever been in his life.

Crown Prince Muqrin retired peacefully to his farm outside of Riyadh where he devoted his time to building his library and fishponds. He was immediately replaced by Deputy Crown Prince Mohammed bin Naif. The new crown prince remained minister of the interior—a powerful position in Saudi Arabia where its holder was, until recently, responsible for all matters concerning the kingdom's internal security. Then, in a move that surprised many and angered some, the king named the recently appointed defense minister Mohammed bin Salman, to be the new deputy crown prince. Many older and more experienced grandsons of King Abdulaziz were passed over in the process. Although the appointments of Muqrin and Mohammed bin Naif had been conventional, that of Mohammed bin Salman to be deputy crown prince was not. It broke decisively with the Al Saud's long-established leadership qualifications of age and experience.

On paper, the new crown prince Mohammed bin Naif remained minister of the interior and chairman of the CPSA; in reality, his authority was being steadily eroded. The crown prince's Diwan, or executive office, had long provided a measure of political and financial independence to previous crown princes, including Salman, Naif, Sultan, and Abdullah. In 2015, the crown prince's Diwan was abolished. Shortly thereafter, Mohammed bin Naif's longtime chief of staff, who had been given cabinet rank, was summarily dismissed by the king. In 2016, the CPSA, which Mohammed bin Naif chaired, was marginalized by the creation of a new National Security Council to which he did not even belong. At the Ministry of the Interior, new men loyal to Mohammed bin Salman were installed to head intelligence and police work. For those who were watching, the writing was on the palace wall.

Meanwhile, the old guard of Saudi politics continued to gradually fade away. Saud al-Faisal and his father had controlled the Ministry of Foreign Affairs almost continuously since its creation in the 1930s. Prince Saud had been foreign minister for forty years. He retired in 2015 and was replaced by a professional technocrat: the articulate former ambassador to Washington, Adel al-Jubeir. Many senior diplomats, including those closely related to Saud al-Faisal, moved on. The same was true at the Ministry of Municipal and Rural Affairs (MOMARA). MOMARA had long been a royal cabinet seat, in which Prince Mit'ib bin Abdulaziz had been succeeded by his son,

Mansour. He too was replaced by a technocrat, who had once led the Riyadh Development Authority. Even at the Ministry of Defense, which had been dominated by Prince Sultan, his sons, and his sons-in-law for decades, a new management team was put firmly in place and all of the service chiefs were changed. The last technocrats from King Abdullah's cabinet, Finance Minister Ibrahim al-Assaf and Petroleum Minster Ali al-Naimi, both of whom had held their posts for more than twenty years, retired in 2016.

By late 2016, it had become clear that King Salman intended to make Mohammed bin Salman king. It was equally apparent that many in the royal family still opposed both the young prince and the quite radical economic reform program that he was proposing. Over the next six months, King Salman persuaded and co-opted many Al Saud family members to support his succession plans. There were thirty-four voting members of the Allegiance Council, which technically would vote on a new crown prince. In a long series of new appointments, Salman made certain that twenty-nine of the Allegiance Council members had sons or brothers serving as provincial governors, deputy governors, or ambassadors to important countries. Like any politician, Salman lobbied for the support he needed by taking care of the voters.

On June 4, 2017 the well-respected Mohammed bin Naif was dismissed from his positions as crown prince and minister of the interior. This time no one doubted that he had been fired. He was summoned to the king's palace for what he believed would be a routine meeting; asking his armed guards to remain behind and surrendering his cell phone were not unusual measures before an audience with the monarch. On this occasion, however, Mohammed bin Naif did not meet his uncle. King Salman and his close aides had contacted most members of the Allegiance Council in person or by phone to obtain their support for Mohammed bin Salman replacing him as crown prince. It is not clear who presented Mohammed bin Naif with a statement that the Allegiance Council supported his removal, or a prepared resignation letter for him to sign. It is, however, clear that, unlike Crown Prince Muqrin two years earlier, Mohammed bin Naif did not immediately accept his dismissal.

What changed his mind was not, as some rumors had it, being deprived of his pain medication or being made to stand in an uncomfortable position for six hours. It was the intervention of the Governor of Mecca, Khalid al-Faisal. The venerable Prince Khalid pointed out to his cousin that he was twenty years older than Mohammed bin Naif; the son of a king, not a crown prince; and had been a provincial governor or cabinet minister for forty-five years, not three. He was by far the most senior and most experienced grandson of King Abdulaziz, and yet he had not protested when the king had passed him over and named his younger cousin crown prince because, "In this family, we obey the King and that is what you will do now." And so, Mohammed bin Naif did.[1]

Having signed his resignation letter, Mohammed bin Naif was immediately greeted by the new crown prince, who had clearly been waiting in the wings. The entire world saw Mohammed bin Salman kneel and kiss the hand of his deposed cousin, promising to always seek his advice. Mohammed bin Naif shook his younger cousin's hand and swore allegiance to him as the new crown prince. No new deputy crown prince was named then—or has been since.

The following month an entirely new security agency, reporting directly to the king through the Royal Diwan and Mohammed bin Salman, was created. Known as the Presidency for State Security, it took over nearly all police and internal intelligence work—leaving the Ministry of the Interior responsible largely for traffic fines, passports, and the administration of the kingdom's fire departments. The new agency was led by General Abdulaziz al-Howairini, who was well and favorably known to the Western intelligence services that had long worked closely with Mohammed bin Naif.

The final consolidation of power in what some were calling the Fourth Saudi State took place on November 4, 2017. Until then, the partnership created by the sons of King Abdulaziz had rested not only on self-interest and consensus, but also on a very real division of physical force—in effect, a balance of power ensuring that no single prince could challenge the central authority of the king, and no king could disinherit his brothers. For decades, the Ministry of Defense, Ministry of the Interior, National Guard and Royal Guard had each reported to a separate senior prince; of these, the National Guard was the most independent. Fiercely loyal to Abdullah bin Abdulaziz, who had led it for forty years, the Guard had more than once firmly resisted efforts for its dissolution or merger into the kingdom's Ministry of Defense.

On November 4, 2017, while Riyadh slept, King Salman issued a royal order dismissing King Abdullah's son, Mit'ib, from command of the National Guard. Mit'ib had spent his career in administrative positions rather than physically leading troops, and had commanded the National Guard only since 2010. He had not earned the loyalty that the Guard accorded his father, and there was no resistance to his dismissal.

Mit'ib was replaced by Prince Khalid al-Muqrin, a long-serving officer from a cadet branch of the Al Saud family whose father had helped found the Guard. A competent and respected professional soldier, Prince Khalid posed no political threat to Mohammed bin Salman. Nonetheless, even he lasted barely a year before being replaced by Prince Abdullah bin Bandar, a grandson of King Abdulaziz with no military experience. Gone, with the stroke of a pen, was the balance of military power that had ensured consensus government in Saudi Arabia. Gone, too, was the last serious threat to the ascension of Mohammed bin Salman. After November 4, 2017, only King Salman or MBS had direct control over any of the kingdom's security forces.

Fate had made Salman king. Had he not outlived Crown Prince Sultan (d. 2011) and Crown Prince Naif (d. 2012), he would have retired as governor of Riyadh. However, unlike the other accidental king, Khalid, Salman was neither reluctant nor unprepared. Those who know him believe he began planning his reign not when he became defense minister in 2012 or crown prince in 2013, but in 2008 when the serious health problems of his two older brothers became known within the family.

Having been in government all his life, Salman was well aware of Saudi Arabia's structural economic problems and administrative inefficiencies. He had watched Qatar and the United Arab Emirates develop more rapidly than Saudi Arabia. He saw talented, educated young Saudis moving to Dubai, New York, and London. Above all, he recognized that the long-running partnership of brothers managing the kingdom could not last much longer. Preserving the dynasty would require a powerful and determined king who could both engineer the transition to third-generation leadership and diversify the country's economy. Intending to rule as a reforming autocrat, Salman was looking for ideas—and his younger son, Mohammed, seemed to have some.

Mohammed bin Salman was the eldest of King Salman's sons by his third wife, Fahada bint Falah bin Sultan al-Hethlayn, whose family are Nejdi aristocrats and the hereditary *sheikhs* of the Ajman tribe. MBS was married to a cousin, Sara bint Mashoor bin Abdulaziz. He had four children, including an eldest son named Salman. His own elder half-brothers had been prominent for many years and appeared more qualified than MBS. One was the Arab world's first astronaut, who flew in the Space Shuttle *Discovery*; another was a respected deputy minister of petroleum; and a third the governor of Medina, with a D.Phil. from Oxford. In contrast, Mohammed bin Salman had held only minor posts while his father was governor of Riyadh, but his youth, ambition, capacity for work, and close relationship with his father set him apart from his siblings.

The rise of Mohammed bin Salman was as remarkable as it was unexpected. In 2010 he was the unknown younger son of the governor of Riyadh; by 2019 he was arguably the single most prominent leader in the Arab world. In 2010 he had held no official position, spending his time trading stocks, developing real estate, and buying expensive cars—most notably, a multi-million-dollar fire-engine-red Bugatti. Stories circulate in Riyadh that some of the firms with which he was associated manipulated stocks on the local Tadawul exchange. Their activities were certainly very profitable. More troubling was a story widely heard in Saudi circles that he had become angry with a notary who delayed a land transaction and mailed the notary an envelope containing a bullet. Far from being intimidated by an unknown prince, the notary went to the minister of justice who, in turn, complained to King Abdullah. The king called in his half-brother, Prince Salman, and shouted "*Imsak Thawrak!*" ("Get control of your bull!").

Chastened by this sharp rebuke, MBS sought an appointment with his uncle, King Abdullah, at which he pleaded that he was a loyal son of the kingdom who asked only to conduct a study on how to improve the nation's economy. That was an inspired request. King Abdullah agreed, and Mohammed bin Salman began his first official project. Over the next two years he worked with various international consulting firms on what became known as the Riyadh Competitiveness Council. In truth, no functioning council was ever created but Mohammed bin Salman learned a great deal about management consulting and economic development in other countries—most notably, the UAE.[2] He also began to develop ideas for improving the Saudi economy and government structure. The *New York Times* columnist, Thomas L. Friedman, would later write, "It's been a long time, though, since any Arab leader wore me out with a fire hose of ideas about transforming his country."[3] Those ideas did not appear spontaneously, or all at once.

Like his father, Mohammed bin Salman received a local education. He graduated from King Saud University's law school where, unlike many other princes, he actually attended lectures and did the work. Characteristically, he subsequently hired the classmate who had been first in the class. This local education left him deeply influenced by Islam and the Arabian traditions of his homeland. He spoke little English, and his worldview differed from the more cosmopolitan and international outlook of his many foreign-educated cousins.[4]

Again, like his father, MBS was a student of Arab and Islamic history who looked to successful historical figures in designing his own strategies. He was particularly impressed by the careers of the first two Muslim Caliphs, Abu Bakr (AD 573–634) and Omar (AD 584–644). The former put down rebellious tribes and successfully unified the Arabian Peninsula under a powerful, centralized Caliphate in Mecca. The latter expanded Muslim influence by seizing Iraq from the Persian Sassanian Empire and Egypt from the Byzantine Empire. In them, the young prince saw leaders with successful domestic and foreign policies.[5]

The core of Mohammed bin Salman's vision was political, not social or economic. A monarchist like his father and grandfather, he was determined to strengthen the Al Saud dynasty, but at the same time he sympathized with youthful Arab Spring protesters who rejected corrupt elites. He was personally pious but did not want a government dominated by obscurantist religious scholars. MBS believed that recent Saudi kings had ceded too much authority to the *ulama*, technocrats, and tribes. He intended to reassert centralized control. He also intended to maintain Saudi predominance in the Arabian Peninsula and increase Saudi influence in Syria, Iraq, and Egypt. Implicitly, this meant competing with Iran while exploring co-existence and cooperation with Israel.

MBS saw himself as the anointed leader of Saudi Arabia's youth, the 50 percent of the population younger than twenty five years old. He understood that King Abdullah had sent thousands of young Saudis to study in the United States, Canada, and Great Britain. Many had been accompanied by wives, children, or siblings. Thus, tens of thousands of young Saudis had learned to speak English while experiencing religious tolerance in open societies. They created a strong, natural constituency for the change he sought to lead. Mohammed bin Salman intended to speak to young Saudis with a voice that was modern and pious, but not democratic.

To some, MBS would become a visionary and a hero; an autocratic reformer fighting to make long overdue changes against a firmly entrenched, reactionary opposition. To others, he was an aspiring dictator—an arrogant, unstable bully, no longer constrained by any individuals or institutions—who was determined to eliminate all opposition and dissent by whatever means necessary. In fact, Mohammed bin Salman embodied a striking combination of both great strengths and real weaknesses.

The young crown prince was openly inspired by this famous grandfather, and frequently refers to continuing King Abdulaziz's legacy. Websites showing the face of Abdulaziz morphing into that of MBS echo the point. Like Abdulaziz, Mohammed bin Salman is physically large and imposing. He works sixteen hours a day, comes to meetings well prepared and, unlike some of his predecessors, he is not prone to drinking, gambling, or womanizing. The crown prince is certainly ambitious, focused, forceful, charismatic, disciplined, and decisive.[6] He believes that years of avoidable indolence, entitlement, and corruption created many of Saudi Arabia's current problems—a situation that he is now determined to tackle. Within Saudi Arabia, the economic and social reforms that he has implemented are widely seen as bold, even courageous.

On the other hand, MBS is inexperienced and focused on princely entitlements. He can be impetuous, impatient, and impulsive.[7] His campaign against corruption and government waste was undermined by his own purchases of a luxury yacht, a multimillion-dollar painting, and a French chateau. His vision of megaprojects, luxury resorts, and futuristic cities may feed inspiring long-term dreams, but critics protest that people need better education, housing, and healthcare today. They protest that one of the first buildings constructed at his entirely new city, NEOM—near Tabuk, in the kingdom's extreme northwest—was yet another palace for himself.[8]

Both Saudi officials and foreign businessmen complain of being summoned on short notice to late-night meetings and then kept waiting five hours for the crown prince to turn up, whereupon he asks good questions, sometimes talks more than listens, and leaves after forty-five minutes. He can be abrupt and critics also note the bad feeling created when, in September of 2018, MBS postponed a state visit to Kuwait twelve hours before he was scheduled to arrive. Many observers believe that the crown prince is surrounded by a circle composed of both competent advisors and a good many self-serving myrmidons who seek to isolate him from bad news. His detractors point out that no one can work sixteen hours a day indefinitely, and that doing so reflects an inability to delegate or organize. Yet even his critics largely agree that such a comprehensive program of social and economic reform would probably not have been under taken by either of the two previous crown princes.

Within four years of ascending to the throne, King Salman has thus dramatically changed Saudi Arabia's political structure. The old system—in which various senior princes ran independent, uncoordinated ministries and where senior technocrats were allied to one senior prince or another—has been dismantled. The king has appointed dozens of new judges and replaced every minister and military service chief, some more than once. Across the Saudi government, all senior technocrats now owe their position not to a variety of princely patrons but solely to the patronage of King Salman and Crown Prince Mohammed bin Salman. The new king and his son have further concentrated royal authority by ending the financial independence that some ministries had long enjoyed.

King Salman appears to have engineered a peaceful handover of power from the sons of King Abdulaziz to his grandsons. Third-generation princes now serve not only as crown prince but in nearly all provincial governor, deputy governor, and royal cabinet positions. Like the young team of brothers that King Faisal assembled in the 1960s, the grandsons of King Abdulaziz installed by King Salman and MBS expect to govern Saudi Arabia into the foreseeable future. This was no small achievement. Many observers would have predicted a far more difficult transition.

However, in making this shift, King Salman has torn up most of the policies that had long underpinned Saudi stability. He has ended the careers of many long-prominent or long-hopeful royal players. He has ended the collective rule of half-brothers that had characterized the Third Saudi State since the reign of King Faisal. He has diminished respect for age within the royal family and dispensed with the requirement for experience in holding senior government positions. He has dismantled the military balance of power among princes that had confirmed consensus rule. Finally, he has used physical force against some members of the royal family and allowed their disputes to become public.

The long-term succession in Saud Arabia is, in fact, now less predictable than it was in 2015. Should King Salman pass away today, organized opposition to the accession of Mohammed bin Salman within the royal family is unlikely because MBS has consolidated control over all civilian and military elements of government. Royal family resistance is nevertheless more conceivable than it was, for example, when Khalid succeeded Faisal—or when Fahd succeeded Khalid.

An even more destabilizing development would be the death of Mohammed bin Salman before his father. MBS certainly has enemies, both foreign and domestic. Saudi Arabia is at war with Yemen's Houthis and, by extension, with Iran and Hezbollah—both of which have sponsored deadly terrorist attacks in the kingdom. Saudi Arabia remains a target of terrorist attacks from al-Qaeda and ISIS, both of which are sworn enemies of the Al Saud. Mohammed bin Salman's aggressive reform agenda within Saudi Arabia has angered many beneficiaries of the old status quo—particularly social conservatives who are potentially violent. It is also possible that a dramatic failure of the crown prince's economic reform program or a series of foreign policy missteps could cause his father to remove the sitting crown prince, something that King Salman has already done twice before with previous incumbents.

So Saudi Arabia now suffers from what investors call "key man risk." Too much is riding on one person. Should Mohammed bin Salman leave the scene for whatever reason, all bets would be off with regard to Saudi stability. There is no obvious replacement. No deputy crown prince has been named, but many princes would like the job and believe that they are well qualified to do it. Dozens of would-be contenders and factions could emerge among King Abdulaziz's hundreds of grandsons. This is precisely what King Salman strove to avoid by orchestrating the rise of MBS in the first place, but it is unclear whether the ageing king would have the physical and mental resilience to repeat that effort. In the absence of Mohammed bin Salman, the kingdom's ongoing economic and social reform programs would certainly lose an energetic champion. However, it is the political consequences of his departure that could be most damaging—as competitive discord over succession could then once more prove to be the Al Saud's Achilles' heel.

Figure 10. King Abdulaziz's 1945 meeting with President Franklin Roosevelt onboard the *USS Quincy* anchored in Egypt's Great Bitter Lake.

Figure 11. Prince Faisal bin Abdulaziz led the Saudi delegation to the opening of the United Nations General Assembly in 1947. He was accompanied by his brothers Talal, Abd al-Rahman and Abd al-Mohsin as well as his young son the future Foreign Minister Saud al-Faisal. Left to Right Talal, Saud, Faisal, Abd al-Rahman and Abd al-Mohsin.

Figure 12. King Saud's 1957 meeting in Washington with Speaker of the U.S. House of Representatives Sam Rayburn (D-Texas).

Figure 13. King Faisal's 1975 meeting with Secretary of State Henry Kissinger in Riyadh. The 1973 Saudi oil embargo had strained bilateral relations.

Figure 14. King Khalid's 1981 meeting in London with Prime Minister Margaret Thatcher.

Figure 15. King Fahd's 1985 meeting with President Reagan at the White House when cooperation in Afghanistan was strong.

Figure 16. King Abdullah's 2007 meeting with Pope Benedict XVI at the Vatican. The first meeting between a Saudi monarch and the Pope.

Figure 17. King Salman's 2015 meeting with President Barak Obama at the White House.

Figure 18. Mohammed bin Salman unveiling his social and economic reform plan Vision 2030 in April 2016.

Part III

Balancing Stakeholders

In a volatile region, the Al Saud have kept their realm surprisingly stable for nearly ninety years. They have benefited from massive oil revenues, Wahhabi doctrines that make political obedience a religious duty, and their own determined will to power. Their rule, while authoritarian, has never been based solely on fear. The Saudi people have accepted and, in fact, supported the Al Saud monarchy, not simply because of its historic role in creating the nation, but also because of the broad coalition of stakeholders that it has built. Although the physical unification of Saudi Arabia was achieved largely through military force, its political unity has been maintained by an ongoing process of give and take among important stakeholder groups managed by the monarchy.

The principal stakeholders in the Al Saud's coalition include the tribes, clerics, merchants, technocrats, and members of the royal family itself.[1] Each stakeholder group has its own interests and leadership elite, with whom the Al Saud have developed mutually supportive relationships. Saudi kings have traditionally consulted these leadership elites in order to build public consensus before setting policy. Government choices favoring one group have often been balanced by subsequent moves favoring another, in a process that emphasized stability more than rapid social or economic change. For many years this slow-motion balancing act successfully avoided both an explosion of accumulated frustrations and a backlash against disruptively rapid change.

King Salman has shifted these priorities. Since his accession in 2015 and the rise of his son, Mohammed bin Salman, the pace of change in Saudi Arabia has accelerated dramatically. Now, rapid social and economic reform take precedence over consultation and consensus building. There is more change, but less stability. Some stakeholders now feel threatened or ignored. The influence and cohesion of the stakeholder elites who long supported the regime has declined. Dissent is now repressed more forcefully, and fear of the government has become more prevalent. The following chapters examine the kingdom's long-established and well-recognized stakeholder groups—who they are, what they want, how their leaders have interacted with the government in the past—and, most importantly, how new policy approaches have affected their relationships with the monarchy.

The Tribes

Saudi Arabia is a large country without a unifying geography or homogeneous population. King Abdulaziz wove its political fabric together from two dozen tribes and a few hundred prominent urban families. When he combined Arabia's geographic regions—the Nejd, the Hejaz, Asir, Al Hasa, and Al Jawf—he also unified tribes with different histories, customs, and traditions of hereditary leaderships. From pre-Islamic times until well into the twentieth century, these tribes were the fundamental social units of the Arabian Peninsula. Before the Al Saud, tribes fought each other for water and pasture. Even today a Shammar tribesman differs significantly from a Shahrani, and tribal quarrels can lead to fatalities—as they did at the 2018 camel beauty contest, an annual month long festival held outside of Riyadh involving thousands of camels.

Most urban Hejazis are the descendants of pilgrims and traders from across the Muslim world who came on the Hajj and stayed. In Mecca there are many Saudis of Indian, Pakistani, and Indonesian origin. Medina is home to many with ancestors from Syria and Turkey. Outside of the Hejaz, however, most Saudis have tribal roots. In the central Nejd region—where large tribes such as Utaibah, Qahtan, and Mutair have hundreds of thousands of members—more than 80 percent of the population identify themselves with some tribal group.[1]

These tribes once had their own armed forces, legal systems, and tax regimes. They controlled the desert and were influential among the settled populations with whom they traded. Today, although no longer providing independent governance structures, they continue to play an important role in maintaining Saudi Arabia's cohesion and stability. The Al Saud's strategy of first destroying the political independence of tribes and then using the remaining tribal structures as a tool for governing a centralized state, has proved highly successful.

If you ask a Saudi whether tribes are still important, the answer you receive will depend very much upon whom you have asked. The Jeddah dentist whose grandfather came from Egypt will tell you that today tribes are finished, totally irrelevant. The National Guard colonel with an American engineering degree will tell you: My family has been Saudi for ninety years. We have been Muslim for 1,400 years. We have been Shammar for 3,000 years. The first thing I did when I joined the Guard was buy a house in my village. That is where I will go if there is ever trouble in the kingdom. Then, even the surgeon will leave his hospital and rejoin his tribe.

That is precisely what happened during the al-Qaeda terrorism campaign of 2003–2007. With law and order breaking down and police officers being killed on an almost weekly basis, tribal loyalties began to re-emerge. Single-tribe camel beauty contests rekindled tribal identities to the point at which the Saudi government banned them and instituted state-sponsored, multi-tribe events judged not by a tribal *sheikh* but by an Al Saud prince. At the same time, Shammar tribesmen began to identify themselves by putting pictures of F-15 aircraft on their cars, and other tribes adopted similar logos until the traffic police stopped the practice.

Tribalism emerged again during the municipal elections of 2005, 2011, and 2015—making clear that the religious establishment and the tribes were by far the most organized political forces in Saudi Arabia and the ones most likely to sway elections. Conservative religious candidates polled best in urban areas, while tribal candidates won in rural communities. Electoral procedures sought to bar the establishment of political parties, but religious voters circulated informal lists of approved candidates. Tribesman needed no lists, they just looked at the candidate's last name.

Tribes mobilize their members through shared ancestry and economic interests rather than religion or political ideology. Blood ties create an indissoluble moral bond so that members share mutual honor and a collective reputation. For the *Bedu* of early twentieth-century Arabia, the tribe was everything: his job, his social club, and his life insurance policy. From the tribe he received economic welfare; social prestige; and, above all, physical protection. Group cohesion, not individual freedom or personal identity were what mattered, since, under the harsh conditions of Arabia, an individual could not survive long without the group.

Before oil, the central economic fact of life in the Nejd was a lack of rain. In the eastern United States, precipitation, including snow, averages between 40 and 50 inches per year. London gets 29 inches of rain per year and even arid Phoenix, Arizona averages 7.5 inches. In the southern Nejd, the average annual total is a mere 4 inches. The rains arrive in autumn as temperatures begin to drop, but some years no rain comes at all. As the British explorer, Wilfred Thesiger, wrote, *Bedu* life revolved around this prospect:

> A cloud gathers, the rain falls, men live; the cloud disperses without rain, men and animals die. In the deserts of Arabia there is no rhythm of the seasons, no rise and fall of the sap, only empty wastes where the changing temperature marks the passage of the year. It is a bitter desiccated land which knows nothing of gentleness or ease.[2]

If it rains, winter provides the best grazing. This is when the *Bedu* disperse into small family groups, taking their herds deep into the desert. In spring the grass begins to wither, and by summer it is dead. Then the *Bedu* must congregate at deep, permanent wells where their camels will live off the fat that they have stored in their humps during the winter. If you are a tax collector or other government official, summer is when you know where to find the *Bedu*. Spring is the time of sandstorms and the harbinger of hot, dry, hard times to come. Whoever coined the term "Arab Spring" was clearly not a Saudi, for to the *Bedu,* autumn, not spring, is the season of hope, rain, and renewal.

Before World War I, the economic life of the *Bedu* depended on the camel – that remarkable combination of a cow, sheep, horse, and dog that provides milk, meat, wool, transportation, and companionship. The camel's ability to survive for long periods without drinking is primarily due not to an ability to store water but to a highly efficient cooling system that allows it to withstand high temperatures with a minimum of fluid loss. In the hot, dry Arabian summer, a camel needs to drink every four days. By comparison, in 115 degree heat (45 degrees celsius) a horse cannot survive without water for more than twenty-four hours. In spring or autumn, a camel must drink roughly every ten days—while in winter, if the grass is good, it will not need to drink at all. Camels can eat desert shrubs that sheep and goats will not touch, and drink alkaline water that is fatal to sheep. Camels can use brackish water, which is unpotable for humans, and turn it into very drinkable milk.

Camels can travel fast and far. When grazing they may move only a few miles every few days, but a migrating herd can cover forty miles a day. Before motorized vehicles, camels were the primary source of transportation in the Middle East and breeding them was one of the Saudi *Bedu's* principal occupations. The *Bedu* sold or traded camels, camel wool, cheese, and butter for the dates, barley, salt, coffee, cloth, and arms that they could not produce themselves.[3]

The second major source of *Bedu* income was raiding. This was an ancient and accepted means of acquiring livestock, arms, slaves, and chattel. Raiding was voluntary, but avoiding it was deemed cowardly. Successful raiders were celebrated and most *Bedu* married only after a few successful raids. By tradition, tribal leaders received one fifth of the loot. Speed, mobility, and surprise were the *Bedu's* weapons. Like a tenth-century Viking longship appearing off the Northumbrian coast, they could travel long distances, appear unexpectedly in overwhelming numbers, strike hard, and quickly disappear before effective resistance could be mobilized. Before motor transport and automatic weapons, camel-mounted tribesmen had no military rivals in the desert except other *Bedu*. They were at the top of the Arabian food chain. Semi-nomadic shepherds riding donkeys, travelers, pilgrims, and settled farmers were at their mercy.

However, raiding (*ghazzuw*) differed from war (*harb*), which was fought over water rights or pastureland.[4] Like the North American Sioux or Cheyenne, the *Bedu* were usually after horses—or, in their case, camels—not scalps. Until the rise of the fanatical Ikhwan, there was an etiquette to raiding: women and children were seldom harmed, nor were men who did not resist. Even in the most heated battles, women were not dishonored and whatever happened to their menfolk they were entitled to a camel for milk and transport to reach their relatives.[5]

The third pillar of the *Bedu* economy was *khuwa*. This was essentially tribute or protection money paid to strong tribes for security by weak tribes, travelers, and settled communities. Once you paid *khuwa*, you had a sponsor in the tribe who was obliged to protect you. *Khuwa* was generally paid collectively rather than by individuals—for example, the Rushaid tribe paid the Mutair, and the Awazim paid the Ajman.[6] The Ottoman authorities paid numerous Hejazi tribes for the safe passage of pilgrim caravans to Mecca, but strong tribes paid no tribute or taxes to anyone and viewed this as a sign of their importance and independence.

Tribes provided, and still provide, social status. 100 years ago, there was a caste system in Arabia. The camel-breeding tribes looked down on the shepherd tribes, who looked down on settled farmers, who looked down on craftsmen, who looked down on slaves. Even today, a Saudi man's social standing depends largely on his tribe or family and his occupation. Within the past decade, Saudi courts have annulled marriages made between women of "noble tribes" and men found to be of lesser status. Despite their teaching that a Muslim's position in the community should depend on his character and piety, the Wahhabi *ulama* have been unable to eliminate tribal prejudices that remain deeply rooted in Saudi society.

The collective reputation of a tribe or family is important to the status of each individual member and is closely associated with the honor of the tribe's women. *Ird*, as female honor is known, has a large sexual component. In the view of conservative Saudis, this honor can be lost not just by promiscuous sexual activity but also by flirting, mixing with men, or just going to a shopping mall unveiled. A woman must avoid not only evil but even the appearance of anything that could be construed as compromising the reputation of her tribe or family.

Although it may seem odd to a Westerner, the loss of a woman's honor reflects poorly on her birth family and tribe. A woman's father and brothers can be more upset about an adulterous affair than her husband because it is their family honor, not the husband's, that she has tarnished. This essentially tribal value has had much to do with Saudi opposition to gender mixing. Tribal honor may seem quaint, something for anthropology textbooks, but it is still very real when foreign embassies have to deal with divorce or child custody cases involving one of their citizens and a tribal spouse.

Finally, the single most important function of the tribe was physical security. In the empty desert where there was no central government, much less a policeman to call, your safety depended on membership in a group. The understanding that your relatives would pursue those who had harmed you for the next several generations was once all that protected you and your family from violence. Westerners can identify with many *Bedu* values such as courage and generosity, but their glorification of vengeance often strikes Western observers as excessively harsh. Yet pursuing revenge was not an option, it was an obligation—and those who failed to seek it lost respect because they endangered the security of their entire tribe.[7]

Membership in the tribe is transferred by the father.[8] A tribal man can marry the girl he met in California and their children will still be part of the tribe; if his sister did the same, her children would lose their tribal status. More importantly, she could damage the chances of marriage of her sisters and even female cousins as some tribal men would not want a non-tribal brother-in-law. If a woman marries outside the tribe, the tribe will be diluted and its ability to provide economic welfare, social status, and physical security reduced. As a result, daughters of tribal families seldom marry non-tribal men.[9]

There are roughly twenty-five major tribes in Saudi Arabia. Among the most important are the 'Anizah, Dawasir, Ghamid, Harb, Mutair, Qahtan, Sbaie, Shahran, Shammar, and Utaibah. Anyone who has spent much time in Arabia will have met people called Utaibi, Mutairi, or Harbi, whose names indicate membership in these tribes. Most tribes have, or at least had, their own geographic region, distinctive attire, songs, poetry, camel brand, customs, dialect, and hereditary leaders. Even today if a man tells you his name is Balawi, Yami, Ghamdi, or Harthi, you will have a very good idea of which part of Saudi Arabia his family comes from and how they fit into Saudi history.

The basic idea of a tribe is that all members are related to a common ancestor; from that ancestor, the tribe divides into smaller and smaller recognized units. The size, number, and even name of these units varies considerably from tribe to tribe, but in general you are a member of a family, which belongs to a clan, which belongs to a section, which belongs to a division, which belongs to a tribe. The family and the clan are economic and residential units. The clan is also the three–five generation unit responsible for seeking revenge or paying blood money to avoid it. The section, division, and tribe become political units. For those interested, the Arabic for family, clan, tribal section, tribal division, and tribe are respectively: *Khams* or *A'ila, Bait, Fakhdh, Butun, Ashira,* and *Qabila.*

This structure closely resembles that of North American Indian tribes. The Iroquois tribe has divisions such as the Mohawk and Oneida, while the Apaches have the Chiricahua and Mescalero. The Sinjara, Abda, Tuman, and Aslam are divisions of the Shammar tribe. Sitting Bull was a Hunk-papa, Lakota, Sioux; Geronimo was a Bedonkohe, Chiricahua, Apache; and the Ikhwan leader, Faisal Daweesh, was a Dushan, Illawa, Mutair. Some tribes are, in fact, more like tribal confederations. The large 'Anizah tribe, for example, has forty-eight major divisions, each with a recognized *sheikh.*[10]

In theory the tribe acts as a unit, but not always. In the same way that some Iroquois fought with the British and some with the Americans during the American Revolution, so the Barga Utaibah under Ibn Bijad joined the Ikhwan Revolt in the 1920s, while the Rouwg under Ibn Rubayan remained loyal to Abdulaziz.[11] This distinction is not entirely forgotten in Riyadh today, by either side.

What would be called nepotism in the West is normal behavior for a tribesman in Arabia. If you belong to a tribe, you help your fellow tribesman and expect him to help you. When you get married, your fellow tribesmen will contribute funds to set up your household. In the workplace, some may disregard a supervisor who comes from a lowly tribe or listen to the son of their *sheikh* even though he is not the boss. Your actions will have consequences. If you do not hire the applicant from your tribe, perhaps no one will show up at your daughter's wedding or your father's funeral. The Arabian tribesman views the Western citizen, with his aggressive emphasis on personal freedom, as selfish to the point of being irresponsible. For him it is the communal group, not the individual, that counts.

A tribal *sheikh* makes important decisions only after consulting the leading men of his tribe. He has limited coercive power and leads not by fiat but by building a consensus. He represents his section to other parts of the tribe or, if he is the paramount *sheikh,* to other tribes. A *sheikh* thus rules by personal influence and persuasion, maintaining his leadership role by successfully resolving internal conflicts and effectively defending the tribe's interests in the outside world.[12]

A *sheikh* must entertain important guests with great hospitality that reflects well both on himself and the entire tribe. As "a father to his people," he knows their problems and renders judgments in their personal disputes. As "a river to his people," he sends food and clothes to the poor, runs an open house for coffee drinking, and frequently hosts tribesmen for meals. There are clear similarities between the behavior of a traditional Arabian *sheikh,* the behavior that Saudis expect from their national leaders, and, until recently, the actual behavior of senior members of the Al Saud family. One thing that a tribal *sheikh* never did was impose a VAT tax on his kinsmen.

Tribal leaders are chosen from recognized families within each tribe or tribal section based on their experience, skill, luck, wealth, and generosity. The position can pass within noble families from father to son or to brothers and nephews. Such families are well known—for example, the Sha'lan of the Ruala, the Dushan of the Mutair, the Hethlayn of the Ajman, the Sharaim of the Murra, or the Jabra of the Sinjara Shammar. Each tribe has, or at least had, a paramount chief known as *sheikh al-shaml. Sheikhs,* at various levels, are officially recognized by the kingdom's Ministry of the Interior with an official warrant and stamp. Today, the most junior government-recognized *sheikh* will be responsible for between 600 and 1,000 families.[13]

Tribes have their own customary law, known as *urf,* which predates Islamic *sharia* law and resembles the notion of "convention" in Anglo-Saxon Common Law. One significant difference between *urf* and *sharia* concerns vengeance. Tribal law accepts the notion of collective revenge while Islamic law holds that a crime is the personal responsibility only of the criminal, not his entire family. Enforcing *sharia* law has been a central element of the Al Saud's nation-building project, which has greatly expanded the organized *sharia* court system and the role of professional judges. Yet even today, tribes often prefer to settle disputes among themselves rather than involve the government. As they point out, "[i]f your son hits your daughter, you do not call the police. You discipline him. It is a family matter and the tribe is your extended family."[14]

Though often poorly understood in the West, the Saudi state and Wahhabi mission have both been profoundly anti-tribal and anti-*Bedu* enterprises.[15] King Abdulaziz was determined to eliminate the political independence of the peninsula's tribes, and while uniting Saudi Arabia's urban centers he systematically divided its tribes. He very deliberately marginalized hereditary paramount *sheikhs* who led entire tribes, and increased the political role of lesser, section chiefs. In northwestern Saudi Arabia the forty-eight divisions of the "Anizah have no single, officially recognized leader. Down on the southern border, the Yam have only three major divisions, but no paramount *sheikh.*

Abdulaziz imposed the *zakat* alms tax on all tribes to make it clear that they were no longer sovereign or independent. He suppressed raiding and abolished *khuwa,* which continued in Syria until 1958.[16] After Abdulaziz signed the 1922 Uqair Agreement, tribesmen who had migrated with their herds since time immemorial and never thought of themselves as anything but Ajman or Mutair, now discovered that they were Saudi citizens, forbidden to cross an invisible line in the sand. In 1953, Saudi land reform legislation effectively ended exclusive tribal use of any grazing areas.[17]

King Abdulaziz's Lebanese-American biographer, Ameen Rihani, wrote, "The Bedu of Central Arabia are as uncontrollable and inconstant as they are superstitious."[18] Theologian Mohammed Abd al-Wahhab would have agreed. He considered the *Bedu* little better than pagans and saw no way to build a proper Islamic state while they resisted central authority, refused to pay religious taxes, and followed their own tribal law rather than the *sharia.* His Wahhabi reform movement sought to suppress superstition, Sufism, and Shi'ism but also to unite the warring towns and tribes of Arabia into one realm where the *imam,* who was always a religious as well as political leader, would enforce *sharia* law. Consequently, Abd al-Wahhab's clerical successors have steadfastly supported the Al Saud's concentration of political power as a necessary step towards building a truly Islamic state.

Unlike the Al Rasheed of Ha'il, who based their state on the strength of the large Shammar tribe, King Abdulaziz was not affiliated with any large tribe. Instead he assembled a multi-tribe coalition and promoted a state based on unifying Islamic beliefs rather than tribal allegiance. His missionaries, like Saudi schoolbooks today, constantly emphasized the Quran's rejection of socially divisive tribal identities. This explicit avoidance of promoting one dominant tribe distinguishes modern Saudi Arabia from many postcolonial African and Middle Eastern states in which tribal or sectarian affiliations still dominate politics. This was not inevitable. When the Sanusi Islamic reform movement spread across Libya in the nineteenth century, it worked exclusively through tribally based lodges to educate the *Bedu*.[19] Unlike the Al Saud's religiously based national ethos, the Sanusi movement promoted the tribal identities that contribute to Libya's current chaos.

King Abdulaziz avoided not only a state dominated by one powerful tribe but also a loose confederation of many tribes, such as has often existed in Yemen or Afghanistan. He wanted a unified state, firmly controlled by the Al Saud family. Through his multiple tribal marriages, that family eventually became a "super-tribe," the one national institution to which many tribes belonged and to which all could eventually express loyalty without a complete loss of their own status. Thus, although King Abdulaziz broke the independence of the Arabian tribes, his sons have consciously made tribal structures a central feature of Saudi Arabia's economic, political, and security architecture.

The traditional *Bedu* lived lives of extreme hardship, carrying all that they owned on their camels and praying for rain. Theirs was a world of poverty, illiteracy, and insecurity. They were often sick, ill-fed, and ill-clothed. There is nothing romantic about hunger, thirst, or sitting in a tent during the Arabian summer. As T. E. Lawrence wrote, "Bedu ways were harsh even for those born to them. For the stranger they were a death in life".[20] While you may still see traces of this life in Syria or the Sinai, in Saudi Arabia it is gone.

Under the Al Saud, life for the ordinary tribesman has greatly improved. There is no more killing for water or pasture. Government wells now provide water for all tribes, and government subsidies make imported barley affordable for every camelman. Mercedes trucks bring water and carry women and children to new grazing grounds. Hired Sudanese herd the camels, which are nowadays fitted with GPS locators and are often more a hobby than a livelihood. *Bedu* tents have generators for large-screen TVs. You can even see *Bedu* children riding their camels wearing a bicycle safety helmet. Things change, even on the lowest rungs of society, when oil revenues are bountiful.

Most of the tribe's traditional economic functions are also gone, long since taken over by the central government. It is the oil-funded Saudi state that provides housing, education, healthcare, security and jobs—not the tribe, and not some extra-governmental organization such as Hamas, Hezbollah, or the Muslim Brotherhood.[21] Still, tribal sensitivities have been carefully respected in the development process. If you examine the enrollment in Saudi elementary schools, you will find that nearly a third of them have less than 100 students. This is a highly inefficient concession to rural tribes, which each wants its own separate school. Unlike in pre-Islamic Republic Iran, where the Shah's White Revolution forcibly consolidated villages in order to provide better public services, in Saudi Arabia tribal preferences took precedence over development efficiency.

There is no National Union of Tribal Sheikhs, which collectively lobbies for tribal interests. Instead, there are roughly 3,000 officially recognized tribal chiefs who have specific political and economic responsibilities for which they are paid both directly and indirectly. Especially in rural areas, these *sheikhs* provide a valuable interface between the bureaucracy and the people.[22]

The northern province of Al Jawf provides a good example of the role played by tribal *sheikhs* in the political and economic system of rural Saudi Arabia. Although some regions have only one tribe, there are three major tribes in Al Jawf: the Shammar, Ruala, and Shararat. Each has a paramount *sheikh* but the governor, a grandson of King Abdulaziz, tries to work through the leaders of each section and clan. In the entire province of Al Jawf, there are roughly 130 *sheikhs* who are technically employed by the Ministry of the Interior. They bring their people's problems directly to the governor, and play an important role in the distribution of government resources.

Each *sheikh* tries to ensure that his tribe gets more than its fair share of roads, schools, clinics, and government jobs.[23] Each *sheikh* knows where to find wanted criminals from his tribe, as well as which families are most needy. He distributes wheat, rice, dates, and cash. He needs to know who the divorced women are, because in Saudi Arabia the social security system covers divorced women as well as widows. Some of these women would be embarrassed to collect their stipend from a government bureaucrat, but they will accept it from their *sheikh*.

The royal governors and the tribal *sheikhs* cooperate to create a mutually reinforcing political relationship, with patronage and compromise remaining cornerstones of the system. The *sheikh* helps to distribute not only cash and jobs but even invitations to the governor's parties. For this he is paid perhaps $15,000 a year, but much of his compensation comes from the prestige of his job. A good *sheikh* is expected to be generous and spend heavily from his own purse. For this reason, some of those eligible actually do not want the job. The governor officially appoints a new *sheikh* upon the death of his predecessor, but he will usually do so only after he has received a signed document from the tribe confirming the man that it wants.

In recent years, the traditional marriage ties between the Al Saud family and the tribal nobility have been augmented by strong business relationships. The government frequently contributes to a local *sheikh's* wealth, and thus his ability to do his job, by supporting his private business interests with government contracts. This can be useful, as it supports the *sheikh* but makes him look less like the government employee that he has, in fact, become. Outside of the kingdom's three main urban areas of Riyadh, Jeddah, and Dammam, members of the local tribal nobility usually play leading roles in the Chambers of Commerce. When the "customary ownership" laws underlying collective ownership of tribal lands ended in 1968, the tribal nobility often ended up among the largest private landowners—again, reinforcing their local status, interest in stability, and support for the Al Saud.[24] Tribal leaders consequently remain loyal to a system that grants them prestige and access to political power, as well as both direct and indirect funding.

Meanwhile, the Saudi Arabian National Guard (SANG) has become the single most important economic institution in rural Saudi Arabia. SANG is a light infantry force that evolved from those elements of the Ikhwan armies that remained loyal to King Abdulaziz. It was formally founded in 1956 and reported directly to King Saud until 1962 when Prince Abdullah became its commander.[25] Nearly all of its members still come from those Nejdi tribes and tribal sections most loyal to the House of Saud. Several of its seven regular brigades are still commanded by Al Saud princes with long military careers. This combination makes SANG the most dependable branch of the Saudi military.

Most of the major Nejdi tribes also have SANG reserve units, known as the *fowj*, with the Utaibah providing the most. Rather like the British officer class before World War I, many *fowj* officers still come from hereditary, noble families—in this case, those of the tribal troops that they command. The only non-royal Saudis still permitted to use the title *amir* (prince) are the *fowj* commanders. Through both its regular and reserve units, the SANG provides training, jobs, and pensions to tens of thousands of tribesmen. Its extensive healthcare system covers a guardsman's extended family—and, thus, a significant portion of the Saudi population.

The Ministry of the Interior also has a specifically tribal security force, known as the Mujahideen. The Mujahideen, who know their local territories intimately, constitute a civilian police force that patrols the thousands of miles of oil and water pipelines that wind though the Saudi desert. King Abdulaziz organized the Mujahideen by *fakhdh* or tribal section. Depending on its size and relationship to the king, each *fakhdh* was designated as having a certain number of rifles, each representing a job. These jobs are valued for both income and prestige and are still passed down from father to son. When demand on the Mujahideen exceeds its capacity, new tribal recruits are allowed to join—but their positions are not hereditary. Tribesmen who once protected travelers for *khuwa* now protect them for the government, but today they get healthcare and a pension as well.

Rural, tribal people are the Saudi stakeholder group least affected by the dramatic social and economic reforms of Vision 2030. What the tribes want most from their government, in order of importance, are security, respect, and financial benefits—and, by and large, they are still receiving all three. Before the Al Saud, tribesmen feared raids and the theft of their livestock. They could not graze in another tribe's area or even move across it to find grass and water. Now their flocks and families are safe; they can graze anywhere they choose and use almost any well in the kingdom. Now, as they say, "the sheep graze with the wolf."

Crown Prince Mohammed bin Salman has been very careful to show respect for the tribes, their *sheikhs,* and tribal culture. This is not what he talks about with *The New York Times* or CNN, but it is a standard part of his stump speech in Saudi Arabia. Both the King and Crown Prince attend Bedouin festivals, camel races, and camel beauty contests. The kingdom's new Entertainment Commission, which is changing Saudi social life, sponsors falcon races as well as rock concerts—and the rock concerts are not being held in conservative, rural areas, at least not yet. The *sheikhs* have access to a private office for tribal affairs in the Royal Diwan, and they are consulted and treated with great respect by provincial governors. No important tribal leaders were detained in the November 2017 anti-corruption campaign.

Like most Saudis, rural tribesmen were not happy when gasoline and electricity prices rose sharply in 2016. However, subsidies for animal feed, diesel fuel, and agricultural water, which are even more important to them, did not change. Nor did the stipends paid to the *sheikhs* who lead tribes and *muariffeen* who adjudicate tribal membership; except where they actually increased. Much to the satisfaction of rural communities, subsidies for growing wheat were reintroduced in 2019. Many of the poorest tribesmen benefited from the new Citizens Fund, which provides means-tested assistance to compensate for increased gasoline and electricity prices. Though no longer led by King Abdullah or his sons, the National Guard remains the principal social safety net for tribal Saudis.

What is currently challenging the importance of tribes in Saudi Arabia is no longer a deliberate government policy to marginalize them but security, prosperity, education, and rural-urban migration. In places where travel to the next village was once dangerous, exposure to foreign cultures through travel abroad has become common. In 1950, roughly half the Saudi population lived in tents with no permanent mailing address while only 5 percent lived in urban centers. Today, 85 percent of Saudis live in towns or cities .[26] The nuclear family is becoming more important than the extended family, and personal choice more relevant than tribal unity. Education is becoming as significant as tribal status in selecting a marriage partner. Even rural tribal couples are marrying later and having fewer children.

As the economy develops and society becomes more open, tribal solidarity—what the fourteenth-century historian and political thinker, Ibn Khaldun, called *asabiyyah*—is gradually eroding, and with it the ability of the *sheikhs* to deliver the support of their followers to the Al Saud.[27] Therefore, the relevance of tribes for Saudi stability lies not in the reemergence of tribes as independent political actors; though that could very well happen should the current political order breakdown. Rather, it is the possibility that in an increasingly modern kingdom the social value of tribal structures and the political importance of their leader's support for the monarchy declines.

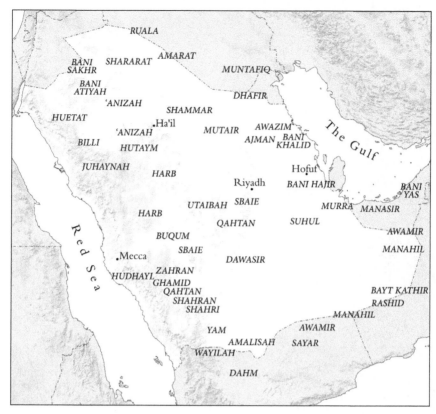

Map 3. Major Tribes of Saudi Arabia.

The Clerics

Early in 2017, the Minister of Islamic Affairs, Saleh Al al-Sheikh, hosted a dinner at his home in Riyadh for the Committee of Senior Scholars, during which Mohammed bin Salman outlined his plans for economic and social reform. The prince told the religious scholars that economic development was crucial to the kingdom's future but could not advance without social liberalization. He assured them that Islam and their role as its guardians would always be respected in Saudi Arabia but insisted that some things would have to change and that their support was both needed and expected.

Like his grandfather, King Abdulaziz, and his uncle, King Faisal, Mohammed bin Salman hoped that by showing the clerics respect he could induce them to agree to his policies. As Faisal had done, he pointed out that social and economic progress are Islamic values. The *ulama* had listened to King Faisal because they respected him. If they did not respect the young crown prince, they at least trusted his father, who had dealt with them firmly but respectfully for fifty years. Moreover, they understood what King Faisal had told them: "People swore allegiance to me, not to you."[1]

King Faisal had used this combination of respect and royal authority to obtain clerical support when he introduced television, ended slavery, and opened schools for girls—none of which the *ulama* had initially wanted. In each instance Faisal had compromised; television stations broadcast more sermons than sitcoms, slave owners were well compensated, girls' education was not compulsory.[2] Even when King Faisal insisted that women needed photographs in their passports, he allowed them to wear a headscarf—but not a veil. Fifty years later, Mohammed bin Salman would open cinemas, end the ban on women driving, and tame the religious police. Like King Faisal, but unlike Mustapha Kemal Ataturk in Turkey or Reza Shah Pahlavi in Iran, Mohammed bin Salman would also make an effort to preserve the dignity, influence, and incomes of the clerics.

There is always a fist in the royal velvet glove. King Faisal imprisoned hundreds of dissidents, including coup-plotting air force officers and pro-Nasser Arab Nationalists. Although King Faisal had welcomed Muslim Brotherhood members fleeing persecution in Nasser's Egypt, King Abdullah had declared them terrorists after the Brotherhood toppled Egypt's president, Hosni Mubarak.[3] Mohammed bin Salman was no exception to this pattern, and in September 2017 he, too, initiated a major crackdown on dissident clerics. Ten mid-grade government clerics, a number of judges, and some thirty non-government clerics who had supported the Muslim Brotherhood or questioned Mohammed bin Salman's reforms went to prison. Those arrested included popular clerics such as Salman al-Auda, Awad al-Qarni, Ali al-Omari, and Abdulaziz al-Tarefe.[4] A very clear message was sent to the kingdom's religious establishment that you were either with the crown prince and his agenda or you supported the Muslim Brotherhood, in which case you were a terrorist.

To appreciate why any of this matters to Saudi stability, it helps to understand that in Saudi Arabia public administration is a religious act. There is no separation between church and state. Government agencies administer religious activities, and most clerics are civil servants working in government offices or state-funded mosques. By law, shops and restaurants close several times a day for prayer. Judges are religious scholars who uphold canon law. The Quran is the Saudi constitution[5] and enforcing *sharia* law is an important government function.[6] The protection and promotion of Islam is a core value that provides part of the Saudi government's legitimacy, much as the promotion of democracy and human rights does in the West. Although these arrangements may seem odd to a twenty-first-century Westerner, they would have been perfectly understandable to an Elizabethan Englishman who recognized very well how heresy and treason could be one and the same.

The followers of Mohammed Abd al-Wahhab do not call themselves Wahhabis. This is a derogatory term coined by their Muslim and non-Muslim critics. They refer to themselves simply as "the Muslims," with the implication being that other Muslims who do not subscribe to their views are nonbelievers. A more neutral term, *muwahideen* or unitarian, is sometimes used because of Abd al-Wahhab's extreme monotheism.[7] At times they are referred to as Salafis because, like several other Islamic movements, they seek to follow the practices of the early Muslims known as the *Salaf*. For the sake of consistency, I have followed the Western practice of calling them Wahhabis.

Whatever you call them, they are evangelists. They practice what is known as the *Dawah*, meaning the Mission or Calling. In the Saudi Basic Law of Governance, "carrying out the duty to spread Islam," as well as the duties of the religious police— "propagating virtue and preventing vice"—are both explicitly spelled out.[8] The Wahhabis are on a mission: they have a gospel to preach, souls to save, heathen to convert, and erring Muslims to reform. They are both ideologically and politically motivated.

Mohammed Abd al-Wahhab's ideas were not particularly original, and he never claimed that they were.[9] He was, in fact, heir to a long tradition of Islamic scholarship that can be traced back to Ahmad bin Hanbal (780–850) and, more specifically, to his much later disciple Taqi Adeen bin Taymiyyah (1263–1328) who lived during the Mongol invasions of Syria. The Mongols destroyed the Abbasid Caliphate in Baghdad in 1258; ten years later they sacked Ibn Taymiyyah's hometown of Harran in modern Turkey. This was a time of great physical destruction, political upheaval, and civil unrest in the Islamic world—what the Arabs call *fitna*, meaning catastrophe or chaos.

Thirteenth-century Islamic doctrine considered fighting fellow Muslims an act of apostasy, and the Mongols claimed to be newly converted Muslims whom their co-religionists should not resist. Ibn Taymiyyah did not see it that way and sought to justify resistance to the Mongols. In his view, anyone who did not behave as a Muslim or failed to follow Islamic rituals correctly was actually an infidel or *kaffir*—and the Mongols' failure to follow *sharia* law annulled their claim to be Muslims.[10] Put succinctly, Ibn Taymiyyah's "big idea" was that not everyone who claims to be a Muslim may be. He went on to argue that *jihad* was obligatory against all those who did not behave as true Muslims. Much as Ibn Taymiyyah deployed this doctrine rallying resistance to the Mongols, the Wahhabis would one day use it against the Ottoman Turks.[11]

The Wahhabi movement's foundation text, Abd al-Wahhab's *The Unity of God*, is a finely argued theological attack on Muslims who associate other deities with God by, for example, calling upon saints or praying at the shrines and graves of holy men. For Abd al-Wahhab these people were polytheists, and his attitude toward them was uncompromising: their lives and property were forfeit. Any person who had accepted Islam and still practiced polytheism should be denounced as an infidel and killed.[12] In early twentieth-century Arabia, the Ikhwan would use this duty to wage *jihad* against non-Wahhabi Muslims to justify their "proselytizing, murdering and plundering."[13]

Mohammed Abd al-Wahhab taught that in an ideal Islamic society, the most important function of the *imam* was the enforcement of *sharia* law. This required a strong, stable government, which made obedience to the ruler essential for a functioning Islamic society. The Quran clearly stated, "O you who believe, obey God, obey the Prophet and obey those in authority among you."[14] Moreover, Ibn Taymiyyah drew a clear distinction between a ruler's private conduct and his upholding of *sharia* law. He believed that "[s]ixty years with a tyrannical imam are better than one night without him." In other words, civil disorder should be avoided at nearly any cost; even an unjust or ignorant *imam* should be obeyed unless he ordered his subjects to contravene the Quran or *sharia* law. Thus, heretics needed to be suppressed not just because they were in religious error but because they ultimately threatened law and order.[15] For the Wahhabis, maintaining political stability became a religious duty.

Abd al-Wahhab taught that a true Muslim must swear and abide by a religious oath of allegiance to an established Muslim ruler if he is to expect salvation on the Day of Judgment. Breaking this oath or *bay'ah* constituted a serious sin. This concept is enshrined in Saudi Arabia's Basic Law of Governance which states: "The citizens shall pledge allegiance to the King and obedience in times of hardship and ease, fortune and adversity."[16] Moreover, Abd al-Wahhab believed that giving advice to a Muslim ruler should be done privately, and only by senior community leaders. He forcefully discouraged public demonstrations as they could lead to civil disorder. He essentially instructed true believers to submit to the will of their rulers, respect their judgment and accept their policy choices without protest.[17]

These two core Wahhabi beliefs—a willingness to declare other Muslims nonbelievers and a deep, religiously based loyalty to the existing political order—were central to the creation of the Saudi state. They have been part of the kingdom's school curriculum for decades and promoted a popular mind-set deeply deferential to political and religious authority. The unregulated condemning of individuals as infidels has often caused serious problems for Saudi society, and has now been made a criminal offense for any but senior scholars.[18] On the other hand, religiously inspired political loyalty to any leader who promotes Islam has been a tremendous source of stability, and helps explain the *ulama's* consistently strong support for the Al Saud.

The Al Saud's alliance with the Wahhabis has always been controversial. For some Muslims, Mohammed Abd al-Wahhab remains a great scholar and theologian. The man who ended centuries of corruption and misguided innovation by restoring the true teachings of the Prophet Mohammed.[19] For others, he was an ignorant eccentric who joined forces with an ambitious, local politician to create a deviant sect that persecuted fellow Sunni Muslims, destroyed historic shrines, and divided Islam with relentless attacks on the Shia and Sufis.[20]

The latter assessment was certainly the view of the nineteenth-century Ottoman sultans. Their campaign against the First Saudi State, between 1811 and 1818, was in fact directed more against what they viewed as the Wahhabi heresy than against the fledgling Al Saud dynasty. As noted in Chapter 1, this campaign—led by the Governor of Egypt, Mohammed Ali Pasha, and his sons—culminated in the destruction of the Wahhabi mission and the Al Saud capital at Dir'iyyah.

That disaster led directly to some very problematic Wahhabi doctrines. As Mohammed Ali's son, Ibrahim Pasha, advanced across Arabia, tribe after tribe and village after village abandoned the Wahhabi movement. Faced with desertions, the Wahhabi scholars became focused on political loyalty and grew even more exclusionist. Mohammed Abd al-Wahhab's grandson, Suleiman Al al-Sheikh, declared that "those who show any approval of the polytheist's religion are unbelievers like them."[21] His extreme xenophobia continues to poison some Saudi classrooms and mosques to this day. It also makes the positive comments about religious tolerance in the Saudi press during Pope Francis's February 2019 visit to Abu Dhabi a noteworthy change.[22]

Meanwhile, in an effort to rally support against the Ottomans, Suleiman's cousin, the future Grand Mufti Abd al-Rahman bin Hasan Al al-Sheikh, developed the view that it was not necessary for *jihad* to be declared by a recognized leader. He argued that it was not the *imam* who made *jihad* but rather the *jihad* that made the *imam*. This was an open invitation to freelance holy war. Many of the *ulama*, then and since, have firmly rejected this view as a threat to the unity and stability of the state—but a dangerous seed had been planted.[23]

The Ottomans wrought terrible retribution on the Al al-Sheikh family as well as the Al Saud. Mohammed Abd al-Wahhab's son and successor, Abdullah, along with hundreds of family members, were marched to Yanbu on Arabia's Red Sea coast and sent into exile in Egypt where most, including Abdullah, died. The fortunate only had their distinctive long beards forcibly shaved off, but many of the Wahhabi *ulama* were executed. Abd al-Wahhab's grandson, Suleiman bin Abdullah, was tortured and then, in the full knowledge that Wahhabis hated music, forced to listen to a one-string Bedouin violin, the *rababa,* before being shot.[24] The chief judge of the Wahhabis had his teeth pulled before being beheaded.[25] With the primary lines of both Mohammed Al Saud and Mohammed Abd al-Wahhab extinguished, the destruction of the Wahhabi mission seemed as complete as that of Carthage at the hands of the vengeful Romans nearly two millennia before.

Yet by 1824, the Al al-Sheikh were back with a new alliance formed among the grandsons of Mohammed Abd al-Wahhab and Mohammed Al Saud. Turki bin Abdullah bin Mohammed Al Saud was the new *imam* of the Wahhabi movement. Mohammed Abd al-Wahhab's grandson, Abd al-Rahman bin Hasan, returned from Cairo to become the chief judge of Riyadh and founder of the most prominent branch of the Al al-Sheikh family.[26] Like the Al Saud, the Al al-Sheikh learned two important lessons from the collapse of the First Saudi State. First, without the Al Saud there would most likely be no Wahhabi mission; second, the movement's political survival may well require theological compromise. Neither lesson is ever far from the minds of the *ulama* today.

Throughout the Second Saudi State (1824–1891) the Al al-Sheikh lent legitimacy to the Al Saud *imams*, who in turn suppressed non-Wahhabi clerics. By the time the Second Saudi State collapsed, Wahhabi Islam—with its extreme monotheism, conservative social values, and emphasis on correctly performed rituals—had become the distinctive religion of the Nejd.[27] The Al Rasheed, who replaced the Al Saud in Riyadh, were themselves devout Wahhabis and governed as such. The Al al-Sheikh did not follow the Al Saud into exile in 1891 but welcomed Abdulaziz's return to Riyadh in 1902.

Under King Abdulaziz, spreading the Wahhabi movement advanced in tandem with the political unification of Saudi Arabia—often by coercion as much as by conversion. In the Nejd, the Al al-Sheikh were restored to a privileged social and political position. Non-Wahhabi clerics were again suppressed and, on at least one occasion, executed.[28] In the desert, Wahhabi missionaries preached to the *Bedu*. These evangelists were men from settled communities who had some basic religious training and went among the people to teach and enforce the correct performance of religious rituals.

During the first two Saudi states, these missionaries had been part-time volunteers.[29] King Abdulaziz transformed them into full-time, salaried employees. Known as *mutawa'a*, they taught people how to make ablutions, how to pray, and what verses of the Quran to read during different prayers. They "flogged all persons caught smoking, wearing fine adornments or procrastinating in their religious duties."[30] They collected the Islamic wealth tax, *zakat*, and kept part of what they gathered.[31] By enforcing the new state's ideology and collecting its taxes, these forefathers of today's religious police helped to consolidate Al Saud authority.

In the Eastern Province, the Shia were required to convert or pay a special tax. They had to renounce specifically Shia holidays and stop making pilgrimages to Shia shrines in Iraq.[32] Some Shia notables were forced, at least nominally, to convert to Sunni Islam, and Wahhabi judges took control of the courts.[33] Some Shia mosques in Saudi Arabia were destroyed, and the building of new ones was prohibited.[34] When the King's advisor, H. St. John Philby asked Abdulaziz if he might consider allowing his Shia subjects to visit their shines in Iraq, the king replied, "I would not object if all their shrines were destroyed and I would do it myself if I had the chance."[35]

The Wahhabis did not limit their religious discrimination to the Shia. In Sunni Islam there are four schools of canon law: Hanafi, Hanbali, Maliki, and Shafi. They have much more in common with each other than any of them do with Islam's Shia or Sufi traditions and can be compared with the different denominations of Protestant Christianity. Most people in the Hejaz were Shafi and all four schools were present in Mecca before 1924. The Wahhabis follow the Hanbali School, and after Abdulaziz took control of the Hejaz, non-Hanbali mosques, judges, and clerics slowly faded away.[36]

In Jeddah, Mecca, and Medina there were many religious shrines and tombs that pilgrims visited. To the local people these shrines were revered places of worship, and important sources of income from pilgrims. To the Wahhabis they were pagan, polytheistic innovations, which focused on individuals rather than God. Like the Taliban in Afghanistan, ISIS in Syria or—centuries earlier—Oliver Cromwell's New Model Army troops during the English Civil War, the Ikhwan set about systematically destroying shrines and religious monuments that they found offensive. Among those they demolished were the birthplace of the Prophet Mohammed; the house of his wife, Khadija; the house of the first Caliph, Abu Bakr; and the mythical Tomb of Eve, which women had traditionally visited in the belief that it aided conception.[37] In Medina they were only just prevented from destroying the Tomb of the Prophet Mohammed, and guards were posted to stop people kneeling in front of it.[38] Like Sunni shrines and Shia mosques, some Sufi lodges were destroyed.

Outside of the Shia communities, Wahhabi Islam gradually became the closest thing that Saudi Arabia has to a national ideology. Although this replacement of local traditions by Nejdi orthodoxy helped to create a unifying culture, it also created lingering resentment among the Eastern Province Shia and urban Hejazi populations.

King Abdulaziz set in place a strategy for dealing with the *ulama* that his sons have largely followed ever since. He showed them great respect, met with them regularly, and consulted with them on religious and social matters. He gave them considerable control over public morals as well as the kingdom's legal and educational systems. As Ameen Rihani wrote, "[t]he *ulama* are the guardians of the Sunna and the Book; and the Sultan is the first to acknowledge this right and accept, within their province, their decrees."[39] The king often gave in to the *ulama* on issues that he considered unimportant. For example, he canceled National Day celebrations, which they found un-Islamic, and cut his robes, which they considered ostentatiously long.[40]

At the same time, King Abdulaziz made it clear to the clerics who the senior partner in their relationship was. Again, as Rihani wrote, "[t]he *ulama* are the power that holds the Sultan and his people together- the medium of control. But they seldom meddle in politics."[41] King Abdulaziz gave them very little say in foreign or economic policy. He introduced important innovations like automobiles, aircraft, the telegraph, and telephones over their protesting heads. He maintained good relations with the infidel British, and invited non-Muslim American oilmen to develop the kingdom's natural resources.

The *ulama* accepted a subordinate position because their involvement in the daily affairs of politics would have violated the ancient division of labor between the *imam* and his religious scholars.[42] Senior clerics also understood that the Wahhabi mission remained controversial outside Saudi Arabia, and that its survival was heavily dependent on the success of the Al Saud dynasty. As long as fundamental Islamic tenets were upheld, social change remained gradual and financial resources continued to flow their way, the senior *ulama* remained firm supporters of the regime.

In 1964 it was the grand mufti, not the new King Faisal or one of his brothers, who announced King Saud's abdication. Yet King Faisal then went on to create a bureaucracy that absorbed and subordinated many of the institutions that the mufti controlled.[43] The king, not the grand mufti, began to appoint judges, religious police chiefs, and members of various scholarly bodies. Secular schools were established. Oil revenues replaced religious endowments and religious taxes as the principal sources of funding for religious activities. Most clerics became civil servants, no longer dependent on their congregations for their incomes. On the other hand, these new bureaucracies greatly expanded the religious establishment's size and authority. Now the *ulama* acquired regular salaries and control over new institutions with generous budgets, large staffs, and formal regulations to enforce.

In 2020 the Saudi state continues to legitimize itself with explicitly religious justifications, and its religious establishment remains a vast, well-organized and well-financed network of individuals and institutions. The most senior clerics are the grand mufti and other members of the Committee of Senior Scholars. They are appointed by the king. There are approximately twenty of them, and they operate through three sub-organizations whose convoluted names are perhaps best translated as the Presidency for Proselytization, the Commission for Religious Rulings and the Council of Senior Scholars. The first two bodies are devoted entirely to religious matters, but the third sometimes becomes involved in political issues as well.[44]

The Council of Senior Scholars is composed of all the members of the Committee of Senior Scholars. It meets roughly twice a year or when called into an emergency session. It makes rulings by majority vote and has dealt with major social issues such as marriage, divorce, inheritance, organ transplants, abortion, and capital punishment. When asked to do so by the king, it also addresses political issues. The council authorized the use of force to retake the Great Mosque in Mecca in 1979 and the arrival of non-Muslim military forces in 1990. It excommunicated both Ayatollah Khomeini and Saddam Hussein, declared al-Qaeda terrorists to be heretics, condemned suicide attacks as un-Islamic, and forbad Saudis from fighting in Iraq during the American occupation of that country.[45]

At no point in the past sixty years have the senior *ulama* openly opposed the Al Saud on matters of political importance. They ultimately ratified the use of force against the Ikhwan, the introduction of the telegraph and radio, the deposition of King Saud, and the education of women.[46] All of these political rulings were intended to support and legitimize existing government policy. Some rulings were unpopular and led to criticism that the *ulama* were not fully independent. That charge is true. The Council of Senior Scholars' budget and agenda are both controlled by the king; those clerics who oppose the government can be, and have been, removed from the council.[47]

However, it is equally true that were a consensus ever to form that royal behavior was truly irreligious, the king could not fire the entire council without seriously jeopardizing his own legitimacy. Moreover, according to the most fundamental Wahhabi doctrine, the *ulama* should, in nearly all instances, support strong, stable government for, as we have seen, they believe that without stable government there can be no *sharia* law or Islamic society. The scholars back the government not just to keep their jobs and salaries but also because the Saudi government gives them independence to promote conservative Islamic social values that maintain their influence in society.[48] Ultimately, the senior *ulama* support the Al Saud because they see themselves not as leaders of an opposition but as respected and influential members of the establishment.

Immediately below the senior scholars come the senior judges who, in Saudi protocol order, outrank cabinet ministers. They are followed by hundreds of lesser scholars and judges, and hundreds of thousands of religious bureaucrats and functionaries. These last-named include the religious police, who are still present throughout the kingdom, as well as the prayer leaders, prayer callers, and sextons who work in some 80,000 government-funded and government-monitored mosques. Among this religious hierarchy is a special body known as the Presidency for the Two Holy Mosques, which administers the large budget and staff of the great mosques of Mecca and Medina, and which began to employ women for the first time in 2018. In addition, there are the faculties of several large religious universities and tens of thousands of religious studies teachers in all the kingdom's schools.

There are explicitly Islamic ministries, such as those for the Hajj and Islamic Affairs. There are also ministries, such as Justice and Education, which, although not specifically Islamic in purpose, have long been bastions of the religious establishment. Finally, there are the Saudi employees of various international Islamic organizations based in Saudi Arabia, such as the Muslim World League, the Organization of Islamic Cooperation, and the Islamic Development Bank.

The most prominent members of the religious establishment remain the direct descendants of Mohammed Abd al-Wahhab, who, over the past two centuries, have produced significantly more senior scholars and judges than any other family in the kingdom. The Al al-Sheikh remain second in influence only to the Al Saud and, like the Al Saud, many receive monthly government stipends from birth.[49] They have often held senior government positions outside the religious establishment, including the Ministries of the Interior and Defense. Al al-Sheikh family members have served as minister of agriculture and chairman of the capital markets authority. Under King Salman, they include among their number the grand mufti, the president of the *Majlis al-Shura* or consultative council, the minister of Islamic affairs, the minister of education, a minister of state, the president of the entertainment commission, the treasurer of the Royal Diwan, and a recently retired minister for municipal and rural affairs.

The Saudi religious establishment is almost exclusively Wahhabi and, in addition to the Al al-Sheikh, has its own partially hereditary leadership elite. Its senior ranks are nearly all Nejdis, and often come from a religious aristocracy that is every bit as real as its royal or tribal counterparts. It consists of families with long traditions of religious scholarship, such as the Al Angari, Atiq, Al Baz, Al Mubarak, and Al Salim. There are usually several self-made men on the Committee of Senior Scholars, whose presence is in part intended to dilute the power of the hereditary religious elite. Since 2009, there have also been representatives of the other three Sunni schools of religious law—but, as of yet, no Shia.[50]

The foot soldiers of the religious establishment include not only the pious but also the poor, the blind, the non-tribal, and the less academic. Admissions standards in the kingdom's religious schools and universities are lower than in their secular counterparts, and for most people it is easier to get a degree in Islamic history than chemical engineering. Like joining the military in the West, joining the Saudi religious establishment is a way out for those lacking connections or other avenues of advancement.

All of these people, from the grand mufti to the *muezzin* at the neighborhood mosque, form an organized, unified interest group. Because modern secularism threatens their livelihoods, social status, and political influence, they have a strong vested interest in maintaining the importance of Islam in Saudi society. *Sharia* law is a principal feature of Islamic civilization that records God's commandments and governs all aspects of life. With ordinances regulating everything from religious rituals, politics, and criminal justice to table manners, divorce, and sickroom conversation, it is the "core and kernel of Islam itself."[51] Thus, defending *sharia* law is one of the religious establishment's most effective means of maintaining its own influence.

The Ottoman Empire was an Islamic state whose legitimacy depended on the upholding of *sharia* law. In theory, the law and the scholars who interpreted it placed a check on the sultan's executive authority. The abolition of the Caliphate in 1924 marked the end of the political system that had governed the Arab world for over a thousand years. First colonial governments and then newly independent, republican Arab regimes sought to replace Islamic institutions with foreign concepts such as elected legislatures, written legal codes, and secular court systems. Nearly everywhere in the Arab world, the *ulama* were marginalized. They became minor officials with no real political authority.

Everywhere, that is, except Saudi Arabia—where there never was a colonial government or secular Arab Nationalist regime, and where the classical Islamic constitutional order in which executive power was counterbalanced by the scholars is to this day preserved in a still-recognizable fashion.[52] The independence of the Saudi *ulama* has clearly been compromised by the king's control of great oil wealth, but they nevertheless remain semi-independent, junior partners in the political establishment and cannot be ignored with impunity.

The *sharia* is based on the Quran, which is the revealed word of God; the Sunna, which is a collection of traditions about the life of the Prophet whose example should be followed; and the Hadith, which records his sayings and judgments. Saudi judges rely on these texts to determine God's will. For them, attorneys, elected legislatures, codified laws, and recorded precedents are at best unnecessary and at worst a dangerous threat to their independence, influence, and status.

It should come as no surprise to learn that Saudi religious scholars have firmly and consistently opposed the introduction of any written constitution other than the Quran, for this would undermine their authority as the sole interpreters of the kingdom's basic legal document. For the same reason, they have opposed the creation of commercial, labor, or criminal legal codes, as well as the establishment of appeal courts that could question the independent interpretations of the *sharia* that each judge feels entitled to make.[53] They prefer *sharia* law with its un-codified body of doctrines, opinions, rituals, and values, which they alone interpret and need not always interpret the same way.

Education is the second pillar of the *ulama's* influence, and they have fought hard to maintain control of it. Traditional Wahhabi education was designed to transmit eternal truths, not to promote critical thinking. The curriculum focused on memorizing the Quran, *sharia* law, Arabic grammar, and enough arithmetic to calculate religious taxes. It was taught by scholars in their homes or in mosques. In many villages the quality of teaching was poor, school attendance low, and literacy limited.[54] Books were very rare. There was no printing press in King Abdulaziz's realm before he conquered the Hejaz.

Riyadh did not get a printing press until 1953. In such circumstances, education remained overwhelmingly religious in format and outlook.

Yet another example of the unusual role of religion in Saudi Arabia is afforded by the kingdom's religious police. Formally known as the Committee for the Promotion of Virtue and the Prevention of Vice, and informally called the *mutawa'a*, theirs is no minor function. The role of the state in "promoting virtue and preventing vice" comes directly from the Quran. Members of the committee are charged with enforcing public morals such as those associated with prayer, fasting, gender segregation, and modest dress. As with its courts, schools, and universities, Saudi Arabia has essentially developed two police forces—one secular and one religious.

For many years the *mutawa'a* constituted a fiefdom of the Al al-Sheikh and were sometimes known as the "Sword of the Al al-Sheikh." Under King Saud they had the authority to arrest, flog, and detain prisoners in their own jails. King Faisal reduced their power and independence. He limited their right to make arrests, incorporated them into the civil service, curtailed their recruitment, and ended the policy of hereditary membership. He unified their previously independent branches in the Nejd and Hejaz and insisted that this unified force report directly to the king rather than the grand mufti. This elimination of an independent police force that did not report to the king was a major step in the political evolution of Saudi Arabia. Yet even King Faisal accepted that, along with consultation and consensus building, patience was a fundamental principle of Al Saud rule—especially when it came to the *ulama* and their control of social mores.

King Faisal played cards each evening with a group of advisors and their sons. One evening in the summer of 1969, the conversation turned to the recent American moon landing. A young man who had just graduated from an American university asked how the king could allow a senior cleric like Abdulaziz bin Baz (1910–1999) to claim that the Earth was flat and that the Americans could not have landed on the moon because it was guarded by angels—surely such nonsense made Saudi Arabia a laughing stock? Shocked by such impertinence, the card players fell silent. The king continued peeling his orange, an act that always took him precisely two minutes. Then Faisal replied. "If I, Faisal bin Abdulaziz, contradict Bin Baz, half the people would believe me and half the people would believe the *sheikh*. It is better to wait."[55] And so King Faisal often waited, and always balanced his reforms with concessions to the *ulama*. Worried about claims that the Al Saud were not Islamic enough, Kings Khalid, Fahd, and, for the most part, Abdullah, did more waiting than reforming as Saudi Arabia's cultural norms slid increasingly out of line with the rest of the world.

King Salman, on the other hand, did not wait. He recognized that times had changed, and with them the demographics, opinions, and aspirations of most Saudis. By 2015, most Saudis were under thirty years of age, and very few thought the Earth was flat. Most thought that women should be allowed to drive. Traditional cultural values, which had provided a valuable stabilizing force during the social upheaval of oil booms, were becoming a liability for an economy that needed to improve its productivity and labor force participation rates. Mohammed bin Salman was very clear about the need for change. Referring to conservative changes that had occurred after the Mecca mosque siege, he stated that "the post 1979 era" was over,[56] adding, "We do not want to end up like North Korea." He spoke about restoring the true, more tolerant and more open Islam: the "Islam of love not fear." He implied that there would be more tolerance for Shia and Sufi Muslims.

King Salman and Mohammed bin Salman have not set out to destroy the *ulama*, dismantle the religious establishment, or diminish the legitimizing role of religion in Saudi Arabia. On the contrary, the king firmly believes in the mutually reinforcing roles of the monarchy and the mosque. He does, however, fully intend to reassert the fact that the Al Saud are the senior partners in this relationship, and that the time has come for more rapid social liberalization. Like his father, King Salman intends to drag the *ulama* along with him rather than push them aside. Unlike some of their predecessors, today's official clerics, the *ulama al-hakim*, support the king because they have learned from history.

To summarize, the Wahhabi Reformation began at a time of widespread violence, political chaos, and pre-Islamic superstition in Central Arabia. The movement sought to restore both religious orthodoxy and political stability, which it continues to regard as two parts of one mission. Its original leaders believed that success required uncompromising orthodoxy. When the Wahhabi army seized Mecca in 1803, the *ulama* sought to prevent non-Wahhabi Muslims from making the Hajj.[57] In 1870, some Wahhabi scholars condemned Imam Abdullah al-Saud for using infidel Ottoman Turks in the struggle with his brother Saud.[58] This doctrinal purity and political inflexibility cost the *ulama* the first two Saudi states. They do not intend to make that mistake a third time.

Instead, the senior clergy supported King Abdulaziz against the Ikhwan in 1928. Unlike the Arab Nationalists, they did not complain to King Saud about the American airbase at Dhahran. During the years of rapid and disruptive oil-fueled economic growth, they provided comforting continuity to people whose lives were changing beyond recognition. Respected senior clerics such as Abdulaziz Al al-Sheikh, Saleh al-Fawzan, Abd al-Rahman al-Sudais, and Saleh al-Luhaidan did a great deal to discredit al-Qaeda in 2003 and ISIS in 2011. They repeatedly preached against terrorism, suicide bombing, unmonitored charitable giving, and government corruption. They strongly condemned terrorist attacks on Saudi Shia citizens despite their continuing doubts about Shia theology.

More recently, their argument for allowing women to drive was both clear and accommodating. Unlike adultery, female driving had never been something absolutely forbidden in the Quran. It had once been discouraged as harmful to society, but circumstances had changed and now the benefits of women driving outweighed the disadvantages. Very similar logic had been used in the past for accepting paper money, bicycles, and even smoking the water pipe or *narghileh*, which Mohammed Abd al-Wahhab had preached should be punished by forty lashes of an ox-sinew whip.

However supportive the senior *ulama* may be, their pragmatism has often been challenged by more junior clerics demanding stricter enforcement of Wahhabi tenets and far less accommodation with Western values. In 1928, some clerics supported the Ikhwan.[59] In 1990, clerics associated with the Awakening, or Sahwa Movement, such as Safar al-Hawali and Salman al-Auda, vehemently opposed the arrival of non-Muslim troops for Operation Desert Storm, and eventually went to prison for their preaching.[60]

In 2003, prominent clerics Ali al-Khudair and Nasir al-Fahd supported al-Qaeda; they are still in prison.[61] Since 2015, several dozen clerics have been arrested on terrorist-financing charges—specifically, for aiding the Al Nusra Front in Syria, which was essentially an offshoot of al-Qaeda.

In 2020, much of the religious establishment and a sizable minority of the general public are opposed in varying degrees to the social liberalization introduced by Mohammed bin Salman. Some want no change at all; some believe that change is simply coming too fast. Many feel that it has been imposed by foreigners and threatens the core values of Saudi society. At the same time, Islam with its emphasis on equality and social justice remains the vocabulary of politics and political dissent in Saudi Arabia. Any future opposition movement to the government—regardless of whether its complaints are political, social, or economic—will probably mobilize support with calls for the restoration of religious orthodoxy, not demands for Jeffersonian democracy.

The most highly organized opposition to the Saudi monarchy today comes from the Muslim Brotherhood. This is an organization that some regard as a terrorist operation and others see as a pro-democracy opposition movement. It has, in fact, at various times and in different places, been both. Although the Brotherhood's clandestine, cell-based membership and many informal sympathizers share the Wahhabis' conservative religious views, they also possess distinctly anti-Western, anti-monarchical orientations that are alien to the traditional, quietist Wahhabi acceptance of the existing political order. Thus, the Muslim Brotherhood, which is outlawed in Saudi Arabia, is potentially a more destabilizing force than the traditional Wahhabi religious scholars.

The question for Saudi stability remains not whether the official government scholars will desert the monarchy—that prospect remains unlikely—but whether they will be outflanked by more radical clerics who incorporate Muslim Brotherhood ideas into their preaching and contest the *ulama al-hakim's* authority to lead the Wahhabi mission. In a kingdom where social media and mosque sermons are closely monitored, it is difficult to gauge the extent of religious opposition. There are certainly some Saudis who believe that the official *ulama* have been un-redeemably co-opted by the state. Whether their numbers grow into a critical mass will depend on both the pace of social change and the success of economic reforms. As the rulers of Iran, Egypt, and Turkey have discovered, there are limits to how fast a Muslim government can push social liberalization without provoking a religious backlash. Likewise, if economic reform falters, restoring a more conservative Islam will certainly become the political platform for those demanding greater economic and social equality.

9

The Merchants

The Third Saudi State has been from its inception, an urban enterprise aimed at settling nomads, protecting settled populations, and promoting trade. Under King Abdulaziz, the political and religious unification of Saudi Arabia was closely associated with, and facilitated by, its economic integration. When Abdulaziz captured Riyadh in 1902, trade between Arabia's regions was limited, primitive, and dangerous. The Indian rupee, British sovereign, Ottoman lira, and Maria Theresa thaler all circulated in Arabian markets. There was no common commercial code, much less a central bank. By the time he died in 1953, all of this had changed. A common currency was introduced in 1928 and a national commercial code created in 1931. The tax code was unified in 1933, as were customs duties throughout the kingdom in 1944. A central bank, the Saudi Arabian Monetary Agency, was established in 1952.[1] For the first time, most of the Arabian Peninsula became a unified market operating with a single currency under a single legal and tax structure. Even before oil, the Al Saud were good for business.

As we have seen, Abdulaziz relied on the *sheikhs* for tribal loyalty and the *ulama* for ideological legitimacy. In his early years, he also relied on prominent merchant families for financing. In many parts of his newly unified kingdom, where the established political and religious elites were swept away by the Al Saud, the traditional landowning and commercial families remained as vestiges of the old order. These urban families formed an identifiable and relatively closed set of Hijazi and Nejdi merchants with long-established wealth and social status. Abdulaziz made alliances with them, and they became part of the Al Saud's national coalition.

Unlike most postcolonial governments in the Arab world, the Al Saud were not socialists or revolutionaries. They never sought to destroy an existing class of landowners and capitalists. On the contrary, they supported the merchant class, relied on them for funding, used their managerial skills, and eventually went into business themselves. Nevertheless, in November 2017, several dozen prominent merchants found themselves detained in Riyadh's Ritz Carlton Hotel on charges of corruption. Others found their bank accounts frozen or their travel restricted. Some are still detained or wearing GPS tracking devices. Some have lost control of their businesses. Many now regard the ongoing economic reforms initiated by the crown prince as part of Vision 2030 to be a direct threat to their business models. Some have begun to question their place in the Al Saud's coalition. How did this come about and how much does it matter for Saudi stability?

100 miles north of Riyadh lies the province of Al Qasim, whose principal towns—
Unayzah, Buraydah, and Ar Rass—lie along the Wadi Rimah, a dry riverbed providing
caravan routes to Iraq. In 1900 the principal towns of Al Qasim were autonomous
emirates, each with its own leading families.[2] These local notables, the *aiyan*, were
always of tribal origin and included leading merchants, major landowners, and local
rulers, who were often one and the same.

The merchants of Al Qasim lent Abdulaziz funds for his early campaigns because
he protected trade.[3] He put an end to raiding, highway robbery, and the payment of
khuwa, the protection money traditionally paid by travelers. Caravans no longer
needed to negotiate with numerous Bedouin tribes in order to cross their territory or
pay fees to do so. In his realm, Abdulaziz provided security and was the only one you
needed to pay for it. The Qasimi *aiyan* became some of the Al Saud's earliest and
strongest supporters. Many of their families—such as the Al Sulaim and Al Zamil of
Unayzah, Abalkhail and Al Sultan of Buraydah, and Al Assaf and Al Malik of Ar Rass—
remain prominent to this day.

Further north lay Ha'il, an important oasis on major trade routes between Mecca,
Iraq, Persia, and India. Its Al Rasheed rulers collected taxes from passing merchants
and often operated their own caravans.[4] However, unlike Al Qasim, Ha'il's fall to the Al
Saud in 1922 initiated its economic decline. When the Al Rasheed lost power, new
judges and rulers arrived from Riyadh. Many Al Rasheed supporters left to join the
Northern Shammar tribe in Iraq. Yet even in Ha'il, the leading merchant families
maintained their status and influence—particularly those families, such as the Al
Subhan, that had helped to negotiate the Al Rasheed's departure.

In contrast to Ha'il, Jeddah and the Hejaz prospered under Abdulaziz. Before World
War I, the Hejaz was the most economically developed region of the Arabian Peninsula.
As the Swiss traveler, Johann Ludwig Burckhardt, wrote in 1814, "[t]he people of Jeddah
are almost entirely occupied with commerce. They are all either seafaring people,
traders by sea or engaged in trade with Arabia."[5] The Hejazi merchant class consisted
of a few families who controlled large-scale international trade as a specialized
profession. They imported rice, sugar, grain, tea, and textiles from India, and their
homes were by far the grandest in Arabia.

When Abdulaziz conquered the Hejaz in 1924, the region lost its political and
religious independence but not its commercial importance. The first foreign bank in
the kingdom, the Netherlands Trading Society, opened there in 1925. It was soon
followed by the locally owned National Commercial Bank. In 1946 the first Saudi
Chamber of Commerce was founded in Jeddah, and a second in Mecca in 1948. By
contrast, the Riyad Bank was not founded until 1957 and the capital did not establish
a Chamber of Commerce until 1961.[6]

Although Abdulaziz eliminated the Hashemite political elite, the non-Wahhabi
religious establishment, and the powerful local guilds that had monopolized the Hajj
industry, he did not destroy the Hejazi merchant class.[7] He maintained Jeddah's role as
the kingdom's business center, and the established merchant families such as the
Alireza, Zahid, and Al Nasseef became prominent members of the Al Saud's coalition.
As in Al Qasim, Abdulaziz borrowed from them to fund his local administration and
often repaid them with land that has become the basis of many modern Hejazi fortunes.[8]

Throughout his realm, leading merchant families furnished King Abdulaziz with more than loans and taxes. They provided local leadership. Many *aiyan*—such as the Alireza in Jeddah, Al Zamil in Unayzah, Al Quraishi in Ha'il, Al Assaf in Ar Rass, or Al Gosaibi in Al Hasa—helped to secure local allegiance to the new ruler just as the tribal *sheikhs* had done with the *Bedu*. These leading merchant families also helped to manage the new Saudi state. They provided talented, well-traveled, relatively well-educated men who served as Abdulaziz's agents abroad. These men were usually not full-time diplomats; they were Saudi merchants with business in foreign cities. In Cairo, King Abdulaziz used Fawzan al-Sabek of Buraydah, who helped to arrange the first geological survey of Saudi Arabia. In Kuwait, Abdullah al-Naffisi was the King's agent for many years. In Bahrain, it was the Al Gosaibi family that Abdulaziz relied on as purchasing agents and diplomats. Under later Saudi kings, the Alireza family provided ambassadors to Washington, Cairo, and London as well as two ministers of commerce.

As the Saudi bureaucracy expanded, the urban, business elite provided much of its senior administrative talent. Kings Fahd and Abdullah appointed Ghazi al-Gosaibi minister of health, minister of industry, and minister of labor as well as Saudi ambassador to Britain and Bahrain. One of the Al Quraishi became governor of the Central Bank, while an Al Gosaibi and an Al Malik became deputy governors. The Al Zamil provided a minister of industry, a deputy minister of commerce, and a chairman of the Riyadh Chamber of Commerce. Outside of a few security-related or religiously focused ministries, neither the kingdom's tribal aristocracy nor its religious elite ever played such a major role in the civilian bureaucracy as the leading merchant families.

The arrival of the oil company, Aramco, liberated the Saudi state from its financial dependence on customs duties, agricultural taxes, pilgrim fees, and merchant loans. Between 1938 and 1958, the Saudi government's annual revenue rose from $7 to $180 million.[9] As King Abdulaziz became less dependent on their loans, the political influence of established merchant families inevitably declined. The state became the most influential feature in the Saudi economy, but as the business community lost independence it also became more open. Along with government employment and state-funded foreign education, government support for the private sector economy has been the greatest source of social mobility in Saudi Arabia over the past forty years. It has provided opportunities that allowed countless Saudis to move up the economic ladder.

Becoming an Al Saud prince, a tribal chief, or a senior religious figure is not something to which the average Saudi can realistically aspire. Access to these elite positions is still largely hereditary. The modern business community is different, and much more accessible than the pre-Aramco merchant class. Although it still includes the traditional Hejazi merchant families, and the established Nejdi *aiyan*, today the Saudi business community also includes many men from humble backgrounds. Prominent business families such as the Olayan, Al Subaie, and Al Rajhi were not plutocrats before oil. Suleiman Olayan spent the first nine years of his career as a storekeeper for Aramco.[10] Mohammed al-Subaei's first job was as a water carrier in Mecca.[11] Osama bin Laden's father started his working life as a porter in Jeddah carrying luggage for pilgrims.[12] Political stability, economic policies designed to circulate oil revenue through a rapidly growing market economy, and commercial talent allowed all of their families to join the business elite.

Today's Saudi business community includes immigrants, young entrepreneurs, and a growing number of women. The immigrants have come from across the Arab and Islamic world. They often arrived as pilgrims or refugees, took up Saudi nationality, and found a place in the kingdom's expanding business community. The most notable were the Hadrami community, which moved to the Hejaz from South Yemen in the first part of the twentieth century. Based largely in Jeddah, families like Bin Mahfouz, Bin Laden, and Bin Zager became leaders in banking, construction, and retail. Others arrived from Palestine, Syria, Egypt, India, and Iraq.

Unfortunately, many government policies designed to distribute oil revenue through the private sector were not genuinely productive. Government procurement contracts became the most common means of supporting private sector business, and after the 1973 oil boom, many fortunes were made selling the government everything from aircraft engines to school desks—but nearly all of these goods were simply imported. Traders, not producers, profited. Maintenance and operations contracts for everything from managing hospitals to cleaning city streets were given to private Saudi contractors, but nearly all the labor was imported. Profits were often based on cheap labor, not capital investment.

Construction boomed; Bin Laden became the largest Saudi contractor and many others, like Saudi Oger, Al Seif, and Al Mabani, grew rapidly. But here again, most of the work was done for the government and most of the workers came from South Asia. With more money in the system, the money-changing and banking families such as Al Rajhi, Bin Mahfouz, and Kakki saw their businesses expand exponentially, operating Islamic bank accounts that paid no interest to customers and placing their deposits in interest bearing government development bonds, or simply taking a cut of the wages sent home by millions of foreign workers. Real estate prices soared, creating still more Saudi millionaires whose wealth was based on speculation and trading rather than on production.

The Saudi government also supported the business community with a series of restrictive commercial regulations designed, not to promote competition and efficiency, but to provide easy incomes for Saudi citizens. Agency and sponsorship laws became the modern version of *khuwa*. Under the new sponsorship system, foreigners no longer paid Saudis for physical protection but for the right to work in the kingdom. Likewise, any foreign firm wishing to sell products in Saudi Arabia had to have a local agent. Many Saudis outside the traditional network of merchant families now became agents for foreign manufactures or sponsors of foreigner workers. Those fortunate enough to become agents for prominent automobile, electronics, or retail franchises made fortunes.

For those who did want to manufacture, the government provided generous financial support through the Saudi Industrial Development Fund (SIDF). Established during the period of the country's first five-year plan, the SIDF supported hundreds of successful private-sector factories with low cost loans—each dependent on the owners putting in some of their own capital and procuring additional funding from private banks.[13] Heavily subsidized power and water, free healthcare for their foreign labor force, and tariff protection all helped to launch Saudi industrialists—primarily in plastics, food processing, and building materials.

Petrochemicals was by far the largest new industry in Saudi Arabia. Initially, the sector was monopolized by the state-owned Saudi Arabian Basic Industries Company

(SABIC). Later, in the 1990s, private-sector firms were encouraged to participate and given access to the very low-cost feedstocks that practically guaranteed their profitability. However, the government insisted that these private petrochemical companies float shares on the local stock market. The founders profited from these initial public offerings (IPOs), and the Saudi public became shareholders in the nation's largest industry after oil. Eventually, even one third of SABIC was sold to the public.

Then there was agriculture. During the 1980s and 1990s, farming became a recognized method of supporting rural populations and rewarding loyal government service. Some large farms were owned by prominent individuals while others were public, joint-stock companies. All benefited from huge government subsidies providing low-cost agricultural loans for equipment and seeds, free access to limited groundwater, subsidized animal feed, subsidized diesel fuel for water pumps, and subsidized electricity for everything from air-conditioned chicken coops to cold showers for dairy cows.

In the early 1980s, government grain silos were paying local farmers ten times the world wheat price. In the first years of the program, it became profitable to buy wheat in Rotterdam, ship it to Yemen, smuggle it across the border, and deliver it to the unsuspecting Saudi grain silos. The Ministry of Agriculture eventually realized that it was buying, at exorbitant prices, far more wheat than Saudi Arabia could possibly produce and initiated a system of quotas. Unfortunately, by 2000 it was clear that in some regions of the country groundwater was being badly depleted. Wells dried up and lakes disappeared. The Ministry of Agriculture began to encourage water conservation methods and dry land crops. By 2016, the formerly extravagant wheat subsidies were gone and the kingdom was once again a major grain importer, though for small farmers the subsidies have now been reinstated.

These policies were neither pointless nor poorly thought out. Although economically inefficient, they were politically powerful. Whether in trade, industry, construction, or agriculture, the Saudi government supported the private sector with contracts, loans, subsidies, and protective tariffs, all in a deliberate effort to rapidly transfer oil revenue to the public. As Professor Mordechai Abir noted, "[u]nlike the Shah's Iran where only a small, self-indulgent upper middle class monopolized the country's oil wealth, the Saudi regime prudently channeled it, however unevenly, to all Saudis."[14] These policies secured the allegiance of the country's business community. They also created an economy that, although nominally based on market principles, was badly distorted, more distributive than productive, and could seldom compete globally in anything other than hydrocarbon-based products.

When the first oil boom arrived in 1974, Jeddah was the unquestioned commercial capital of Saudi Arabia. It was home to all the kingdom's foreign embassies, bank headquarters, and most large foreign corporations. As Abdulaziz's viceroy, King Faisal had lived in the Hejaz for much of his life. He was sympathetic to the Hejazis and recognized that they were still his best educated subjects. Accordingly, he used many of them to staff his new ministries.

In the Eastern Province, Aramco adopted a policy of enlightened self-interest. It built schools and hospitals for its employees, improved their nutrition, and eradicated malaria. The oil company drilled water wells, paved roads, and built power plants. It initiated successful programs encouraging employees to buy their own homes and start their own businesses; many did both. Private firms sprang up to supply Aramco with everything from drill pipe to bread. Sulaiman Olayan went from employee to entrepreneur when he won a contract to unload pipes. Aramco and Bechtel gave him the contract for 25 percent more than he had bid because they wanted to encourage local businesses.[15]

King Khalid and Crown Prince Fahd saw that compared with the peninsula's two coasts, the central provinces of the Nejd were poor and falling behind economically. Of the seven largest private firms in Saudi Arabia, only the General Motors dealer, Al Jomaih, was based in Riyadh.[16] The king and crown prince no doubt also recognized that the Nejd, with its tribal orientation and religious conservatism, had been, and probably always would be, the region of the kingdom most devoted to the House of Saud. They began an affirmative action program for Nejdis in both business and government.

The results were unmistakable. Whereas in 1971 there had been 14,000 registered businesses in the Nejd and 26,000 in the Hejaz, by 1983 there were 73,000 in the Nejd and 57,000 in the Hejaz. Between 1975 and 1981 the size of industrial zones in Jeddah grew from 1 to 4 square miles, while in Riyadh they swelled from 0.45 to 5 square miles.[17] The traditional terraced farms of the southern Asir Province, bordering on Yemen, were abandoned as mechanization, irrigation, and subsidies moved farming to the new wheat fields of the Nejd. In 1980, Jeddah was the largest city in Saudi Arabia; by 1990, Jeddah and Riyadh were roughly the same size. In 2020, Riyadh, with over seven million people, is the most populous city in the Arabian Peninsula, and distinctly larger than either Jeddah in the Hejaz or the Dhahran-Dammam conurbation in the Eastern Province.

Foreign embassies all moved from Jeddah to a new purpose-built, architectural award-winning diplomatic quarter outside of Riyadh. Most major corporations followed and even the large Jeddah trading families recognized the need to open offices in Riyadh, which was the source of government contracts. All of the banks except National Commercial Bank (NCB) moved their headquarters to the capital. The government switched its very profitable accounts from NCB, which was owned by Jeddawi money-changers, to Riyad Bank, which was primarily owned by Nejdi *aiyan*.[18] The business elites of the Hejaz and the Eastern Province were not ignored—they remained prosperous and influential—but preference often went to Riyadh.

A similar pattern of change occurred within the Nejd as Riyadh became Saudi Arabia's dominant economic center. The caravan trade that had supported provincial cities disappeared, and all of the kingdom's newly paved roads led to the capital. Cash payments in a single national currency replaced barter, and the banks were now headquartered in Riyadh. As low-paid foreign workers replaced Saudi shopkeepers and sharecroppers, thousands left Al Qasim to find better paid, but not necessarily more productive, jobs in Riyadh's rapidly growing government bureaucracy. Those who remained behind usually found some form of local government employment.

Today, places like Ha'il and Unayzah are provincial market towns, peripheral to the national economy. The last camel caravan left Unayzah in 1945. Local merchants no longer import or export directly. Nearly everything available in their shops is manufactured abroad, and imported by large commercial establishments in Jeddah or Riyadh.[19] These towns benefited from joining a large, integrated, national economy but—like the tribes, clerics, and large merchant houses—the kingdom's provincial economies became increasingly dependent on the oil-funded, centralized Saudi state.

Unlike the Al Rasheed of Ha'il, the Al Saud had not traditionally engaged in commerce. Abdulaziz sought to promote the merchants' prosperity because he relied on them for taxes, customs duties, and loans. He did not compete with them and instructed his sons to stay out of business. King Saud continued his father's policy, and in 1956 and 1959 issued royal decrees prohibiting princes and civil servants from engaging in private business.

King Faisal, however, recognized the need for change. With more and more princes coming of age, they could not all be given large stipends or senior government positions—nor could they be prohibited from earning a living. King Faisal's own son, Abdullah, had served as minister of the interior but wanted to go into business. When a new decree was issued in 1976 allowing members of the royal family to engage in commerce, Prince Abdullah al-Faisal became Saudi Arabia's Sony dealer.[20] This fundamental legal change ensured that the Al Saud would eventually join the kingdom's commercial, as well as its social and political, elite.

There are now thousands of Al Saud family members, both men and women, engaged in commerce.[21] Some are silent partners in private firms in which they have contributed influence and government connections more than capital or effort. Some have certainly used their political status to advance their business interests. Others have diligently built their own businesses, though they too have benefited from government contracts, low-cost loans, and subsidies. Among the most well-known have been Faisalia Group CEO Mohammed bin Khalid al-Faisal, Khalid bin Abdullah bin Abd al-Rahman, who launched the Mawarid Group, Sultan bin Mohammed Saud al-Kabeer, who established Almarai Dairy, and the founder of Kingdom Holdings, Waleed bin Talal bin Abdulaziz. These commercially successful princes have aligned their personal interests with those of an essentially market-oriented business community. They have a clear vested interest in protecting and developing the private sector rather than in attacking it.

All stakeholder groups in the Al Saud's coalition have access to the political leadership, but the business community has the most formalized and democratic structures for this purpose. Businessmen and women are represented by the Chamber of Commerce and Industry, which, unlike many countries, is not a voluntary organization. There are more than twenty-five local Chambers of Commerce, and membership in one of them is legally required in order to be allowed to open a business. Each Chamber has a board of directors, one half of which is elected by the membership while the other half is appointed by the minister of commerce. The first female board members were government appointees, and other women have subsequently been elected. Each Chamber has a permanent staff under the direction of a secretary general. The local Chambers are united under the umbrella of the Council of Saudi Chambers, which is located in Riyadh.

The Chambers of Commerce work through sector-specific committees and subcommittees. In a large Chamber, like that in Riyadh or Jeddah, there may be sixty committees. In the past these were dominated by traders, contractors, and industrialists, but today accountants, lawyers, and consultants are among the more important sectors represented. New committees for photographers and graphic designers have sprung up alongside the more traditional ones for jewelers and fishermen. Local sector committees elect a chairman who serves on their national sector committee. Sector committees generally have a secretary on the Chamber's permanent staff and must meet at least quarterly.

The views of these committees are routinely sought by the government on most commercial, and some social, issues. Chamber of Commerce committees review new legislation, and are free to lobby for changes to existing laws and government policies.[22] Their national leadership meets formally with the relevant ministers; *Majlis al-Shura* committees; senior members of the *ulama*; and the Committee of Experts at the Royal Diwan, which supervises the technical drafting of legislation. Should they need it, they also have access to the king. The business community certainly does not get everything that it asks for, nor does it always present a unified position. Nevertheless, the Chamber of Commerce has helped to consolidate the business community's support for the monarchy because it assures that its members' views are heard and considered by economic policymakers.[23]

What the Saudi business community has wanted most is political stability—followed by government-funded economic growth, abundant subsidies, limited regulation, and no taxes. It has generally supported trade liberalization, intellectual property rights protection, and privatizations, while resisting pressure to employ Saudi nationals or provide health insurance to foreign workers. On some issues, like anti-trust legislation, it has been divided. When oil prices are low and the government is not spending, the business community complains loudly about a lack of new government contracts and delayed payments for existing ones.

Recently, what the Saudi government has sought most urgently from the business community has been well-paid jobs for Saudi nationals. Foreign workers fill 90 percent of non-oil, private-sector jobs in Saudi Arabia. Saudi employers like it that way because expatriates are usually less expensive, better qualified, and more easily dismissed than Saudis. Of roughly four million working Saudis, only 600,000 are employed in the non-oil private sector; the rest are civil servants. But the public sector can no longer absorb the majority of Saudi jobseekers. Although this problem has been recognized for many years, the collapse of oil prices in 2014 made finding a solution urgent.

The Saudi government's solution was Vision 2030, about which there will be much more later. Vision 2030 is a wide-ranging social and economic development plan created by Mohammed bin Salman and a small army of international consultants, who have earned billions of dollars preparing and implementing it. Introduced in 2016, the program's principal economic objectives are balancing the government's budget, reducing the budget's dependence on volatile oil revenue, and increasing the non-oil private sector's share of GDP from 40 to 65 percent within fifteen years. At least initially, the impact of Vision 2030 on the Saudi business community was largely negative.

As the government cut spending to balance the budget in 2016 and 2017, many Saudi businesses saw sales decline by 20 to 30 percent. Real estate prices fell, and numerous projects were canceled or delayed. Government contracts became harder to find, and margins were squeezed. Payments were often delayed and "Saudization" quotas were enforced more firmly. Subsidies for electricity, water, gasoline, and petrochemical feedstocks were sharply reduced, and existing business fees increased. New property and value-added taxes came into effect—as did more demanding ways of calculating the Islamic *zakat* tax.

Vision 2030 specifically sought to eliminate middlemen in government contracts. Many influence peddlers and commission agents, who had grown fat on cost-plus contracts, found their services no longer needed. Some major business figures discovered that their close ties to senior princes no longer guaranteed projects or prompt payments. For the first time in Saudi history, the new VAT tax meant that businesses would have to keep accurate records that tax authorities could monitor. Some owners, wishing to avoid such scrutiny, simply shut up shop.

Residence fees for foreign workers rose sharply, and as a result nearly two million expatriates left the kingdom between 2017 and 2020—taking with them demand for everything from apartments to groceries.[24] As skilled foreign workers departed, labor productivity declined and labor costs rose; by the end of 2017 the kingdom was in recession. All of these difficulties fell most heavily on the owners of small and medium-sized businesses, who had come to rely heavily on cheap expatriate labor and who lacked the economies of scale or access to capital needed to compensate for their loss. Thousands of these businesses failed.

Nevertheless, like the *ulama al-hakim*, the Chamber of Commerce usually supports government policy. The Chamber's leadership endorsed Vision 2030—specifically, its goals of privatizing state-owned enterprises, private-sector growth, and regulatory reform. As we have seen, some clerics have considered senior *ulama* too accommodating of the Al Saud's social reforms. Likewise, some small businessmen believe that their leaders have been too accepting of Vision 2030. They had played by the accepted rules for thirty years, only to have the rules changed overnight and their business model destroyed. Unlike senior business leaders with close ties to the government who tended to support Vision 2030, these smaller businessmen were more likely to face bankruptcy. Some complained bitterly that they were being ignored, while young people, women, and civil servants became the new preferred stakeholders in the Al Saud's coalition.

In November 2017, the anti-corruption campaign saw major business figures, both royal and non-royal, detained for two months and forced to repay large sums of money. Members of long-prominent merchant families, who had once considered themselves nearly as untouchable as princes, were jailed and stripped of their property holdings. Important commercial families across the country worried that they, like the Bin Ladens, could lose control of their businesses. Some prominent businessmen remained in prison because they could not or would not repay what the government said they owed. The Jeddah merchant class felt that King Salman had singled them out for especially harsh treatment. Some prominent royal family entrepreneurs lost control of their business empires, as well as their equestrian stables and charitable institutions.

New accusations, investigations, and arrests continued long after most of the original guests at the Ritz Carlton went home. Although many businessmen welcomed what they saw as a long-overdue leveling of the commercial playing field, they were disturbed by allegations of torture and a lack of transparency surrounding the financial settlements. Some major businessmen left the country, moved their families abroad, or bought second passports—with EU member Malta being a favorite destination. Others simply parked their cash and refused to make new investments, expand their business, or hire Saudi nationals.

It is impossible to know how much private money left Saudi Arabia after the Ritz Carlton round-up because capital flow data includes government as well as private funds. Some private money certainly left and, although there are still no formal capital controls in Saudi Arabia, capital flight reached a point at which informal restrictions had to be implemented. Large transfers abroad are now likely to require an explanation about where you got the money in the first place. Transfers by royal family members are being scrutinized with particular care and, in some cases, reportedly denied.

The merchants have always been an important part of the Al Saud's coalition and today Crown Prince Mohammed bin Salman needs them to help fund Vision 2030 every bit as much as his grandfather needed their loans in the 1920s. The Saudi government needs private-sector investment in order to solve its most pressing problems—job creation, affordable housing, power generation, and economic diversification—but in 2020, members of the country's business community may no longer be interested in helping. They are unlikely to take their protests to the streets, but they may well take their cash to Geneva—leaving Saudi Arabia poorer and less stable.

The Technocrats

Much has been written about the Al Saud's alliance with Wahhabi clerics, urban merchants, and rural tribes whose leaders form Arabia's traditional elite. These prominent individuals dominate the kingdom's religious, judicial, commercial, and rural affairs. Their status is based on tradition, heredity, wealth, and religious scholarship.[1] Many come from families that supported King Abdulaziz in his campaigns to create Saudi Arabia in the early twentieth century and their enduring alliance with his sons has enhanced political stability.

Less often appreciated is the Al Saud's relationship with a new technocratic elite based on merit and a secular higher education that was often acquired abroad. The kingdom's technocrats do not have large, easily mobilized followings like the *sheikhs* or the *ulama*, nor are they characterized by their ownership of land and capital like the *aiyan*. They are bureaucrats in development-oriented ministries that provide healthcare, communications and transportation. They are chemical engineers at petrochemical giant, SABIC; pilots at the national airline, Saudia; petroleum engineers at Saudi Aramco; mining geologists at Ma'aden; and electrical engineers at the Saudi Electric Company. They are professors, research scientists, bankers, journalists, and newspaper editors; some serve as military officers. Many are pious, tribal, or from established merchant families. What makes them a distinct group is their technical education and expertise in various secular fields. They constitute the newest, fastest-growing and least-organized stakeholder group in Saudi Arabia.

Like nearly all institutions in Saudi Arabia, the technocrats are the result of both oil wealth and deliberate policy choices made by the Al Saud. King Abdulaziz valued any technology that could help him unify the country. In 1902 he rode a second-rate camel from Kuwait to retake Riyadh, carrying a sword and wearing chain mail that you can still see in the National Museum. Twenty-six years later, he rode in a Chevrolet to his final showdown with the Ikhwan at Sabillah, where his British machine guns quickly decided the battle. Nineteen years after Sabillah, he flew to Dhahran in his own DC-3 to witness the first shipment of Aramco oil. The most interesting room in King Abdulaziz's palace is not one of the grand salons where he conducted court business but the small, easily overlooked, closet-like radio room. From here, the king rapidly communicated with his realm at a time when his opponents communicated with each other at the pace of a camel. From here he also remained in touch with the outside world, keeping several translators busy as he listened to the news from London, Berlin, and Moscow.

Although obtaining the latest technology, especially in security-related fields, has long been part of the Al Saud playbook, even King Abdulaziz's relatively simple equipment required skills not then found in Arabia. The king did not hesitate to use foreign experts, but from the outset he was determined to train Saudis to use new technologies. In 1929, Abdulaziz sent four young Saudis from the Mecca Post Office for a ten-month radio operator training course with the Marconi Company in England.[2] In the 1940s, he sent Saudi students to study secular subjects in Egypt.[3] This was a very different policy choice than that made by some of King Abdulaziz's contemporaries in the Arabian Peninsula. For example, Imam Yahiya Hamid al-Din of Yemen (ruled 1904–1948), Sheikh Shakhbut al-Nahyan of Abu Dhabi (ruled 1928–1966), and Sultan Said al-Said of Oman (ruled 1932–1970) all vigorously sought to isolate their people from outside influence and technology; in the case of Sultan Said even banning the use of sunglasses. Perhaps not surprisingly, all three rulers were either deposed or assassinated.

Despite the *ulama's* opposition and a brief hiatus during the reign of King Saud in the 1950s, the Saudi policy of sending students abroad to study technical fields, at government, corporate, or personal expense, has now been in place for three generations. Whether trained at home or abroad, technically educated Saudis have become the core of a new, Saudi upper-middle class. Although religious scholars guard the kingdom's Islamic values and tribal leaders promote its internal cohesion, it is largely Western-educated technocrats who have transformed oil revenue into economic development.

King Abdulaziz ruled through a set of loyal regional governors and a small permanent staff.[4] Outside of the Hejaz there was no professional civil service—the *Bedu* depended entirely on their *sheikhs* to interface with the king—and, in fact, most Saudis lived their entire lives having very little contact with the government. But at that time Saudi Arabia was still poor, even by the standards of the Arab world. When Abdulaziz died in 1953, schools and hospitals did not exist outside of major cities and no paved roads connected those cities. Except in the oil sector, salaried employment was rare; rural electricity was unheard of; and, outside of Jeddah, desalinated drinking water was unknown.

Within fifty years, all of this had dramatically changed. By 2000, most Saudis lived in cities, drove imported cars on paved roads, read newspapers, watched television, carried mobile phones, drank desalinated water, and worked for the government. Government-provided medical care, education, and electricity were available in even the most remote village. All of this was the result of massive, oil-funded government intervention into the lives of ordinary Saudis, which gave them all a stake in the state's survival.[5]

The instruments for this change were the newly created ministries such as agriculture, communications, commerce, health, transportation, labor, and social affairs. The technocrats who administered them, along with foreign experts and expatriate labor, built the roads, railroads, power lines, pipelines, and a domestic airline service that bound together an area the size of Western Europe, excluding Scandinavia. In a parched desert, they provided wells, dams, irrigation systems, and desalination plants that assured both access to water and dependence on the state for it.

The evolution of the Saudi bureaucracy often had as much to do with balancing competing royal family factions as it did with development priorities. Many of these new ministries had ill-defined or overlapping responsibilities. Their bloated, inefficient staffs sometimes obstructed free enterprise.[6] Yet at the same time, they distributed government contracts, business loans, agricultural subsidies, healthcare, education, social services, and virtually free utilities. In short, they delivered economic development to most Saudis. As the immense financial resources of the central government eroded the independence of traditional elites, the Saudi bureaucracy's role in promoting national cohesion became second only to that of the monarchy itself. The creation of a cabinet, or Council of Ministers, in 1953 was a major step in this evolution.

King Abdulaziz had no need of a cabinet to coordinate government policy since for most of his reign he had only two ministries; foreign affairs and finance. The Council of Ministers he created at the end of his reign brought all ministers together for the first time in a single body. It began the transition from King Abdulaziz's personal rule to the more institutionalized government of his sons. Another Royal Decree in 1958 gave the Council of Ministers explicit legislative, executive, and administrative authority. That decree stated, "[t]he Council of Ministers shall be responsible for drawing up internal and external policies, including financial, economic, educational and defense policies, and supervising their implementation. It shall have executive authority and be the final authority in financial and administrative affairs of all ministries and government agencies."[7] This second decree transformed the Council of Ministers from a largely advisory role into the formal policymaking body that it is today.[8]

Princes held 60 percent of the seats in King Saud's first cabinet. Since then the number of technocrats in the cabinet has risen steadily. Kings Faisal and Fahd replaced princes with non-royal technocrats as ministers of education, finance, communications, and agriculture. King Salman replaced a prince with a technocrat at the Ministry for Municipal and Rural Affairs but did the opposite at the new Ministry of Energy. Nonetheless, by 2020, princes held only 15 percent of cabinet posts—primarily, in the security-related ministries of defense, the interior, foreign affairs, and the National Guard. The non-royal technocratic ministers now have more influence on the day-to-day management of government than tribal leaders, urban notables, or even the clergy. Their views carry weight and their support is important to effective government.

Chaired by the king, the Council of Ministers meets weekly. It makes most routine decisions by majority vote, with the king and crown prince not always being in the majority. The king usually acts in his capacity as President of the Council of Ministers, and issues new policies as Council of Ministers Decrees. He can, and sometimes does, bypass the Council, issuing Royal Orders in his own name. The most common example of this is his appointment and dismissal of ministers, who have no more independent political authority than American cabinet secretaries.

Ministers who perform poorly or disagree with the king too loudly, are fired.[9] One well-known example was Health Minister Ghazi al-Gosaibi who in 1984 opposed the tendering process for hospital construction that often favored the king's associate Rafik Hariri and his construction firm Saudi Ojer. Al Gosaibi published a poem entitled "The Pen Bought and Sold," which, while modeled on classical Arabic verse, was a thinly veiled criticism of King Fahd and Defense Minister Prince Sultan. The next day al-Gosabi read in the newspaper that he had resigned.

Thirty-five years later, King Salman and Mohammed bin Salman are attempting to rapidly implement major social and economic reforms. To succeed, they will need a much better, coordinated bureaucracy than any of their predecessors ever enjoyed, and they are trying hard to create one. King Salman has, for instance, replaced all of the ministers and many of the deputy ministers who served under King Abdullah. As has been mentioned in Chapter 6, a dozen cabinet subcommittees that rarely met have been replaced by two that meet weekly, the Council for Economic and Development Affairs (CEDA) and the Council for Political and Security Affairs (CPSA). Now, all large projects must be approved by CEDA or CPSA, and will be approved only if they directly support Vision 2030. The Ministry of Finance, which previously played a powerful role in resource allocation and often acted as a bottleneck, has become a mere accounting office for CEDA and CPSA. Furthermore, the independent bank accounts of many ministries have been closed or reduced, and new regulations have centralized procurement processes. Most ministers now maintain independent control of only small projects, salaries, and maintenance programs.

The most significant change to the Council of Ministers has been the creation of Vision Realization Programs. In the past, ministers often worked independently or even at cross purposes. Many had independent agendas, unrelated to the national five-year development plans, and there was no central monitoring authority. All of this has changed since Salman became king. Now, there are twelve Vision Realization Programs for issues such as improving the Hajj, improving housing, and developing the financial sector. Each program is the responsibility of one minister, who has authority over all departments in any ministry involved in his program. These super-ministers sit on a Vision Realization Council, a forum in which they can no longer blame their failures on parts of the bureaucracy that they do not control.

Many Saudi military officers are technocrats with advanced degrees in technical subjects from Western universities. However, in sharp contrast to most Middle Eastern militaries, the Saudi Armed Forces have never become an independent political actor. Unlike what occurred in Turkey, the Saudi officer corps has never formed the vanguard of a new, secular, modernizing elite. It has nothing like the independent economic resources enjoyed by the Egyptian military or the Iranian Revolutionary Guard Corps both of which hold extensive business interests in their own names. Nor is Saudi Arabia like Pakistan, where the army, when not actually running the country, has ignored the political leadership on critical security matters. In Saudi Arabia, the military has always been firmly under civilian control.

Unlike the Shah of Iran, who had no close relatives in government, there are hundreds of Al Saud princes serving in the military as both junior and senior officers. Although princes do not command all major military units, they are found in leadership positions in nearly every unit down to brigade level. The most important branch of the Saudi military, the Royal Saudi Air Force, is commanded by a career officer, Lt General Prince Turki bin Bandar bin Abdulaziz, whose brother, Lt General Khalid bin Bandar, was commander of Saudi Land Forces. Their brother, Faisal, governs Riyadh, while another brother, Abdullah, is the civilian minister for the National Guard.

Senior generals serve entirely at the pleasure of the ministers of defense or the National Guard, who are always civilian princes. In 2020, the minister of defense and his deputy are both sons of King Salman, while the minister of the National Guard is his nephew. Senior commanders are rotated every four or five years, and never remain in place long enough to develop an independent power base. They are often retired with very little advance notice. Since becoming king, Salman has rotated or retired the Army, Navy, Air Force, Air Defense, and Strategic Missile Force commanders. Upon retirement, some generals are appointed to the *Majlis al-Shura*, others go into business. Only after retirement are military personnel of any rank allowed to vote in municipal elections, but they do not form a distinct or organized class in retirement any more than they did while on active duty.

During his own short tenure as minister of defense (2011–2015), King Salman recognized significant shortcomings in the Saudi military and, upon becoming king, appointed his son to replace him. As minister of defense, Mohammed bin Salman hired hundreds of foreign consultants to improve the structure, personnel system, procurement policies, and coordination of the Saudi military. New agencies have been set up to coordinate and regulate government-owned military industries and the joint ventures that they operate with foreign defense manufacturers. A new Procurement and Tenders Law has centralized defense acquisitions across services. For the first time ever, a multi-ministry committee has studied which drones the kingdom should buy instead of each service choosing its own system.

Military reform has faced resistance from vested interests, some of which were notably corrupt. Yet the old guard is retiring and most young officers welcome changes intended to build a more effective fighting force. One thing that could unite them in opposition to the monarchy would be a prolonged, unwinnable war or outright military defeat in Yemen. Recall that it was their disillusionment over the disastrous 1948 war with Israel that united Egypt's officer corps against King Farouk.

In 1990, it took three days for the Saudi press to acknowledge Iraq's invasion of Kuwait, yet in 2020, the Saudi media is even more firmly controlled than it was then. Although Saudi newspapers are privately owned, they are all controlled by members of the royal family or their loyal lieutenants. Some, such as *Sharq al-Awsat* and the *Arab News*, are controlled by the king's immediate family through the Saudi Research and Marketing Group, of which three of King Salman's sons—Ahmed, Faisal, and Turki—have been chairmen. The popular pan-Arab daily, *Al Hayat*, is owned by Prince Khalid bin Sultan, while *Al Watan* is owned by Prince Khalid al-Faisal. Members of the royal family and their associates are—or in some cases were, until the November 2017 Ritz Carlton asset confiscations—major shareholders in four regional television broadcasters: Rotana, Middle East Broadcasting Center, Arab Radio, and Television and Orbit Showtime Network.[10]

Article 39 of the Saudi Basic Law of Governance prohibits publishing anything that might lead to "sedition and division" or that "undermines the security of the State or its public relations or is injurious to the honor and rights of man."[11] According to other regulations, the minister of information must approve the appointment of new editors. The minister of the interior has the authority to fine, suspend or remove editors and to shut papers temporarily or permanently. Government control of the media thus operates not through direct censorship but through well-understood authority which leads to self-censorship. No criticism of the king or ruling family is allowed, and harsh attacks on the religious establishment can result in imprisonment. Ministers may be criticized and government policies debated, but unpleasant domestic news is usually reported only if it promotes the government's agenda.

Such was the case with the 2004 fire at Girls' Middle School Number 31 in Mecca that left fifteen junior high school students dead. Urban legend has it that the religious police forced girls back into the burning building because they were escaping unveiled. Ministry of interior officers who investigated the incident found that what in fact happened was that fire exits were locked and girls were crushed while trying to flee.[12] Wherever the truth lies, the Saudi press—particularly Mecca's *Al Nadwa*, Jeddah's *Okaz,* and the English language *Arab News*—published detailed accounts and critical editorials about the tragedy.

This highly negative coverage was permitted because it had a desired effect. The president of the Presidency for Girls' Education, Dr. Ali al-Murshid, was forced to resign. Two weeks after the fire, his agency was formally abolished and its activities incorporated into the more secular Ministry of Education. Few complained. The clerics responsible for girls' education had been advised to either go quietly and keep their pensions, or protest and face a major investigation that could well result in jail terms for the negligent. The clerics went quietly, and girls began to study the same curriculum as boys—including, for the first time, English. To be clear, King Abdullah used the fire and the press coverage to orchestrate changes that he wanted to make in girls' education, but which would have met fierce resistance if public opinion, led by the press, had not turned against the Presidency for Girls' Education.

Although King Abdullah allowed limited press liberalization, his successor, King Salman, has reversed that trend. Charmed by rock concerts, women driving, and new movie theaters, some have overlooked the fact that under King Salman freedom of speech and freedom of the press has declined. There is less scope for constructive criticism; and editorial policy, like everything else in the kingdom, has become more centralized and controlled. When the United States moved its embassy from Tel Aviv to Jerusalem in 2018, there were none of the usual and anticipated anti-American editorials. There were, in fact, louder protests in the European press than in the Saudi media.

Economic and social reform is happening in Saudi Arabia—but its direction, pace, and means of implementation are not open for debate. Controversy is now avoided, active support for Vision 2030 is mandatory, and social media is closely monitored by the Ministry of the Interior for political dissent as well as terrorist threats. No criticism of Mohammed bin Salman is tolerated. Punishments have become harsher. For example, when journalist Saleh al-Shehi suggested that the Royal Diwan distributed land based on personal connections, he was sentenced to five years in prison.[13] After receiving numerous warnings from the Ministry of the Interior, Essam al-Zamil was arrested "for giving foreign diplomats information about the Kingdom," alleged ties to the Muslim Brotherhood, and criticizing aspects of Vision 2030.[14] In fact, his most serious offense was posting a video in which he claimed that the 2 trillion dollar valuation the Crown Prince had placed on Saudi Aramco would require including the firm's oil reserves in the IPO. Those reserves al-Zamil argued belong to the people and were not Mohammad bin Salman's to sell.

Criticizing the religious establishment has been and remains a red line. In 2003, Jamal Khashoggi was removed from his post as editor of the newspaper *Al Watan* when he allowed criticism of the religious police. Restored to his post in 2007, he lasted only three years when the paper again published articles questioning Wahhabi orthodoxy. Blogger Raif al-Badawi went to prison for blasphemy in 2012—an unusual charge that, although it remained on the books in England and Wales until 2008, had not led to imprisonment there since 1921. In the United States, the Supreme Court ruled in Joseph Burstyn, Inc. v. Wilson (1952) that the New York State blasphemy law was an unconstitutional restraint on freedom of speech. In Saudi Arabia, al-Badawi remains in prison.

As has been discussed, Saudi Arabia's tribal, religious, and business communities all have formally recognized leaders who have regular, direct access to the highest levels of government and are consulted on policies affecting their constituents. Until quite recently, the technocratic, meritocratic, modern middle class lacked such representation. The creation of a Consultative Council or *Majlis al-Shura* in 1993 was an effort to address that deficiency with an appointed body authorized to initiate and approve some forms of legislation.[15]

This seemingly progressive step was highly controversial with Saudi Arabia's religious establishment, which has long opposed the very concepts of elections, parliaments, and democratic change. Even today, the *ulama* remain highly suspicious of any legislative body that might infringe upon God's authority, or their own. For them, all important laws are divinely inspired and have already been written. They accept the need for modern interpretations of religious law but insist that this can only be done by trained religious scholars such as themselves. The idea that men and women, even democratically elected men and women, could make laws that contradict Islamic teachings is anathema.

To reassure the clergy that the kingdom's proto-parliament would not transgress Islamic values, the king has always chosen a senior cleric to lead the *Majlis al-Shura*. It first met under the leadership of the respected Islamic scholar and former justice minister, Sheikh Mohammed Ibrahim al-Jubeir. When he died in 2002, he was replaced by the conservative scholar, Saleh bin Abdullah bin Humaid, who was in turn replaced in 2009 by another former justice minister, Dr. Abdullah bin Mohammed Al al-Sheikh.

Although the *Majlis al-Shura's* leaders have all been locally educated religious scholars, most *Majlis* members are well-respected technocrats. Some are retired military officers, businessmen, or tribal leaders, but most are former bureaucrats, academics, or journalists. More than 70 percent of them hold doctorate degrees in secular subjects, mostly from Western universities. Though this does not mean they all endorse secular attitudes, they are nevertheless a better educated group than the members of either the British Parliament or US Congress. Members are appointed to four-year terms with most serving only one, or at most two terms. In principle, they each represent the entire kingdom, but in practice they are carefully selected so that all regions and important towns are included.

In 2013, King Abdullah appointed the first female *Majlis* members and decreed that in the future women should hold no less than 20 percent of *Majlis* seats. Again, these appointments were strongly opposed by religious conservatives and the king's method for dealing with their concerns was classic Al Saud. Like his father, King Abdulaziz, and brother, King Faisal, before him, King Abdullah argued with his religious opponents in religious terms by using the story of the Prophet Mohammed's wife, Umm Salamah.[16]

All Saudis know that during the early years of Islam, the Prophet Mohammed faced open revolt among some of his followers. It was the sound advice of his wife, Umm Salamah, that helped him to defuse the crisis. King Abdullah argued that if a woman had given such excellent advice to Mohammed, why should he be denied their counsel. Then, as always, the king compromised in a way that gave religious conservatives something without detracting from his primary objective. He agreed that the new women *Majlis* members would have to enter through a separate ladies' entrance and sit in a separate ladies' section of the council chamber. Interestingly, the conservative voices on this occasion were not all male—10 percent of the new female *Majlis* members objected to even sitting in the same hall with unrelated men and asked for a separate room. They did not get their wish.[17]

Since 1993, the *Majlis al-Shura* has grown in size and stature. Its membership has increased from 60 to 150 and, although initially permitted only to comment on legislation proposed by the Council of Ministers, since 2003 the *Majlis al-Shura* has been able to initiate new legislation or changes to existing regulations, which it now sends to the Council of Ministers. If there are differences in the versions of a bill adopted by the *Majlis al-Shura* and the Council of Ministers, then in theory the king makes a final decision. In practice, he generally waits until the two bodies reach a compromise—and this was the case with the kingdom's new mortgage law. The Council of Ministers believed that each bank could independently guarantee that its products were Islamically correct. Some religously conservative *Majlis* members wanted one government board to supervise religious aspects of all mortgages. The *Majlis al-Shura* eventually got its way.

The *Majlis al-Shura* has no budgetary authority and very limited oversight authority, but its annual reports on each ministry have sometimes produced results. For example, the Ministry of Municipal and Rural Affairs (MOMARA) is responsible for the kingdom's partially elected municipal councils. The *Majlis al-Shura's* report to King Abdullah on MOMARA asked why women could not vote in municipal elections. The king acknowledged this concern and replied that not only should women be allowed to vote in the next municipal election but that they should also be allowed to run as candidates. King Salman accepted his predecessor's view, and when the next round of elections took place in late 2015, women participated for the first time as both voters and winning candidates.

In 2020, Mohammed bin Salman needs technocrats and civil servants, not religious scholars or tribal chiefs, to implement Vision 2030. Without their support he cannot succeed, but their support is not guaranteed. Like the kingdom's religious scholars and merchants, many technocrats and bureaucrats are uncertain about the social and economic changes unleashed by Vision 2030. Saudi civil servants have been accustomed to a great deal of autonomy and independence from either royal or public oversight. Now they are being held accountable by an activist crown prince with a long list of key performance indicators and a task force of monitors. Some feel threatened; others simply resent having their routines challenged by 25-year-old foreign consultants.

Many bureaucrats now worry about job security for themselves and their children. They saw that when airport operations were privatized at Medina, the government saved money, but civil servants from several government agencies lost authority; now, they fear and resist further privatizations. Jobs for life appear to be a thing of the past. It had been nearly impossible to fire government employees for poor job performance. Now, for the first time, they are facing serious performance evaluations that include potential penalties, as well as rewards for high achievers.

Like all Saudis, the technocrats have seen taxes increase along with the price of fuel and electricity. Unlike other Saudis, they had some civil-service benefits suspended in 2016. Although their benefits were subsequently restored, this has caused concern about future pay cuts. Many bureaucrats also benefited from established patterns of corruption, overstaffing, and nepotism. Now, the anti-corruption campaign and the installation of anti-corruption watchmen in most ministries has reduced the opportunities to misuse public funds that some had considered a normal part of their income. In 2017, the minister of the civil service himself was very publicly fired after hiring his son for a well-paid position instead of a more qualified candidate; something that would have been considered normal five years ago.

Under King Salman, the technocrats' representative body has lost influence. Major reforms have bypassed it. The *Majlis al-Shura* did not vote on Vision 2030, the Saudi Aramco initial public offering, or the imposition of a value-added tax. Twice, its conservative leadership found procedural ways to block a vote on women's driving. Eventually, the king ignored them and acted on his own. *Majlis* members' allowances have been reduced, and their salary level is not attractive to many. Some of those asked to join the *Majlis al-Shura* have declined and others have stopped attending, though no one has yet formally resigned.

Bureaucratic resistance to Vision 2030 is not only about personal gain. Saudi civil servants have long regarded themselves as the stewards of public funds, defending the people's oil revenue from selfish businessmen who pay no taxes but always demand more subsidies. Some of these bureaucrats now resent what appears to them to be a growing role for the undeserving private sector. Many civil servants remain hidebound and philosophically supportive of government intervention rather than free-market solutions. Other technocrats firmly believe that Vision 2030's urgent implementation is undermining responsible administration.

For example, Saudi banks grant loans to civil servants based on their entire income, including benefits. When these benefits were suspended in 2016 and incomes fell by 30 percent, bankers were simply instructed to change their debt to income limits. Likewise, during the anti-corruption campaign, senior bankers were instructed to transfer funds from some detainee's accounts to the government. When the bankers pointed out that such large withdrawals would impact on their capital adequacy ratios, they were told "just do it."[18] Needless to say, such episodes caused concern and resentment among the kingdom's bankers and financial regulators.

Passive resistance on the part of Saudi technocrats is possible and is nothing new. Twenty years ago, they derailed Crown Prince Abdullah's initiative to invite foreign oil companies to explore for gas in the kingdom. More recently they sought to delay privatizing grain silos, airports, and hospitals. The retirement of oil ministers Ali al-Naimi and Khalid al-Falih were at least partially the result of their lack of enthusiasm for the Aramco IPO, a policy that many Aramco professionals did not welcome. Ultimately, a lack of both commitment and capacity on the part of some Saudi technocrats remains one of the greatest challenges facing Vision 2030 and its efforts to assure political stability through economic reform.

11

The Royal Family

We can imagine Iraq without the Hashemites, Yemen without the Hamid al-Din, or Libya without the Al Sanusi dynasties. It is difficult to imagine Saudi Arabia without the Al Saud. Like the Ford Motor Company, the founder's name is on the door. Theirs is the only nation on earth created by and named after a family, which remains the political glue holding it together. Three times the Al Saud created their kingdom and twice they lost it. Those painful defeats are embedded in their collective memory and make them determined not to lose their patrimony again.

The Al Saud have ruled Central Arabia for most of the past 280 years, since well before the founding of the American Republic or the French Revolution. According to some historians, they were originally part of the powerful 'Anizah tribal confederation of northern Arabia.[1] King Salman has suggested that they were initially part of the much smaller Bani Hauta tribe. In either case, they are a long-established Arabian family that can confidently trace its local roots back ten generations.[2] The same could not be said of the last kings of Egypt or Iraq—both of whose families were recent, foreign arrivals. In sharp contrast, the Al Saud grew from their own soil and that is where they remain. Generations of Al Saud are buried together, and in accordance with Wahhabi practice, in the unmarked graves of south Riyadh's al-Ud Cemetery.

Although the Third Saudi State is a federation of several once-distinct political entities, it is no European Union created by mutual consent. The unification of Saudi Arabia was a prolonged and bloody process. Abdulaziz al-Saud forcibly welded his kingdom together through thirty-two years of nearly continuous warfare and tribal diplomacy. He fought and defeated the Ottoman Turks; the Al Rasheed of Ha'il; the Hashemites of Mecca; the Shaalan *sheikhs* in the north; the Idrisi in the Asir; the Imam of Yemen; and his own former allies, the Ikhwan. He lost a finger to a Turkish bullet and was severely wounded more than once. Unlike the rulers of other Gulf states, he was not peacefully handed his sovereignty by a departing colonial official. Yet military prowess was not enough to hold his disparate kingdom together—especially before the discovery of oil.

In 2020, monarchy remains the principal institution holding Saudi Arabia together, and the preeminent branch of its royal family is largely the creation of one man. For many years, Abdulaziz pursued an active policy of serial polygamy and was quite open about what he was trying to do in bed. He said that he created his kingdom with "a sword of steel and a sword of flesh." He had dozens of wives during his lifetime— though, in accordance with Islamic law, never more than four at any one time. King

Abdulaziz had forty three sons, only thirty four of whom survived into adulthood, and even more daughters. Five of his sons remain alive, though his last daughter, Madawi, passed away in 2017.[3]

In his early years, Abdulaziz carefully chose his wives on political grounds and married into families of tribal, religious, or local political importance as well as into many collateral branches of the Al Saud family itself. King Saud's mother was from the Bani Khalid tribe, which once controlled the kingdom's Eastern Province; King Abdullah's mother was from the powerful Shammar tribe of the north; King Faisal's mother was from the Al al-Sheikh religious dynasty; and King Khalid's mother was from the Bin Jaluwi branch of the Al Saud. Abdulaziz's own mother (Sara), as well as those of Kings Fahd and Salman (Hussa), came from the Al Sudairi family—who are prominent members of the Dawaisir tribe, traditional *amirs* of Al Ghat in Al Qasim Province, and longstanding allies of the Al Saud.[4]

Like the Hapsburgs or Queen Victoria's children, King Abdulaziz's sons and grandsons continued this policy of weaving marriage alliances throughout the kingdom's prominent tribes and families. King Saud, who had even more children than his father, married into the Mutair, Shahran, Qahtan, and Shammar tribes; King Faisal married into the Al Sudairi family as well as the Bin Jaluwi and Thunayyan branches of the Al Saud; King Abdullah married into the Shammar, Bani Sakr, and Ruala tribes as well as the Bin Jaluwi and Al al-Sheikh families; King Salman has married both Al Saud and tribal women. Crown Prince Mohammed bin Salman's maternal grandfather was a paramount *sheikh* of the Ajman tribe.

Over time these marriages transformed the Al Saud family into a unifying super-tribe, and the only institution that encompasses most of Central Arabia's traditional elites. King Abdulaziz's grandsons now govern most of Saudi Arabia's thirteen provinces, and in many instances the governor's mother is from a prominent local family or tribe. Rival tribes and regions that would never agree to be governed by one another accept the unifying role of the Al Saud monarchy.

Of all the constituencies with which King Salman and Crown Prince Mohammed bin Salman must deal, their own family is the most diverse. There are liberal princes and conservative princes, playboys, and pious Muslims. Some have doctorates from Oxford, while others lack even a high school education; some are fantastically wealthy, while others work at middle-class jobs. Most believe that change is inevitable, but some do not. The Al Saud are in effect a hereditary political party in a one-party system with thousands of members, different wings promoting different agendas, and various individuals competing for leadership positions. Balancing rival factions and maintaining consensus within the royal family has always been an important part of a Saudi king's job description.

By now, approaching the twelfth decade of the Third Saudi State, there are many great-grandchildren and even great-great-grandchildren of King Abdulaziz living in Riyadh.[5] A simple calculation shows that if he had eighty children who each had ten children and whose children and grandchildren then had only five children, there would be at least 20,000 descendants of King Abdulaziz—most of whom are still alive. However, this is an extremely conservative estimate. Most of King Abdulaziz's sons had far more than ten children. King Saud had more than 100; King Abdullah and Crown

Prince Sultan each had more than thirty; King Salman has had at least thirteen, though some have passed away.

In addition, King Abdulaziz had more than a dozen brothers and sisters as well as several dozen first cousins.[6] When all the collateral branches of the family are added together, the Al Saud number well over 100,000. Far from being a small, isolated group like the Shah of Iran's family, they comprise at least one half of one percent of the entire Saudi population and a much larger portion of the nation's political, social, and economic elites.

This family is clearly too large to govern collectively. Not every prince can become a minister or a general. Many princes with limited talent, education, or ambition remain largely out of the public eye, and the vast majority hold no formal government position at all. Consequently, there has long been a clear distinction between the large "Royal Family" comprising thousands who play no role in politics and the much smaller "Ruling Family," which is itself divided into two sections. There are those princes who may be consulted on various important issues; these include the surviving sons of King Abdulaziz, his most prominent grandsons, and a few leading members of cadet branches of the family—most notably, the descendants of Abdulaziz's cousin, Saud al-Kabeer, and his brother, Abdullah bin Abd al-Rahman. Then there is a very small group of princes who actually run the country on a daily basis: the king, crown prince, royal ministers, and provincial governors.[7]

There are, in fact, many similarities between the Ford Motor Company and the Al Saud dynasty. For one, they are both cyclical businesses heavily dependent on the global economy and the price of oil. Both are also family-dominated firms. In Detroit, the Ford family still controls their namesake firm through a special category of voting share. In Riyadh, no one outside the Al Saud family makes decisions on major policy issues, senior appointments, or succession. Many stakeholders are consulted, but on important matters only a few family members decide—and even their number has been decreasing of late.

Like many business empires, the Third Saudi State was founded by a visionary entrepreneur who did something better than his competitors. Such a founder is usually the sole proprietor. Involved in all aspects of building the empire, he is the only one who makes important strategic choices; he grows the business and establishes core policies. His children often form a limited partnership and try to follow his guidance. These siblings face difficult questions about the employment of family members, compensation packages, and the diverging interests of the family and the firm, but they usually make collective decisions and resolve conflicts quietly among themselves.

The founder's grandchildren form an even larger, looser, and more complex consortium of cousins. The ties of the original, nuclear family grow weaker and tensions increase; not all branches of the family can remain equal. Enforcing a line between personal and corporate property becomes critical, and the debate over paying current dividends to the family or investing in the firm's future intensifies. Above all, without primogeniture it becomes less and less obvious who should lead the firm or the kingdom.

In business, only a third of family firms make the transition to a second generation; less than ten percent survive into the third generation. The important, North African, political thinker Ibn Khaldun (1332–1406) predicted a similar cycle of rise and decline for all Arab dynasties. In his influential book the *Muqaddimah* or Introduction, with which King Salman is almost certainly familiar, Ibn Khaldun described how Arab dynasties usually last for three generations or 120 years—whichever came first.[8]

The few business families that have beaten these odds all understood the need for structures. They created a unifying family mission, established a family governance system, and often recruited professional managers. In nearly all successful multi-generational family firms, one branch of the family clearly became dominant. The Al Saud have adopted all of these strategies.

Preserving King Abdulaziz's memory and his legacy of national unification has become his family's unifying mission. Rare is the prince, however junior, who is not fiercely proud of his ancestor's role in creating Saudi Arabia and, although often airbrushed or embellished, there is substantial truth to the story. A bold, charismatic, young warrior united his people under the banner of religious reform. With luck and good judgment, he evolved into a capable king who brought security and prosperity to his people.

Tales of Abdulaziz's courage and wisdom are common in Riyadh. Whether true or exaggerated, they are widely believed and often intentionally focus on the old king's role in creating and preserving the nation. One relates to a battle during the Wars of Unification with the Ajman—the tribe that killed his younger full brother, Saad. During a campaign in the Eastern Province, Abdulaziz was severely wounded and carried from the field. As rumors flew that their commander was dying, the ever-fickle *Bedu* began to desert. Suddenly, they were surprised to hear the beating of wedding drums. To their amazement, Abdulaziz had sent to a nearby village for a young woman to marry. He took his new bride into a tent and emerged with evidence of having consummated the marriage. Needless to say, the crowd went wild—what a guy! Shot in the stomach and still a stallion. Desertions ceased, the army rallied, and the Ajman were defeated—at least, temporarily. According to some versions of the story, the king then sent for the American missionary doctors in Bahrain who saved his life.

In another story, during World War II the British Army requested permission to move troops stationed in Iraq through Saudi Arabia to help defend Egypt. Abdulaziz assembled his advisors and asked how he should respond. All agreed that under no circumstances should the British Army be allowed to enter the country. When they had finished the king asked, "If the British Army decides to march across Saudi Arabia, just how are we going to stop them?" He suggested it would be much wiser to welcome them, help them on their way, and get their implicit agreement that the route they proposed taking was part of Saudi Arabia, not of Iraq or Jordan. Abdulaziz actually gathered trucks to help transport British troops to Egypt, but after the British victory at El Alamein in 1942 their presence was no longer needed. Nevertheless, the letter implying that their intended route had been through Saudi territory was later used in border negotiations.

Although preserving Abdulaziz's memory is straightforward, the second part of the Al Saud's family mission—preserving his legacy of national unification—is not. Saudi Arabia is no city state. In area, it is roughly the size of the United States east of the Mississippi River. With more than 10,000 villages and four cities of more than a million people, it contains two dozen major tribes and scores of minor ones. Almost within living memory, it was as fragmented and violent as its southern neighbor, Yemen, is today. Maintaining national unity is a difficult task and, after maintaining their own privileged position and wealth, this remains the House of Saud's foremost priority. This mission furnishes them with a practical task. Combined with the global mission of defending Islam, preserving Saudi Arabia's national unity gives the Al Saud a deeply felt moral purpose greater than simply holding on to power. It also gives them, at least in their own minds, an unquestioned right to rule.

In 2020, King Abdulaziz's legacy of national unification remains in place and his role as founding father is carefully preserved. Although he has been dead since 1953, his portraits still hang everywhere in Saudi Arabia. Although the endless pictures of current rulers found in most autocracies express a nervous, forced loyalty, those of Abdulaziz reflect respect for an historic national leader, rather like the pictures of Ataturk to be seen in Turkey or George Washington in the United States. This close association of the ruling Al Saud family with national unity has been a stabilizing factor for Saudi Arabia. It contrasts sharply with the experience of nations such as Lebanon or post-Saddam Iraq, where political elites have drawn their legitimacy from divisive sectarian agendas.

A shared mission helps to maintain family cohesion, but the Al Saud also need internal governance structures. The king is their chairman of the board and ultimately responsible for dealing with compensation, promotions, and misbehavior. Although the king may initially owe his position to the family, once in place he is invested with authority over it. Likewise, within the royal family direct descent from King Abdulaziz has become increasingly important. As noted previously, King Faisal formalized the distinction between Abdulaziz's descendants through a male line who are referred to as His or Her Royal Highness (HRH) and other family members who are simply His or Her Highness (HH).[9] Many HH princes hold important positions in the civilian bureaucracy and the military, but it is always the HRH princes who hold the most senior posts. When the Al Saud family deposed King Saud in 1964, nearly half of those signing the decree were HH princes. Today, all members of the Allegiance Council, which technically approves a new crown prince, are HRH princes. In the past, collateral branches of the Al Saud often provided provincial governors. Today, nearly all provincial governors and deputy governors are HRH princes.

Respect for age has been another fundamental principle within the Al Saud.[10] It is a clear, easy-to-follow concept, which creates a private family hierarchy that differs from official government structures. The king is the overall leader of the family, but he shows considerable deference to any older brothers who are still alive. As we have seen, even King Abdulaziz always stood behind his father, Abd al-Rahman, at prayers, and the sons of Abdulaziz always showed great deference to their last surviving uncle, Abdullah bin Abd al-Rahman. Even today Al Saud princes know their place, and at family gatherings routinely sit according to age rather than official position.

A Saudi king's most direct means of controlling his family is the allocation or denial of government positions and financial benefits. He may consult with family members on senior appointments, but the final decision is his alone and is usually made by Royal Order without reference to the Council of Ministers. Appointments have generally been based on seniority, experience, and the balancing of family factions. For example, in 1985—when King Fahd appointed his own son, Mohammed, governor of the Eastern Province—Defense Minister Sultan demanded that his son, Fahd, be made governor of Tabuk. King Fahd complied, and then gave additional governorships to the sons of King Saud and Faisal as well.[11]

The Al Saud operate something like baseball's farm team system, in which ambitious young princes start off with relatively junior minor league positions and, if they are talented and fortunate, advance to more senior major league posts. Sultan bin Abdulaziz had been deputy governor of Riyadh, governor of Riyadh, minister of agriculture, and minister of transportation before assuming the very senior role of defense minister and eventually crown prince. The current governor of Asir was for many years deputy governor of Al Qasim, while the current governor of Mecca spent many years as governor of the much smaller Asir Province. The dozen young princes whom King Salman has appointed as deputy provincial governors, are the next generation of Al Saud leadership in training. Their effort and ability are being tested and judged.

King Abdulaziz and his long-time finance minister, Abdullah Suleiman, made no distinction between the king's personal wealth and the state treasury. They simply supported the royal family from government revenues. Today, many, but not all, members of the Al Saud still receive monthly stipends from the government. King Faisal formalized the Saudi Civil List, and while its details have never been made public, the broad outlines are known. Payments vary greatly based on lineage and age, with each generation receiving significantly less than the one before it. In each category a princess receives half of what a prince does, the difference being based on the view that women have husbands and are not the principal breadwinner of their household.

The few remaining sons of King Abdulaziz are paid directly by the Royal Diwan, and each receives several million dollars a year. The vast majority of princes and princesses collect their stipends from a special agency known, oddly, as the Office of Decisions and Adjustments. Within each category payments can vary significantly, but in general a grandson of King Abdulaziz receives roughly $200,000 a year and a great grandson $100,000. There are very few of Abdulaziz's nephews still alive; their sons, however, receive roughly $50,000 a year. The descendants of King Abdulaziz's cousins receive less. Distant branches of the family, such as the Farhan or Thunayyan, may receive nothing at all.

Some male members of the Al al-Sheikh and other prominent tribal and *aiyan* families also receive monthly payments, many of which date back to the founding of the kingdom. Again, the senior members of these families receive substantial sums but most are less than $50,000 a year and some are purely symbolic. Although these payments are not particularly large, they are coveted status symbols that have done a great deal to promote elite cohesion, and thus stability, in Saudi Arabia.

Land grants are far more important to many princes than their modest monthly stipend, and these were until recently very much under the direct, personal control of the king.[12] Outside of the major cities, most land in Saudi Arabia was initially empty desert and now belongs to the government. As cities expanded, this land became increasingly valuable and was often distributed free of charge. The average citizen received a few thousand square meters; a more prominent individual received several times that. Some princes were awarded huge tracts, which they in turn often sold to property developers. Even more profitable was to sell land that the government gave you for free back to the government for a greatly inflated price. This was sometimes repeated several times with the same piece of property, and such shenanigans were by no means uncommon before King Salman came to power.

All large government construction contracts in Saudi Arabia are, on some level, royal favors—and princes, or firms associated with them, have received many of them. Defense Minister Prince Sultan's sponsorship of the German construction firm Philipp Holzmann, King Khalid's son Bandar's partnership with the Dutch firm that built the causeway to Bahrain, and Prince Mishal bin Abdulaziz's involvement in building railroads are well-known examples, but there are many others.[13] Thus, while Saudi kings did not routinely adjust royal stipends, they maintained considerable influence over the financial position of individual princes. Allocating important jobs, land grants, and government contracts have long represented the king's most effective levers of control over his family.

Although the Al Saud created their kingdom largely by force of arms, they have maintained it largely through the principles of consultation and consensus building, *shura* and *ijma*; concepts strongly enjoined in the Quran. This is no desert democracy; it is a client–patron relationship, in which the client has little to say about the final outcome. Nevertheless, the concepts of consultation and consensus building have been as fundamental to Saudi politics as campaigns and elections are to Western democracies.

The Al Saud have developed a range of structures to facilitate consultation. The king meets formally with the Cabinet every Monday, the religious scholars every Tuesday, and royal family members on Friday. As noted earlier, the king's executive office complex, or Royal Diwan, has a specific department for tribal matters; the business community is heard through the Chambers of Commerce; and the middle-class technocrats through the *Majlis al-Shura.*

In addition to these formal structures for consultation, the royal family's large size has made it a very effective, informal instrument for gauging public sentiment. Many princes with no official position have a set of clients for whom they mediate with the bureaucracy.[14] They often hold gatherings in the formal sitting room, or *majlis,* of their homes or palaces. For a senior prince these may be scheduled, public events, while for a more junior prince they are only for invited friends and retainers. Current events are always informally discussed and junior princes feed what they learn of the public mood to more senior princes. In rural areas, it is usually the tribal *sheikh* who gathers public sentiment and relays it to the local Al Saud governor. When he was governor of Riyadh, King Salman was known to ask everyone from wealthy merchants to street sweepers what was on their mind. This informal polling process allows senior princes to make policy choices based on a broad reading of public sentiment. Just as importantly, it has made many junior princes with no official position feel that they are part of the governing system.

A further role for princes with no formal government position is their part in the Saudi distribution system. While the bureaucracy distributes official welfare, the royal family dispenses patronage. Princes often pay for everything from free meals to divorce settlements and foreign medical bills for their retainers. Senior princes fund substantial charitable foundations. The Prince Sultan Foundation builds homes for the poor and operates a major rehabilitation hospital, while the King Faisal Foundation runs its own co-educational university. King Khalid, King Fahd, King Salman, and Crown Prince Mohammed bin Salman all have charitable foundations. These charitable distributions are designed to instill in the public a sense of gratitude and obligation towards the royal family. Making such paternalistic gestures is an important part of a prince's job, and his monthly stipend will depend in part on how much of it he is seen to be giving away.

When King Salman inherited leadership of the royal family in 2015, it was clear that these established mechanisms for conducting family business could not guarantee the monarchy's survival in the twenty-first century. Most of Salman's brothers were dead and a problematic transition to third-generation princes was looming. With oil prices collapsing, the generous entitlements enjoyed by the Saudi people were no longer sustainable. A new royal family team and a less generous social contract would need to be negotiated. These two adjustments were bound to be controversial with princes and the public. Both would become impossible to implement if the fading, collegial governing system of brothers dissolved into a free-for-all game of thrones amongst King Abdulaziz's grandsons; which is precisely what Ibn Khaldun had predicted.

King Abdullah had experimented with tentative steps towards more participatory forms of government, including the *Majlis al-Shura*, municipal councils, and the Allegiance Council. King Salman rejected his predecessor's approach. In Salman's view, preserving the monarchy required not consensus-driven, incremental adjustments but rapid change implemented through more autocratic governance. In addition to reforming the economy, he set out to secure the succession, concentrate power within the royal family, and reduce excessive royal privilege. Although all of the stakeholders in the Al Saud's coalition have experienced the effects of Salman's policy shift, none have felt it more painfully than the royal family themselves.

King Salman knew very well that his father had been in his mid-twenties when he captured Riyadh. Age and experience were not the qualities that had led to this success. What King Abdulaziz had, and what King Salman was looking for, was fire in the belly. Brought up in humiliating exile, King Abdulaziz had been fiercely determined to restore his family's honor. He had combined exceptional ambition with a ruthless will to power. Such vigor and resolve would be needed again in order to manage a generational leadership transition and drive forward much needed, but contentious, economic and social reforms.

These were not qualities easily found among King Abdulaziz's grandsons. Brought up in privilege, many third generation Al Saud princes had spent more time on French beaches or in American universities than in the Saudi desert; many were chronically complacent and out of touch. As a result, King Salman very deliberately replaced age and experience as the criteria for the throne with ambition, determination, and a capacity for hard work.

Recognizing these attributes in one of his younger sons, King Salman rearranged the line of succession and made Mohammed bin Salman heir to the throne by removing first his half-brother, Muqrin bin Abdulaziz, and then his nephew, Mohammed bin Naif, from their positions as crown prince. Had the king only wished to maintain the Al Sudairi branch of the royal family in power, he could have easily left his full brother's son, Mohammed bin Naif, in place. Had he wished only to concentrate power in his own immediate family, he could have easily made one of his older, more experienced sons crown prince. Instead, the king very deliberately engineered the unconventional, complicated, and controversial rise of the young and relatively inexperienced Mohammed bin Salman because, to paraphrase *The New York Times* columnist Thomas L. Friedman, If you think there are another dozen princes in Riyadh with the steel, cunning, and ruthlessness as Mohammed bin Salman, you are wrong.[15]

Age and experience were not the only important principles that King Salman discarded as he reverted to his father's more autocratic style of rule. Since King Faisal's death in 1975, there had been, in effect, a king and several vice-kings—each with his own sphere of political influence, financial network, and set of retainers. These princes were brothers who cooperated to keep the family in power, but they and their sons also competed for positions and resources.[16] Not anymore.

King Salman systematically dismantled the institutional power bases that other senior princes had enjoyed for decades as ministers of defense, the interior, the National Guard, municipal affairs, and foreign affairs. All of these positions had provided the minister-prince with opportunities for patronage and personal financial gain; this arrangement had been part of the royal family's internal revenue-sharing system.[17] Now these positions have been handed over to technocrats or politically impotent junior princes, and new anti-corruption watchdogs have been installed to monitor spending.

Likewise, under previous kings, officers from the Ministries of Defense, the Interior and the National Guard had seldom met in the same room, much less supported each other in military operations. This was a deliberate policy and part of the division of power that existed among the sons of King Abdulaziz from 1953 to 2015. That division, too, is now gone. The new defense, interior and National Guard ministers are all King Salman's men, not his brothers or rivals.

King Salman further recognized that the monarchy could not survive with thousands of princes all demanding deference. The Saudi people will accept paying homage to, or being bumped off airline flights by, a few dozen senior royals. They will not tolerate indefinitely hundreds of great-great-grandsons of King Abdulaziz illuminating their palaces and gardens for free while their own electricity bills tripled. Gaining public support for painful economic reforms required fighting high-level corruption and downsizing the royal family.

Since 2015, the great majority of Saudi princes have lost their sense of immunity, if not their sense of entitlement. One was executed for murder, another jailed for mistreating his servants, and others lost widely publicized court cases. Royal bank accounts have been frozen and royal land grants revoked. Most princes must now go through customs at the airport, some have had their right to travel limited, and others have left the country altogether. With royal electricity and phone bills no longer being paid by the Royal Diwan, princes have begun turning off their garden lights or at least putting them on timers. Most princes no longer receive cost-plus contracts, free airline tickets, or large blocks of work visas that they can profitably resell. The days of Prince Sultan skimming vast sums from the Ministry of Defense or his brother, Mishal, fencing off huge tracts of public land for his own use appear to be over.

Cabinet ministers no longer feel obliged to come to their ministry door to greet any random prince who turns up without an appointment. Every prince no longer has easy access to the king and royal middlemen are being cut out of government contracts. Eleven princes were detained at the Ritz Carlton Hotel as part of the November 2017 anti-corruption campaign; they were questioned, and indeed humiliated, by commoners. Some—including King Abdullah's son, Turki, whose firm, Petro Saudi, was involved in defrauding billions of dollars from the Malaysian sovereign-wealth fund, 1MDB—have remained in prison for months.[18] A further thirty princes were detained in early 2018—allegedly for protesting against their new electricity bills, but in reality for challenging the authority of the Crown Prince. For many Saudis these detentions, especially that of National Guard Commander Mit'ib bin Abdullah, were a transformative moment in Saudi history: the day that Al Saud princes lost their halos.

Another factor in Al Saud cohesiveness has been the fact that aged kings never stayed in power too long. In fairly rapid succession they handed over power to another brother from a different branch of the family. It paid to wait your turn rather than rock the boat. That incentive to cooperate is no longer present as Mohammed bin Salman could easily be king for the next fifty years.

In 2020, the opaque, consensus-driven, power-sharing Al Saud family structure is gone. Respect for elders has been replaced by deference to a 35-year-old crown prince who has solidified his position by deploying both patronage and fear. The average prince has lost some of his political influence and social status. There are now fewer princes in government and fewer with the wealth and power to distribute significant patronage. People have begun to ask: "Why should I attend the *majlis* of a prince who can no longer help me with the bureaucracy"?

Many, probably most, of the Al Saud are privately angry about these changes. There are undoubtedly members of the royal family who would welcome a new crown prince and would quickly pounce should Mohammed bin Salman stumble. King Salman has a plan. He installed the prince that he considers best suited to lead the country and began to reduce the royal family's sense of entitlement. However, in doing so, he has damaged the family's internal cohesion as well as the prestige of many princes. These are potentially destabilizing developments for Saudi Arabia

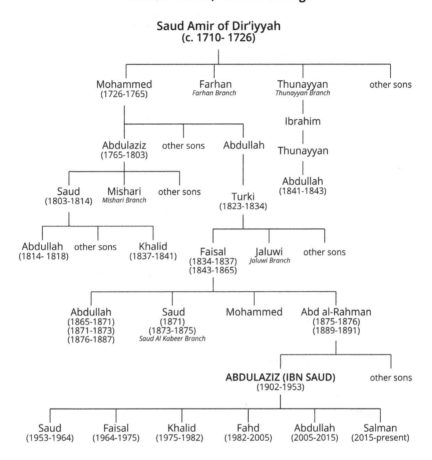

(DATES OF REIGN)   **Al Saud Imams, Amirs and Kings**

Saud Amir of Dir'iyyah
(c. 1710- 1726)

Mohammed
(1726-1765)

Farhan
Farhan Branch

Thunayyan
Thunayyan Branch

other sons

Abdulaziz
(1765-1803)

other sons

Abdullah

Ibrahim

Thunayyan

Saud
(1803-1814)

Mishari
Mishari Branch

other sons

Turki
(1823-1834)

Abdullah
(1841-1843)

Abdullah
(1814- 1818)

other sons

Khalid
(1837-1841)

Faisal
(1834-1837)
(1843-1865)

Jaluwi
Jaluwi Branch

other sons

Abdullah
(1865-1871)
(1871-1873)
(1876-1887)

Saud
(1871)
(1873-1875)
Saud Al Kabeer Branch

Mohammed

Abd al-Rahman
(1875-1876)
(1889-1891)

ABDULAZIZ (IBN SAUD)
(1902-1953)

other sons

Saud
(1953-1964)

Faisal
(1964-1975)

Khalid
(1975-1982)

Fahd
(1982-2005)

Abdullah
(2005-2015)

Salman
(2015-present)

Chart 1. Family Tree of House of Saud.

The Sons of King Salman

Salman bin Abdulaziz Al Saud

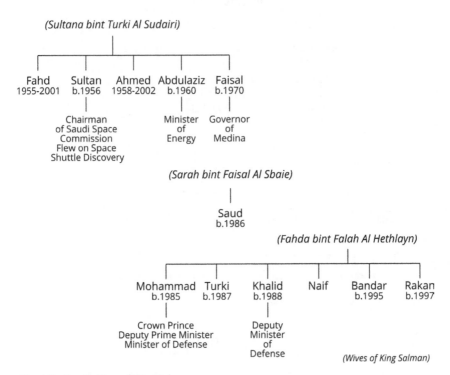

(Sultana bint Turki Al Sudairi)

Fahd	Sultan	Ahmed	Abdulaziz	Faisal
1955-2001	b.1956	1958-2002	b.1960	b.1970

Sultan: Chairman of Saudi Space Commission Flew on Space Shuttle Discovery

Abdulaziz: Minister of Energy

Faisal: Governor of Medina

(Sarah bint Faisal Al Sbaie)

Saud
b.1986

(Fahda bint Falah Al Hethlayn)

Mohammad	Turki	Khalid	Naif	Bandar	Rakan
b.1985	b.1987	b.1988		b.1995	b.1997

Mohammad: Crown Prince Deputy Prime Minister Minister of Defense

Khalid: Deputy Minister of Defense

(Wives of King Salman)

Chart 2. Family Tree of King Salman.

Part IV

Delivering Competent Government

The Al Saud's playbook for successful governance contains many practical guidelines, which they have usually followed. Stay close to the *ulama* and the religious establishment that they lead; stay close to the tribes and co-opt their leaders; stay close to the Americans, but not too close; promote private-sector trade and commerce; pay attention to the needs of the poor and all the kingdom's diverse regions; avoid collective punishment of families or tribes and try not to humiliate anyone; forgive those willing to reconcile, but firmly crush those who will not; and never appear to make concessions under pressure.

These are all tactics that have helped the Al Saud maintain power, but they are not the underlying source of their legitimacy—nor is the historical European concept that kings had a divine right to rule. The Al Saud's widespread support among the Saudi people is drawn from deeper wells. As we have already seen in Parts I, II, and III, they are a local family who unified a nation, found ways to peacefully transfer political power amongst themselves, balanced often competing stakeholder interests, and maintained elite cohesion. Just as importantly, they have provided competent government that has satisfied most Saudis—at least most of the time.

In a society that remains as hierarchical as the Kingdom of Saudi Arabia, community leaders will deliver their followers' allegiance as long as their own leadership status is protected. However, neither stakeholder elites nor their constituents will indefinitely support an incompetent monarch. Just what competent government means in Saudi Arabia is the subject of the current part of this volume. It includes providing domestic and international security, promoting economic development, and encouraging social change at a pace that the majority of the population finds acceptable. To assess whether the Al Saud will be able to deliver these benefits to their people in the future, it helps to understand how they have done so in the past.

12

Providing Internal Security

Older Saudis remember from their own experience or their parents' stories just how violent and unstable Arabia used to be. Every town had a wall and a watchtower; you can still see the city gates of coastal Jeddah and the watchtower at Ar Rass in the center of the peninsula. Gates were closed at night. Watchtowers were manned in case *Bedu* appeared from the desert to steal crops and livestock; if you resisted them, you might well be killed. There were no policemen. Travelers paid for protection. Tribes fought for water and pastures. Banditry was common, even for pilgrims on the Hajj.

Much of King Abdulaziz's initial support came from Arabia's settled populations because he promised to improve security. In 1928 it was *Bedu* lawlessness that finally allowed him to mobilize an army to confront the Ikhwan.[1] He ended protection payments, tracked down brigands, and punished them very severely.[2] Villages no longer feared midnight raids, travel became safer, and trade became more profitable. When the Saudi people consider what the Al Saud brought them and what they continue to deliver today, internal security is at the top of the list—particularly, when they reflect upon current conditions in many neighboring countries.

There have been security problems in Saudi Arabia since the defeat of the Ikhwan. During the 1950s there was violent labor unrest at Aramco.[3] During the 1960s there were attempted coups by air force officers.[4] There have been numerous violent protests by the Shia population of the Eastern Province—most notably, in 1979. The kingdom's worst security breach in recent years was the seizure of the Great Mosque in Mecca in that same year. However, all of these events were limited geographically and contained in a matter of days or weeks. For many years, most people traveled safely throughout Saudi Arabia with more to fear of traffic accidents than highwaymen. That changed in May 2003, when al-Qaeda began a four-year terror campaign that presented the Al Saud with their most serious internal security threat since the Ikhwan Revolt of 1929.

Al-Qaeda intended to overthrow the House of Saud. Its well-planned campaign began in Riyadh with simultaneous attacks on three residential compounds the night before the American Secretary of State, Colin Powell, was due to arrive. Thirty-six people died in those attacks and more than 100 were badly injured. The American Embassy evacuated families and did not bring them back for six years. Those who believe that diplomats spend all their time at cocktail parties or that only military families paid a price during the War on Terror should reflect on the lines of embassy mothers and children leaving for the airport while their husbands and fathers remained behind in increasingly uncertain circumstances.

During the next four years, dozens of people died and hundreds were injured in a seemingly endless series of attacks and shoot-outs that cost the lives of many Saudi police officers. The Ministry of the Interior headquarters in Riyadh and Aramco's vital gas processing facility at Abqaiq were attacked. Foreigners were killed in their offices and homes or as they went about their business. BBC cameraman Simon Cumbers was murdered and British journalist Frank Gardner severely injured. On December 6, 2004, al-Qaeda terrorists crashed a car into the gates of the American Consulate in Jeddah, shot the Saudi National Guardsmen on duty, and took control of the large consulate compound for nearly eight hours. Six employees were killed and a dozen wounded before Saudi National Guard units stormed the compound, killing or capturing all the terrorists—including one who tried to hide in a palm tree.

Within days of the initial attack, the Saudi government made the first of several strategic choices that led to it defeating the terrorists. The civilian police, not the Saudi Army or the National Guard, were charged with eliminating Al-Qaeda in the Arabian Peninsula, or AQAP as the local group was known. The decision not to militarize a domestic security problem sent a message. The country faced a crime wave, not a war: al-Qaeda members were common criminals rather than holy warriors—and criminals were primarily the responsibility of the civilian police, not the army.[5]

The Ministry of the Interior took up its new task with vigor. Checkpoints appeared across the country. Heavily armed units began to guard Western residential compounds. The first of many "Most-Wanted" posters went up across the kingdom. Crown Prince Abdullah announced, "We specifically warn anyone who tries to justify these crimes in the name of religion. We say that anyone who tries to do this will be considered a full partner to the terrorists and will share their fate."[6] Prominent religious scholars, such as Sheikh Nasir al-Fahd, who had given ideological support to al-Qaeda, were quickly arrested and remain in prison to this day.

The beginning of the end for AQAP came in April 2005, when the government's massive public awareness campaign and policy of paying large rewards for information paid off. Young men with long beards and short robes who came and went at odd hours from newly rented apartments were no longer being ignored; now they were reported to the police. Such was the case in Ar Rass, a rural community 150 miles northwest of Riyadh, where residents notified the police of suspicious activity at a local farm. What followed was a three-day gun battle, in which dozens of police officers were wounded and the last of AQAP's leadership died. The organization never recovered from this blow; leaderless, finding new recruits scarce, and clearly lacking broad public support, the remnants of Al-Qaeda in the Arabian Peninsula began fleeing to Yemen.

Saudi Arabia thus contained and then largely eliminated an organization that had proven extremely resilient in other parts of the world. The Saudi counterterrorism program certainly sought to kill or capture active terrorists, but it placed much more emphasis on preventing young men from joining al-Qaeda in the first place or convincing them to quit if they had already signed up. It offered the repentant a well-defined path back into society. In such respects it was a very Saudi approach, which the kingdom's security forces divided into three distinct programs targeting "The Men", "The Money", and "The Mindset".

In 1929, when King Abdulaziz headed for his final showdown with the Ikhwan at Sabillah, he knew his opponents well. Faisal Daweesh and Sultan Ibn Bijad had fought with him to unify the kingdom on many occasions. He knew how the Ikhwan were organized and what motivated them. Seventy-five years later, when three large, residential compounds were attacked in one night, King Abdullah and his brother, Interior Minister Prince Naif, had much less clarity about whom they were fighting. Like their father, who had supported the Ikhwan Movement only to have it turn against him, they discovered that the Saudi government had unwittingly helped to create Al-Qaeda in the Arabian Peninsula.

In its effort to counter the secular, anti-monarchical Arab Nationalism preached by Egypt's president, Gamal Abdul Nasser, the Saudi government had since the 1960s encouraged the idea of Pan-Islamic Solidarity. The Soviet invasion of Afghanistan in 1979 dramatically amplified this Pan-Islamic sentiment in the kingdom and across the Muslim world. With the full support of the United States, the Saudi government and the kingdom's religious leadership had mobilized young men and private funding to fight the godless Soviets in Afghanistan. No one called them terrorists, in fact, many members of the royal family encouraged these *mujahideen* with financial and moral support to go help their Muslim brothers kill Russians. As governor of Riyadh, King Salman had played a prominent role promoting this effort, and many answered his call.

Estimates of how many Saudis fought against the Soviet Union in Afghanistan range from 12,000 to 25,000.[7] When Soviet forces withdrew in 1989, most of these Saudi *mujahideen* returned home to resume normal lives; some even joined the kingdom's police force. These original *mujahideen* were all middle-aged men by 2003. Very few of them joined AQAP, but their example had created a climate of sympathy for oppressed Muslims abroad and justified the use of violence in their defense.

Although the original Saudi *mujahideen* did not return home to start a terror campaign against their own government, a few, like Abdulaziz al-Muqrin, did become hardened "lifestyle jihadists" who went on to fight in, among other places, Bosnia and Chechnya.[8] It was these conflicts—and, later, the Anglo-American invasion of Iraq— that kept a new generation of young Saudis coming to al-Qaeda training camps in Afghanistan long after the Saudi government had stopped encouraging such behavior and stripped Osama bin Laden of his Saudi citizenship and bank accounts.

The second wave of Saudis who went to Afghanistan after 2000 were not impoverished Iraqis willing to kill an American soldier for $100 in order to feed their family. Nor did they resemble the opponents of a repressive, anti-Islamic government like others in Algeria. They did not form a nationalist or sectarian movement like Hamas or Hezbollah. They were not alienated immigrants, as appeared in France and Britain. So, just who were they?

A great deal of original research, much of it conducted by Western-trained Saudi sociologists, went into finding out who joined al-Qaeda and what motivated them. The results, gathered from scores of interviews, made interesting reading. These Saudi terrorists were, for the most part, urban high-school graduates in their twenties from lower middle-class backgrounds. Most were unmarried and had jobs with steady, but modest, incomes. Most had been to Afghanistan or had relatives who had been on *jihad* abroad. They were motivated not by oppression at home but largely by events outside of Saudi Arabia, and what the sociologists called "humiliation rage." Fueled by religious zeal, this boiled down to a three-part agenda: defend foreign Muslims who were being abused by non-Muslims; get the foreigners and their non-Islamic values out of Saudi Arabia; and overthrow the Al Saud, who were clearly aligned with the non-Muslim foreigners.

On September 11, 2001 there were several hundred humiliation-enraged, young Saudis training at al-Qaeda camps in Afghanistan. They had gone there not to fight Russians, like their older brothers, but to support the country's Taliban government. When US Special Forces units and CIA officers organized Operation Anaconda to topple the Taliban regime, these Saudis found themselves on the run. Some were killed; some found shelter in Iran. More than 100 were captured and imprisoned at Guantanamo Bay, the detention camp set up at an American naval base in Cuba. Their leaders, including Osama bin Laden, retreated into Pakistan. Most of the others, between 300 and 1,000 deeply committed individuals, simply went home to Saudi Arabia.[9]

These Afghan veterans provided the bulk of the leadership and many of the foot soldiers for AQAP, which remained largely restricted to the Afghan alumni network, their friends, and relatives.[10] For two years they organized a five-cell structure in the kingdom with military, finance, media, and religious units. They acquired weapons, established safe houses, and recruited new members. Although they did occasionally speak out against government corruption, their most vehement grievance became the presence of foreigners—and, in particular, American military personnel—in Saudi Arabia. Although the original Saudi *mujahideen* had not been particularly anti-American, this became a defining characteristic of those who went to Afghanistan after the Soviet withdrawal, especially after the Anglo-American invasion of Iraq in 2003.

Whatever their motivation, the Saudi Ministry of the Interior made slow but steady progress in arresting or killing those who had made the transition from moral outrage to outright violence. The resources of the Saudi state greatly exceed those of al-Qaeda. The Ministry of the Interior's overall budget rose sharply, reaching $8.5 billion in 2005 and $12 billion in 2007—with most of the increase going to fighting terrorism.[11] A great deal of new equipment was purchased, five new prisons were built, and more than 800 retired prison guards returned to duty. Highly sophisticated electronic and physical surveillance systems were installed across the kingdom. Cooperation with Western intelligence services expanded.[12]

Eventually, more than 150 terrorists were killed by the police and several thousand suspects arrested.[13] However, the Saudi campaign against al-Qaeda was noticeably less vicious and more targeted than the brutal, widespread repression of the Islamists in Syria, Algeria, and Egypt. This was another clear policy choice: Prince Mohammed bin Naif—who, as deputy minister of the interior, directed government counterterrorism efforts—emphasized again and again that although it was important to eliminate existing terrorists, it was much more important not to create new ones with heavy-handed police tactics.[14]

Al-Qaeda needed funding, but where did its money come from? The 9/11 Commission Report found "no evidence the Saudi Government as an institution or senior Saudi officials individually gave money to al-Qaeda," but it did note that al-Qaeda took advantage of a system of charitable giving that was both "essential to the [kingdom's] culture and subject to limited oversight."[15] Islam requires Muslims to give 2.5 percent of their liquid net worth to charity each year. This *zakat* tax constitutes one of the five "pillars of Islam," and Saudis, who pay no income tax, have long been mindful of this charitable dictum. Many Saudis contributed to the kingdom's 200 plus charities—which in 2001 were, indeed, largely unregulated.

In sending funds overseas, most of these charities were doing nothing more than what they claimed; however, it is now clear that a few were diverting funds to terrorist organizations. Although some Saudis knew this, many did not. In some cases, such as that of the Al Haramain Society (a foundation largely dedicated to proselytizing, supporting orphans, and feeding poor Muslims around the world), only part of the charitable organization had been corrupted. In others, like that of the Benevolence International Foundation, the entire operation was nothing but a front for al-Qaeda.

A 2007 study by one of Washington's leading Middle East researchers, Jon Alterman of the Center for Strategic and International Studies, concluded that "thus far, publicly available evidence does not suggest that Saudi charities act as a wholesale front for sowing global terror."[16] What appears more likely is that lax administration, rather than any official plan, allowed some branches of essentially reputable charities to be misused. Such was the case with the branches of the International Islamic Relief Organization in Indonesia and the Philippines, both of which were designated as terrorist organizations by the US government.

After 9/11, Saudi authorities began screening charity staff members, freezing suspect accounts, and closing suspect charities. More than a dozen branches of Al Haramain were closed. Then the entire organization, which had once been very prominent and respected in the kingdom, was shut down. In 2002, charities were required to notify the Saudi Foreign Ministry before sending any funds abroad. Thousands of unsupervised collection boxes that had hung in mosques and supermarkets for years were removed; all contributions were subsequently to be made only to the specific bank accounts of registered charities.

Later in 2002, the Saudi Arabian Monetary Agency issued detailed regulations criminalizing money laundering and terrorist financing. The Ministry of Interior established a Financial Intelligence Unit to investigate suspicious bank transactions. The FBI and the US Treasury sent agents to work with the Saudis on terrorist financing. Arrests were made. Saudi Arabia joined the EGMONT Group and the Financial Action Task Force, two well-respected international organizations designed to set standards and facilitate sharing of financial intelligence related to counterterrorism and money laundering.

In 2004, King Fahd issued a royal decree ordering all Saudi charities to stop sending funds abroad without explicit official permission for each transfer. This was a draconian step. The governor of the Saudi Arabian Monetary Agency received angry calls from respected philanthropists whose half-built schools and half-dug wells could now not be completed. Nonetheless, although supervised exceptions were made for some official organizations, such as the Saudi Red Crescent, this regulation remains in effect today.

Since 2015, King Salman has very deliberately and effectively further centralized control of charities. He founded the King Salman Humanitarian and Aid Center, which, by mid-2017, had become the only Saudi charity sending funds abroad. Even the International Islamic Relief Organization—which was part of the Muslim World League and had once been the largest Islamic charity in the world—ceased sending funds abroad and essentially shut down.

It is difficult to track financial flows in a cash-based economy, and impossible to know how much private Saudi money has gone to support terrorism. Millions of pilgrims who come and go from all over the world generally carry cash, not credit cards. Many Saudis have foreign bank accounts that their government cannot monitor. What can be said confidently is that Saudi citizens have become more careful about making charitable donations, Saudi banks have grown more vigilant about checking suspicious transactions, Saudi charities against which there was evidence of suspicious activities have been closed, and the Saudi government has implemented UN Security Council resolutions concerning terrorist financing—including freezing the assets of individuals on the Security Council's Sanctions Committee list.

These actions reduced the resources available to al-Qaeda. As US Assistant Treasury Secretary Juan Zarate stated, "[t]he targeting actions and systemic reform undertaking by the Kingdom of Saudi Arabia clearly demonstrated its commitment to work with us and the international community to combat the global threat of terrorist financing."[17] In September 2018, the Financial Action Task Force assessment on the country noted, "The Kingdom of Saudi Arabia is achieving good results in fighting terrorist financing, but needs to focus more on pursuing larger scale money launderers and confiscating their assets." It is worth noting that large-scale money laundering was one of the charges made against some of those detained in the Ritz Carlton, and their assets were indeed confiscated.

Changing the mindset of violent jihadists was a more sensitive matter than arresting them or cutting off their funds, because the Saudi government had long endorsed a culture of religious intolerance. Changing the way people thought required admitting that there were serious problems in the kingdom's religious and educational institutions. It meant developing innovative rehabilitation and public awareness programs to address these problems. Most of all, it required the support of the *ulama*. Saudi authorities worried far more that a respected theologian would endorse al-Qaeda than they did about another terrorist bombing for the damaging impact of such an endorsement would have been more widespread and long lasting.

Because religion was the language that resonated most forcefully with the population, the Saudi government always condemned al-Qaeda in theological rather than in political terms. Instead of using the standard Arabic word for terrorist, for instance, Saudi officials referred to al-Qaeda as "religious deviants," or "misguided corrupters on earth." Official internet chat rooms staffed by moderate clerics sought to refute subversive online preaching. Religious rulings by prominent clerics directly countered al-Qaeda propaganda. The kingdom's most senior religious scholars, including the grand mufti of Saudi Arabia and the imam at the Great Mosque in Mecca, repeatedly denounced terrorism in their sermons.

The few respected scholars in Saudi Arabia who supported al-Qaeda had been arrested soon after the group's first bomb attacks. Several less important clerics were killed in shootouts. The remaining al-Qaeda ideologues were not religious scholars and did not command significant popular respect. Then in June 2004, two of Saudi Arabia's most controversial religious leaders came out strongly against al-Qaeda. Safar al-Hawali and Salman al-Auda were serious scholars who in the early 1990s had led non-violent, anti-government protests known as the Awakening or *Sahwah* movement. Now they undermined al-Qaeda's legitimacy with a theologically articulate assault on terrorism. The fact that both of them had once gone to prison for their anti-government views helped to validate their arguments against al-Qaeda. At least one al-Qaeda member, Ali Faqsi, took Safar al-Hawali's advice and went with the cleric to surrender to the Ministry of the Interior.[18]

Condemning al-Qaeda's agenda was not difficult for the Saudi *ulama*. Many of the terrorist group's tenets, such as resisting Western influence or using political violence, came not from Mohammed Abd al-Wahhab but from the twentieth-century Muslim Brotherhood. Although Abd al-Wahhab had nothing to say about confronting secular Western values, he had a great deal to say about avoiding *fitna* or civil unrest. Many Wahhabi texts are direct attacks on anything that disrupts the social order or political stability. These were the arguments that the *ulama* used against al-Qaeda, which they clearly understood posed as serious a threat to themselves as it did to the House of Saud.

Throughout this difficult period, the Saudi *ulama* denounced al-Qaeda; supported the monarchy; and brought with them most of the kingdom's religious policemen, mosque preachers, and religious studies schoolteachers. The theological offensive that they supported helped to prevent al-Qaeda from ever becoming a broadly based, popular movement in Saudi Arabia. In return, the *ulama* benefited from a vigorous crackdown on any criticism of the Wahhabi clergy or the conservative social values that it endorsed.

The battle against terrorism extended into Saudi schools, where a great deal of the day was taken up with religious studies. In 2003, the Saudi government began a full-scale review of school textbooks. These books clearly and repeatedly proclaimed the superiority of Islam over all other religious, social, or political systems. The books explicitly promoted intolerance more than violence, but they created an atmosphere in which violence was more easily accepted. As the then Foreign Minister Saud al-Faisal stated, "Ten percent of what we found was questionable. Five percent was actually abhorrent to us."[19]

The goal of the Ministry of Education's fourteen-man review committee was to remove passages that "directly or indirectly promote enmity, hostility and hatred."[20] Western critics complained that some provocative passages remained. Saudi clerics insisted that the superiority of Islam over other religions was self-evident, and that if the Islamic principles expressed in schoolbooks were compromised to please foreigners, the ruler would lose his legitimacy as Custodian of the Two Holy Mosques. The process was highly contentious, yet despite many protests the textbook revision process continued—and continues today, to the point at which most Saudi schoolbooks have not so much been revised as simply replaced with more modern and less problematic volumes.

Deputy Minister of the Interior Mohammed bin Naif realized that keeping thousands of young men in prison indefinitely would eventually cause a public backlash. He needed a way to identify terrorists who could be reformed, and to re-integrate them into society. Hence, for many al-Qaeda members who had served their prison sentence, the path back to normal life began with six months in the Prince Mohammed bin Naif Rehabilitation Program—an original Saudi creation. Not all prisoners were eligible for the program and not all participants were eventually released, but a regime for rehabilitation was devised for at least some of them.

The program was run jointly by the ministers of Islamic affairs, the interior, and social affairs. Inmates were called beneficiaries rather than prisoners. Accommodations were secure, but comfortable; the food was good. The program included sports; vocational training; psychological counseling; art therapy; and, most importantly, anger management classes. More than 150 Wahhabi religious scholars worked with the beneficiaries to improve their understanding of Islam—particularly, the concepts of holy war (*jihad*) and excommunication (*takfir*). They stressed that only the head of state can call for *jihad*, and only senior religious scholars can declare someone a *kaffir* or nonbeliever.

The program sought to get beneficiaries out of the al-Qaeda cult and back into society with changed views and changed behavior. Counselors worked to defuse the humiliation rage that drove many beneficiaries and helped prepare them to deal with the stigma of having been in prison. Families were involved, and conjugal visits allowed near the end of the program. If necessary, the families of beneficiaries were supported financially because the ministry understood that if it did not help to support an inmate's family, al-Qaeda would shove cash under their door at night.

There were repeated psychological evaluations. If, after six months of rehabilitation, a beneficiary was still judged dangerous he went back to prison. Evaluations were made by psychologists and social workers, but the final decision on release was always taken by a senior police officer. Once released, the ministry tried hard to make sure that the former prisoner had something to lose; they had noticed that not many suicide bombers had a wife, three children, a good job, and a mortgage to pay. The state helped with dowries, car loans, and mortgages as well as employment. Some prisoners were released on the recognizance of their families or tribes, who would face a humiliating loss of face, financial, and even legal consequences if they reverted to terrorism.

One former Guantanamo inmate, Khalid al-Hubaishi, who was returned to Riyadh and placed in Al Haier prison, told the author Robert Lacey, "I had heard horror stories about Al Haier, but when I got there it was like a five star hotel compared to Guantanamo. [...] They let me call my family [...] They bought me a new *thobe* [robe] to wear when I met my mother. They paid for my family to stay in a hotel while they visited me." Al Hubaishi was released after a further fifteen months in prison and secured a job with the Saudi Electric Company. As Mohammed bin Naif told Robert Lacey, "We try to transform each detainee from a young man who wants to die into a young man who wants to live."[21]

Some released beneficiaries did rejoin al-Qaeda. Official reports cite recidivism rates of between 10 and 20 percent depending on the period examined and the set of prisoners involved.[22] However, it was not the course content or recidivism rates that most concerned the Ministry of the Interior; what really mattered was the program's dramatic public relations impact. In the Saudi leadership's view, it did not matter if some of the rehabilitated terrorists returned to al-Qaeda. They could be recaptured and might not be given a second chance to surrender. What mattered was that the Saudi public saw how the police had trusted these young men and tried to help them, only to be repaid with lies and more criminal behavior. No one would feel sorry for them the second time they were arrested for terrorism. For Mohammed bin Naif, losing a few dozen prisoners was less important than keeping millions of Saudi citizens on his side.

A recognition that defeating al-Qaeda required keeping public opinion onside, colored everything that the Saudi Ministry of the Interior did and was fundamental to its success. The police understood that every time a terrorist was killed, his parents would have one of two reactions: either the wicked security forces killed my son, or the wicked Osama bin Laden misled my son and got him killed. Mohammed bin Naif tried hard to encourage the second reaction.

On the night that the Ministry of the Interior was bombed, the then Deputy Minister Mohammed bin Naif called the parents of the dead suicide bombers so that they would not first learn of their sons' deaths on TV. He did not delegate the task to a clerk or a minor official, he telephoned the parents himself. He described the young men who had just tried to kill him as victims and apologized for not having been able to stop them before they began their bombing run. He asked for help in stopping other young men from joining al-Qaeda, and he got the help he asked for. The fathers of dead or incarcerated terrorists began appearing on TV warning parents to avoid a similar fate for their sons. They described signs that indicated that your son might be going off the rails: Was he no longer watching television or praying at the family mosque? Was he avoiding old friends? Was he withdrawing into himself or becoming moody? They told parents where to turn to for help before it was too late.

On other occasions, a senior police officer would call parents to tell them something like: We have good news for you. We caught your son on his way to join al-Qaeda in Iraq; thank God. Now he won't be driving a suicide vehicle into an American checkpoint in Baghdad. Of course, he is going to prison for a while, but he is safe. The parents were in fact usually relieved and grateful. Then there were the young men who did get to Baghdad but did not get what they bargained for. One in particular appeared frequently on television and spoke at various public events. His face was badly burned. He had gone to Iraq to fight the foreign invaders and been asked to drive a truck past a military checkpoint. What he had not been told was that the truck was a bomb, which was remotely detonated as he passed through the barrier. The disfigured young man would go on to explain how the American troops had pulled him from the burning truck and taken him to a hospital. His message was clear. al-Qaeda lied to me and tried to kill me. Now I am scarred for life. Don't let the same thing happen to you.

Soon after the first bombings, checkpoints went up all over Saudi Arabia. They became a part of life, and in many places still are. It soon became clear that some al-Qaeda operatives were moving about dressed as heavily veiled women. The Ministry of the Interior nevertheless made a conscious decision not to start lifting women's veils to look for terrorists. Better by far to let them pass than antagonize hundreds of conservative husbands or innocent women. Better to keep the general public on the ministry's side. Even at the height of the violence, the Saudi government never imposed a curfew.

Torture has certainly occurred in Saudi Arabia. The prisons of King Abdulaziz and even King Faisal were not pleasant places. Yet in 2003, the Saudi leadership recognized that harsh treatment for those many regarded as misguided youths rather than criminals would be unpopular. Mohammed bin Naif specifically prohibited the torture of incarcerated al-Qaeda prisoners, and had hundreds of closed-circuit television cameras installed to monitor both prisoners and guards.

Thomas Hegghammer, who has conducted the most thorough academic study of the Saudi counterterrorism program, wrote,

> To further encourage desertions [from radical *jihadism*], the State attempted to appear merciful and forgiving towards repentant militants. This began with de facto abstention from prisoner abuse. By most available accounts it seems that the police really did not torture captured al-Qaeda militants, at least not with the methods and on the scale it had done in the mid-1990s. In order to counter the lingering torture stories from the 1990s, the authorities sought to create a degree of transparency about prisoner treatment.[23]

As a result of these efforts, the Saudi public accepted that torture of al-Qaeda prisoners in Saudi prisons was rare. Saudi fathers began informing the police of their sons' suspicious activities, and even turning them in to Mohammed bin Naif. They knew that their sons would go to prison; they would probably have been less cooperative had they expected that torture would be part of the process. Announced periods of amnesty and reduced sentences were used to encourage al-Qaeda desertion, and by all accounts these pledges were honored.

The lack of torture was only one strand of a highly sophisticated public relations campaign designed to keep the people on the government's side in its struggle with al-Qaeda. Across Riyadh, billboards went up showing the clasped hands of a citizen and a police officer: "Working together for Security." All Saudi papers echoed the sentiments of *The Arab News* in their editorials: "This battle will take years to finish, but there is no other option. Sitting back and sticking our heads in the sand is the worst thing we can do. The devil is here and he is knocking hard on our door."[24]

Elementary schools held anti-terror poster competitions, in which artwork depicting burning buildings and bleeding bodies made it clear that the children did not see terrorists as heroes. In many public spaces, monuments to police officers killed fighting al-Qaeda were erected. Like war memorials found in the West, their message was clear. These people died to keep you safe. In the small, central Arabian town of Bukahriah, a prize is given every year to the citizen who has done the most for the community and the winner's picture is prominently displayed in the village hall. For years the prize had gone to businessmen and philanthropists; in 2007 it went to a local police captain killed in a shootout with al-Qaeda.

Like the *ulama*, media, and schools, the tribes were brought into the fight against al-Qaeda. Interior Minister Prince Naif bin Abdulaziz was an expert in tribal affairs. He met frequently with tribal leaders, who helped him locate wanted tribesmen and monitor those released. Prince Naif and his son, Mohammed, carefully avoided apportioning collective blame or punishment that could alienate an entire tribe.

Because the *sheikhs* did not want their tribes' names tainted by association with terrorism, the names of those arrested were sometimes not announced.

A respect for tribal values directly affected police operations. When an al-Qaeda cell with members from a particular tribe was uncovered, it would become known informally that police officers from that tribe might want to call in sick on the day on which arrests would be made. Then a prince or senior police officer would meet with the tribe's chief to show him the incriminating evidence against his fellow tribesmen. The prince would guarantee that all the wanted men would be given a fair chance to surrender. He would assure the *sheikh* that the government regarded most of his tribe as loyal, law-abiding citizens with whom it had no quarrel. Then as the *sheikh* was leaving, the prince would say: "Send several young men from your tribe to the Ministry tomorrow morning. We will have good jobs waiting for them. You choose the men to send and next time you're here perhaps we can talk about that new school you keep saying your village needs." Like the religious scholars, the tribal leaders stayed with the government, brought their followers with them, and got something in return.

In their struggle with al-Qaeda, the Al Saud benefited from several favorable circumstances. The Saudi state was highly centralized, backed by important allies, and determined to hold on to power. A state-controlled media and consolidated education system got out the message that al-Qaeda members were criminals, not heroes. Oil prices were high and the Saudi treasury full. The war in Iraq drew militants away from Riyadh. In a very conservative society, revolution was never likely to be popular and, as we have seen, the Al Saud made some deliberate, and ultimately effective, choices.

Unlike the rulers of Algeria, Egypt, Iran, Iraq, Libya, or Syria, the Saudi government did not turn the army on its own people. Torture was officially abandoned. Collective punishment of families and tribes was avoided. Less dangerous terrorists were treated more as prodigal sons than criminals. Police officers attended the weddings of released terrorists to indicate that they were still part of the community. In a deeply religious society, al-Qaeda was delegitimized in religious terms by respected theologians.

It was a very Saudi approach. Many foreign observers found it not only successful but highly original, and experts came from Washington and London to study the Saudi counterterrorism program. Conferences were held to discuss it and books written about it. In reality, the program was not new. Although modified for current circumstances, it came straight out of King Abdulaziz's playbook.

As we saw in Chapter 3, King Abdulaziz sought religious sanction from the *ulama* for his war against the Ikhwan and emphasized their common criminality just as King Abdullah would do with al-Qaeda members eighty years later. King Abdulaziz issued detailed "Most-Wanted" lists for the Ikhwan leadership and made extensive use of amnesty to draw away rank and file Ikhwan, just as Interior Minister Prince Naif would do. At the Ad Dawadimi Conference in 1929, King Abdulaziz specifically spoke about rehabilitation for the former "criminals" just as his grandson, Prince Mohammed bin Naif, would do in 2004.[25] After Sabillah, the families of some dead or imprisoned Ikhwan leaders were given state pensions, much as the Ministry of the Interior would later provide for the families of incarcerated al-Qaeda members.[26] King Abdulaziz married a widow of Faisal Daweesh's son Azayiz, much as the Ministry of the Interior arranged marriages for rehabilitated al-Qaeda terrorists. The Saudi leadership clearly understood these historic precedents, even if most Western commentators did not.

Whatever criticisms may be leveled at the Al Saud for having maintained a religiously intolerant autocracy, they must also be given credit for establishing internal peace and security. In Europe or the United States, it may be difficult to grasp just how important and difficult this was to achieve in a place where constant tribal warfare and sheer banditry had been the norm for centuries. In other tribal or sectarian societies—such as Afghanistan, Iraq, Lebanon, Somalia, Syria, or Yemen—it is not difficult to imagine what people would give for a strong central government that kept their families safe.

Inside Saudi Arabia, people are well aware of what the Al Saud have achieved. Internal security remains a primary reason why the Saudi people did not opt for the violent, dramatic political change that al-Qaeda offered them in 2003, and why they largely ignored the Arab Spring protests of 2011 or their 2019 aftershocks in Algeria, Lebanon, and Sudan. Instead, the vast majority of Saudis—both liberal and conservative—have long chosen to support the family that has provided them with stability since King Abdulaziz created the kingdom in 1932.

The great irony is that in 2020, many Saudis have come to fear their own government more than al-Qaeda terrorists. Saudi Arabia was never an open society, but it was not a police state. Now the highly sophisticated technical apparatus installed to thwart al-Qaeda has been turned on peaceful citizens. Telephones and social media are closely monitored. Saudis no longer feel comfortable making even mild criticisms of their government. They switch off their cell phones or go into the garden to talk. The scope of acceptable debate has narrowed, with both conservative Muslim Brothers and liberal feminists being arrested. The anti-corruption campaign, however popular and even necessary, has cast doubt on the rule of law. Restrictions on travel and the seizing of assets have become more common. Allegations of torture have reappeared, and the dissident journalist Jamal Khashoggi was killed by government agents.

Totalitarian governments are not always unstable. Joseph Stalin died in his bed; North Korea's Kim dynasty has been in place for three generations; it took a foreign army to remove Saddam Hussein from power. What is often destabilizing, as both Mikhail Gorbachev and the Shah of Iran discovered, is attempting to maintain authoritarian political structures while pursuing social liberalization and economic reform. That is the risk that Saudi Arabia now faces—for, in the long run, opening cinemas will not compensate for closing newspapers.

Figure 19. The Great Mosque in Mecca burning after it was seized in 1979 by armed militants from over a dozen Muslim countries who believed the Al Saud were both impious and too close to non-Muslim foreign powers.

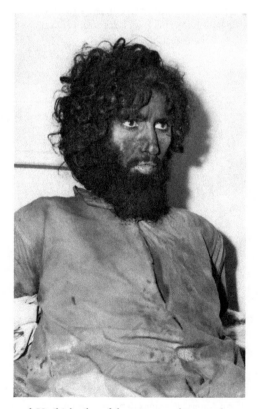

Figure 20. Juhayman al-Utaibi, leader of the mosque takeover who was captured and executed along with sixty six of his followers.

Figure 21. In August 1990 Iraq invaded and occupied Kuwait. Operation Desert Storm was the 1991 military operation to expel Iraqi forces from Kuwait. Here Saudi troops prepare to cross into Kuwait, January 1991.

Figure 22. Half a million American troops deployed for Operation Desert Storm: Here the US Army First Armored Division advances into Iraq.

Figure 23. The ruins of Riyadh's al-Hamra residential compound after al-Qaeda's 2003 attack, which began its four-year campaign of terrorism in Saudi Arabia.

Figure 24. The twenty-six most wanted al-Qaeda terrorists. The Saudi government offered a one million riyal reward (US$ 266,000) for information leading to the arrest of each of them.

Figure 25. The blind Grand Mufti of Saudi Arabia, Abdulaziz bin Abdullah bin Mohammed Al al-Sheikh, leader of the religious scholars, guardian of traditional values and a direct descendant of Mohammed Abd al-Wahhab.

Figure 26. Saudi Arabia's inchoate, appointed parliament, the *Majlis al-Shura* or Consultative Council which is composed largely of secularly educated technocrats.

Promoting Rapid Economic Development

If providing physical security is the first service that Saudis want from their government, then improving economic conditions comes a close second—and in that regard, they had a very long way to go. Saudi life expectancy in 1950 was only 39 years.[1] By 1960 the majority of Saudis still lived at no more than subsistence level, and their nutrition was poor even by developing-world standards. Safe drinking water was unavailable in many places; cholera, malaria, and bilharzia were common; and a majority of the population suffered from trachoma.[2]

There can be no argument that the health, education, housing, and general welfare of the Saudi population has improved dramatically since then. Life expectancy was 75 in 2019.[3] Cholera, malaria, and malnutrition are mostly now a memory. A system of subsidies introduced in the 1970s has kept prices low for water, electricity, gasoline, and basic foodstuffs such as bread, sugar, and rice. Education and healthcare services have improved and are provided free of charge, as are some housing loans.[4] Most Saudis believe that the Al Saud have made their lives better by producing oil efficiently, building high-quality infrastructure, and providing public services at a fraction of their real cost. How did they achieve this in the past and, more importantly, can they maintain that record in the future?

Before oil, King Abdulaziz had no money for basic infrastructure, much less a welfare state, nor was an Arabian *sheikh* traditionally responsible for providing healthcare and education. Yet from the outset, perhaps because of what he had seen in Kuwait during his family's exile there, Abdulaziz tried to improve the services that his nascent government was able to provide. In 1917 he was still a minor, provincial chieftain in what was nominally part of the Ottoman Empire when he summoned American, Christian missionary doctors to Riyadh from Bahrain. Here is what the wife of one doctor wrote about the meeting:

> "While you are here, my house is yours" added the *sheikh*, motioning for Paul to sit down with him. A slave brought in coffee, and while Ibn Saud sipped, he explained that he had asked the doctor to come, neither for his health nor the health of his family, but for the needs of his people. He had already arranged the use of a nearby house for a hospital where he wanted his people treated without any cost to them.[5]

These American missionary doctors from Bahrain continued to come to the Nejd and Eastern Province of Saudi Arabia until the 1950s, when the Aramco Health Department took over from them.

Abdulaziz paid particular attention to the Hajj, where pilgrims were chronically short of water. In 1927 he ordered two distillation condensers from Egypt, which not only helped thirsty pilgrims but also gave him real leverage over his new subjects since he recruited the water carriers to be his tax collectors.[6] He set up quarantine facilities, which helped to reduce the number of outbreaks of epidemics that had long been associated with the Hajj. Then, over the "piteous protests of the camel owning Bedouin," who feared the loss of their livelihood, Abdulaziz introduced motor vehicle transportation for pilgrims.[7]

There is a well-known story about Abdulaziz's early efforts to improve education. Before World War I, the Dabbagh family had been well educated, prominent officials in the Hejaz. When the Hashemites fled, some members of the family had opposed the Al Saud and led the ill-fated Hejaz Independence Movement. Some were killed in the fighting. Hussein Dabbagh fled to Iraq until a general amnesty was offered. Then he set off for home, only to be summoned to Riyadh to meet the king.

Dabbagh was surprised when King Abdulaziz said he would like to apologize for something he had done. Specifically, he had intercepted Dabbagh's letters from Iraq and read them. He was sorry to have intruded on Dabbagh's privacy. Then the king took a letter and read out, "We are now governed by ignorant *bedu* and their ignorant king." "Did you write that"? He asked the now alarmed Dabbagh, who could hardly deny the evidence.

"Well," said the king, "perhaps you are right. Perhaps we are ignorant *bedu* and perhaps you can help us. How would you like to become the Director of Public Education?" The once exiled former revolutionary accepted and served in that post for nearly twenty years. Although his funding was limited, Dabbagh established new non-religious schools outside the Hejaz—including one for the king's sons in Riyadh.[8]

The Great Depression severely reduced Abdulaziz's already limited income. In 1927, no less than 132,000 fee-paying pilgrims came to Mecca.[9] By 1933 their numbers had fallen by 80 percent, and the king's revenues had collapsed from $1,000,000 to barely $250,000.[10] Abdulaziz was hard-pressed for cash and told his advisor, H. St. John Philby, "if anyone were to offer me a million pounds now, he would be welcome to all the oil concessions he wants in my country."[11] Fortunately for the king, someone was interested.

In 1933, when Abdulaziz invited foreign oil companies to bid for a concession, today's Chevron bid only £50,000 and won a sixty-year concession to seek, produce, refine, and market Saudi oil.[12] The Aramco agreement of 1933 promised further annual payments and royalties if oil was discovered. It proved to be a very profitable agreement for both parties. By 1939, the kingdom's oil revenue had risen to $2,000,000 a year.[13] Over the next 60 years, efficiently producing and distributing oil revenue would become as important to Al Saud legitimacy as providing religious orthodoxy and physical security.

Not all national oil companies are the same. Many are corrupt, inefficient, politicized bureaucracies that can barely maintain their current production levels; others are only shells for foreign operators and contractors. Saudi Aramco is different. It is not only the world's largest and most profitable oil company but is also a well-managed, technologically advanced firm operated on a commercial basis, which produces 10–12 percent of the world's oil every day. Aramco is widely recognized as more efficient, technically competent, and meritocratic than most Saudi institutions.[14] Nearly all its senior management and 80 percent of its 75,000 employees are Saudi nationals. Twenty percent of its new hires are women. This was not a fortuitous accident, it was the result of two deliberate strategic choices made at the highest levels of the Saudi government.

The first choice related to the largely consensual manner in which Saudi Arabia took control of the Arabian American Oil Company, or Aramco during the 1970s. There was surprisingly little bitterness towards the Americans who had created Aramco and managed it for nearly half a century. In part, this was because the United States government had never exerted anything like the level of political influence in Saudi Arabia that Britain had in Iraq, Iran, or Egypt; there was no anti-colonial rush to throw the foreigners out. Aramco had faced violent labor unrest in the late 1950s, and its segregation policies between Arab and American workers had at times bordered on racism; many of its more liberal policies had been adopted only after the outbreak of strikes.[15] But Aramco had also provided jobs, electricity, education, technical training, healthcare, paved roads, and independent business opportunities to a local community that had never seen a refrigerator before the company arrived in the kingdom.

Most Saudis still have a positive view of the Aramco experience, whose legacy lingers in the Eastern Province—the only part of Saudi Arabia where the work day starts at 7:00 a.m. and where, until recently, the electricity supply was 120 volts, rather than 240 volts as in the rest of the world outside North America. The fact that the Saudis chose to keep the name Aramco after the firm's nationalization says a good deal about their attitude towards the former owners.

In Mexico, Iraq, and Iran, foreign oil assets were seized by the government. Because the Saudi government wanted to maintain both the efficiency of its oil industry and its security relationship with the United States, Saudi Arabia purchased Aramco from what are today Chevron and Exxon Mobil. Although he would not disclose the final price paid, one of the last American presidents of Aramco, Frank Jungers, made it clear in 2017 that there was no expropriation. It was a takeover that the Saudis negotiated in accordance with accepted business practices.[16] Aramco's ownership was transferred gradually and completed in 1980, although the firm remained incorporated in Delaware until 1988 when it became Saudi Aramco.

Even more significantly, the Saudi oil industry was not subsequently taken over by military officers, politicians, or trade unions. The foreign management structure and American culture of Aramco was left largely in place, and Saudi Aramco maintained technical cooperation agreements with its former foreign owners until 2010. As a retired American Aramco Vice President noted:

> The American partners were well compensated and generally relieved to get out of the act. The employees were unaffected because all of them were Aramco career people. Nothing much changed in terms of day-to-day operations. And the Saudi government was very careful not to impose a new set of managers.[17]

This Saudi approach to nationalization was not limited to the oil sector. It was also applied to foreign banks. In the 1970s, local investors purchased majority stakes in the

Saudi operations of American, British, French, Dutch, Egyptian, and Pakistani banks. They created institutions such as the Saudi American and Saudi British Banks, which maintained management contracts with former owners Citibank and HSBC. Over time, a majority of banking positions, including senior management posts, were filled by Saudi citizens. In 2020, the kingdom's banking system is modern and sophisticated in comparison with that of many other Arab countries.[18] Again, in many developing countries such as Iran, Iraq and Libya, foreign control of the banking sectors ended very differently.

The second choice that Saudi Arabia made differently from many other oil-exporting countries related to Petromin. Initially, Saudi Arabia had two parallel oil companies: Aramco and Petromin. The latter was established in 1962 and had responsibility for all oil and minerals in the kingdom located outside of the Aramco concession area. The clear expectation was that Petromin would eventually replace the American firm. However, unlike Aramco, it was not a private company with a board of directors and its own financial resources. It was a state agency with a government budget whose governor reported to the petroleum minister. Like the petroleum ministries in many socialist oil-exporting countries, Petromin followed a statist approach to development; on a daily basis it operated like a government agency, not a private company.[19]

Petromin saw itself as the national champion and natural leader of Saudi Arabia's industrialization efforts. By 1973 it had 3,000 employees and was negotiating petrochemical, pipeline, and refining agreements with Mitsubishi, Shell, Dow, and Mobil.[20] Although it had expansive plans, Petromin became increasingly viewed as an inefficient organization characterized by delay, waste, and politicized planning.[21] The development of the Rabigh refinery, north of Jeddah, proved an extremely expensive fiasco, tainted with corruption and indicative of all that can go wrong with state-led industrialization programs. As a result, Petromin's mandate was drastically reduced in 1975 and limited to oil refining and marketing. The decision to keep Petromin and Aramco separate confirmed that Saudi Aramco would largely maintain the historic corporate independence and management culture that it had enjoyed under American ownership.[22]

This outcome was, in part, because of lobbying by Aramco's Saudi managers. They were led by Ali al-Naimi, who had joined the firm as an 11-year-old office boy, been educated at company expense in Lebanon because there were no modern high schools in Saudi Arabia, and then sent to study geology at Lehigh and Stanford universities in the United States. Al-Naimi would become both Saudi Aramco's first non-American president and the minister of petroleum before retiring in 2016 aged eighty-one. Like most of Saudi Aramco's employees, he welcomed the firm's gradual nationalization, but he did not want it to be absorbed by either Petromin or the Ministry of Petroleum. Aramco employees wanted the firm's independent identity, commercial culture, and private sector salaries preserved.

In 1986—when the minister of petroleum, Ahmed Zaki Yamani, once again considered merging Saudi Aramco and Petromin—Ali al-Naimi presented him with a letter of resignation for himself and the entire Saudi management team effective immediately upon any Petromin takeover of Saudi Aramco. Yamani backed off.[23] When the minister of finance wanted Saudi Aramco's revenues to be deposited directly with his ministry instead of in Aramco's New York bank accounts, Ali al-Naimi again refused.[24] Saudi Aramco's managers fought hard to maintain their independence, and recognized that the best way to succeed was to operate efficiently and profitably.

Until its 2019 initial public offering, Saudi Aramco was entirely owned by the government of Saudi Arabia. Like any private oil company, it pays income tax and royalties to the government and dividends to its stockholders—who remain, largely, the Saudi government. Together, taxes, royalties, and dividends account for roughly 93 percent of Saudi Aramco's operating profits.[25] Saudi Aramco thus has an incentive to maximize profits because it gets to keep part of them. Petromin, on the other hand, relied solely on Ministry of Finance funding. It was slowly dismantled and finally dissolved in 2005, whereupon its assets were transferred to Saudi Aramco. Had Aramco been abolished instead of Petromin, both Saudi Arabia and global energy markets would look very different today.

Saudi oil revenue increased rapidly after World War II and exploded in the 1970s, rising from $10 million in 1946 to $333 million in 1960 and $104 billion in 1980.[26] Again, unlike many Arab oil producers, the Saudi government did not simply park its cash in US Treasury bonds or create a sovereign wealth fund that invested abroad. Until 2016, Saudi Arabia's Public Investment Fund focused almost entirely on building domestic industries. One of the first revolved around desalinated water.

In a land without a single river, water was once a very scarce resource—a commodity over which people fought. For generations in Central Arabia, the fear of water shortages was nearly as grave and persistent as the dread of *Bedu* raids. King Faisal's son, Mohammed, set out to allay those fears. In 1973 he became the first chairman of the Saline Water Conversion Corporation (SWCC), which began to extract potable water from the sea on an immense scale. Between 1975 and 1980, Saudi Arabia spent as much on building desalinization plants as it did on education.[27] Over the following two decades, SWCC built a nationwide network of more than twenty desalinization plants, pipelines, pumping stations, and storage facilities that brought fresh water not only to major cities but also to most towns in the arid Nejd and mountainous Asir. As a result, the Saudi people became dependent on inexpensive water sold to them at a fraction of its true cost.

The same was true of electric power.[28] In 1975 there was no national electricity grid in Saudi Arabia. Instead, there were fifty-six small, privately owned, unconnected electric companies—each serving its local community. They all lacked the capital and expertise to meet rapidly growing demand. The government unified these companies into the Saudi Consolidated Electric Company, gave the original owners shares in the new firm, and provided the capital for its expansion. Shareholders in the new company were guaranteed a profit of 15 percent, consumers were charged much less than production costs, and the number of Saudis connected to the grid rose rapidly. Overall, between 1969 and 1984 electricity generation multiplied twenty-five times while the production of desalinated water rose from 4 million to 355 million gallons per day.[29]

All ten of Saudi Arabia's development plans have called for economic diversification, which initially meant utilizing natural gas to produce petrochemicals. Saudi Aramco operates hundreds of oil wells that also produce gas—which until the mid-1970s was simply flared. For those flying into Dhahran at night, this flared gas produced a memorable sight of tall flames streaking across the sky in all directions. It was spectacular, but also a huge waste of energy and an environmental disaster.

Collecting and distributing all this gas was neither easy nor inexpensive. Aramco's American owners were not eager to spend the $40 billion (at today's values) that this required.[30] Nevertheless, Saudi Arabia's political leadership went ahead with the Master Gas System, which captured methane to fuel the kingdom's desalination and power plants as well as ethane feedstock for its emerging petrochemical industry. The system proved a great success. Today, while Iraq and Iran still flare most of their gas, the Saudis have been utilizing theirs productively for more than forty years while also developing non-associated gas reserves.

If capturing flared gas was the first step in diversifying Saudi Arabia's economy away from simply exporting crude oil, then building a petrochemical industry was the crucial next step. In 1974, King Faisal approved his technocrats' plan to build new industrial cities at Jubail on the Persian Gulf and Yanbu on the Red Sea. Both were to be connected to the Master Gas System and given special powers to bypass the delays inherent in normal bureaucratic procedures.[31] At Jubail, the American construction firm Bechtel undertook an engineering challenge that *TIME* magazine called a "project of moon-landing proportions."[32] There are many interesting statistics about the venture, but perhaps the most extraordinary is that the volume of process water that Jubail's pumping stations and canals provide is greater than the combined flows of the Tigris and Euphrates Rivers at their mouths.[33] The project at Yanbu was coordinated by Parsons of Pasadena, California, and was nearly as extensive. Dozens of refineries and petrochemical plants have now been built at Jubail and Yanbu. In 2018, these facilities produced 12 percent of Saudi Arabia's total GDP.[34]

The largest of these petrochemical companies is the Saudi Arabian Basic Industries Corporation (SABIC), which was spun off from the Ministry of Industry and Electricity in 1976. Unlike Petromin, which had attempted to operate complex industrial facilities without Western investors, SABIC always sought joint-venture partners. However, attracting partners to a desert country lacking good highways, ports, or telecommunications was a challenge. What Saudi Arabia offered its prospective partners in return for their technology and marketing networks was low priced petrochemical feedstock and crude oil delivery commitments. For every million dollars invested, the joint-venture partner could purchase 1,000 barrels a day of Saudi oil. There would be no discount on the market price, but if there were 200 companies lined up to buy oil, the joint-venture partners would go to the head of the line. Shell, Exxon Mobil, and many others found this an attractive proposition.[35]

SABIC was a "runaway success."[36] It had started in a rented apartment with seven recently graduated Saudi engineers and, by 2019, was the fourth largest chemical company in the world.[37] Unlike Petromin, its plants were immediately profitable and, within twenty years, SABIC and its partners operated sixteen of them. Thirty percent of SABIC shares were eventually sold to individual Saudi citizens and foreign investment funds. Until the Aramco IPO in 2019, it was the largest publicly traded company in the Arab world.

Initially there were no private Saudi petrochemical companies—only the largely state-owned SABIC and its foreign partners. Then in 2002, Chevron Phillips joined with the Saudi Industrial Investment Group to build the kingdom's first entirely privately owned petrochemical plant. Within a few years, other privately held firms such as Sahara, Tasnee, and Sipchem were together producing nearly as much as SABIC. All of these new firms were allocated feedstock from Saudi Aramco on the condition that they offered shares to the public.[38] They all currently trade on Saudi Arabia's Tadawul, which is the largest stock exchange in the Islamic world. The Tadawul was opened to foreign investors in 2015 and has been included in both the FTSE and MSCI emerging market indexes.

Building physical infrastructure principally requires spending money. A country can hire Americans to design refineries, bring in Koreans to build them, and call it development. At some point, however, more than physical infrastructure is needed. Institutional structures must also be put in place. The Saudi government recognized this and began building a series of free market regulatory institutions. In 1999, a cabinet subcommittee known as the Supreme Economic Council was established to expedite these reforms, and over the next five years guided the implementation of many pieces of new economic legislation in the kingdom.

The new legislation covered, among other things: patents, trademarks, telecommunications, power and water production, capital markets, health insurance, general insurance, mining, port operations, and the practice of law.[39] A Saudi Capital Markets Authority and Saudi Food and Drug Administration were created. A new telecom regulator oversaw the sale of one third of Saudi Telecom to the public, and a second mobile phone license was awarded to Etisalat from the UAE. Internet service provision was opened up to competition. Port operations were privatized. The Water and Electric Company was created to purchase power and water from new private producers. Although electricity and water continued to be sold to the public for much less than their true cost, the production costs—and, thus, the amount of government subsidy required—decreased with private production.[40]

Saudi Arabia's entry into the World Trade Organization (WTO) in 2005 was King Abdullah's most ambitious regulatory reform initiative. The king used the discipline of joining the WTO to drive forward numerous unpopular, but necessary, economic reforms.[41] On more than one occasion when negotiations reached an impasse, Saudi chief negotiator, Dr. Fawaz al-Alamy, consulted the king, who instructed him to close the deal. In doing so, Saudi Arabia negotiated new, mostly lower, tariffs on 7,000 products and adopted forty-two new commercial laws. The kingdom opened its markets to foreign agricultural products, financial services, telecommunications providers, and retail sales. The Saudis secured foreign markets for their petrochemicals, fertilizer, cement, and aluminum. In joining the WTO, Saudi Arabia took a long step away from economic isolation and became part of a rules-based international trading system that reduced costs and increased investor confidence.[42]

Because their own material circumstances have so clearly improved, most Saudis have long accepted that the Al Saud manage the kingdom's economy competently. However, at the same time, their once modest expectations mushroomed even as the economy was becoming more distributive than productive. New and very costly government institutions provided utilities, housing, education, healthcare, and employment—all of which came to be regarded as entitlements.[43] Like their compromise with the *ulama* over education, the Al Saud's oil-funded economic bargain with the public laid the foundation for potential stability problems.

By 2015, Saudi Arabia's generous, distributive economic system was beginning to unravel, and a failure to deliver material benefits was becoming the monarchy's most immediate problem. Young Saudis were growing increasingly concerned about the very issues mentioned above: affordable housing, healthcare, education, and employment. They began to wonder if they would ever be as well-off as their parents. Tensions grew between the entitlements they had come to expect and the productivity improvements needed to maintain their standard of living.

Unlike his predecessors—who had done a good job building infrastructure, diversifying within the hydrocarbon sector, distributing benefits, and providing government jobs—King Salman faced the much more difficult problems of expanding economic diversification beyond hydrocarbons, creating private-sector jobs, and maintaining public support while shrinking the kingdom's welfare state. Chapter 17 will look in detail at how he has addressed these challenges.

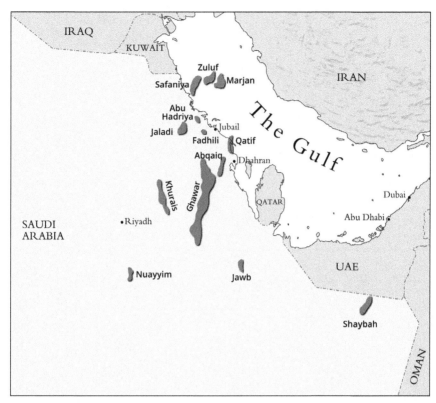

Map 4. Major Oil Fields of Saudi Arabia.

14

Foreign Policy: Keeping Powerful Friends

The first Western diplomat assigned to Riyadh was the colorful Captain William Shakespear. Born the son of a colonial official in India, he joined the Bengal Lancers and drove a single-cylinder Rover from Iran to England when most of the roads were horse tracks. In 1914 he was sent to convince the Amir of Riyadh that he should support Britain in the coming conflict with Turkey.[1] Shakespear spoke good Arabic, got on well with Abdulaziz, and was favorably impressed with the future king's leadership potential. He reported that Abdulaziz did not display "a trace of the fanatical spirit which might have been expected from the ruling Wahhabi family"; that he was "particularly intelligent" and "a broad minded man who could probably be trusted."[2]

Shakespear might have become a household name like Lawrence of Arabia had he not been killed in January 1915 at the Battle of Jarab in Central Arabia. Abdulaziz urged Shakespear not to join him in fighting the Ottoman-allied Al Rasheed and suggested that if the Englishman insisted on coming along, he should at least take off his British uniform. Wearing his khaki uniform of the 17th Bengal Lancers, Captain Shakespear insisted on joining the battle in which his failure to heed Abdulaziz's advice cost him his life. The Al Rasheed gave his helmet to the Turks, who hung it outside the gate of Medina as proof of the Al Saud's cooperation with infidels—something that they are still routinely accused of by adversaries as diverse as al-Qaeda, ISIS, and the Muslim Brotherhood.[3] Without Shakespear in Riyadh, British support drifted towards Abdulaziz's long-standing rival, Sharif Hussein of Mecca, who was much easier to find than the wandering Amir of Riyadh and much better positioned to attack the Turks in Mecca and Medina. It was the Sharif's son, Faisal, whom T. E. Lawrence eventually joined in the Arab Revolt against the Turks.

Captain Shakespear left behind the earliest photographs of Abdulaziz, his family, and his army. He also provided one of the earliest and clearest statements of what motivated the future king:

> Ibn Saud was motivated first by intense patriotism for his country, by which he meant the Al Saud dynasty, and then by a profound veneration for his religion as well as a single-minded desire to do his best for his people by obtaining for them lasting peace and security.[4]

There's nothing here about building schools and roads or securing the blessings of liberty. What his people needed most was "peace and security." In a badly war-torn region, Abdulaziz and his sons have been remarkably successful at bringing "peace and security" to the Saudi people. Since the official founding of the kingdom in 1932, very few Saudis have lost their lives in foreign wars. Until 2015, modern Saudi Arabia had seldom been attacked by foreign enemies. Iraq's brief cross-border assault in 1990 and Egyptian air raids in 1962 were minor events in comparison to the war damage and casualties experienced by Egypt, Iran, Iraq, Jordan, Lebanon, Syria, Yemen, or even Morocco.

The Al Saud maintained their security not through great strength but despite well recognized military weakness. They have a large area to defend, but only a small population with limited technical skills. More than once the long-serving Defense Minister, Prince Sultan bin Abdulaziz, called for conscription and a unified military command structure to strengthen the Saudi armed forces. Each time, other senior princes resisted any concentration of military power that might weaken the balance of political power among King Abdulaziz's sons.[5]

Nor can their achievement be attributed to the routine maintenance of long-standing, stable security relationships with trusted neighbors. On the contrary, the Saudis live in a very dangerous and volatile part of the world where their relations with neighbors have been anything but stable. Egypt has gone from being a revolutionary regime bombing the kingdom in 1962, to an ally in 1973, to an outcast following the 1978 Camp David Accords and back to an ally under General Abdul Fattah Sisi's government in 2013. Iraq has gone from being the home of the Al Saud's Hashemite rivals to a bulwark against Iran, to an invading army lobbing missiles at Riyadh. Iran, once a partner in promoting a pro-Western security structure in the Persian Gulf, has evolved into what most Saudis today consider their most significant foreign threat.

Maintaining national security and political independence in such difficult circumstances with limited military resources required skillful diplomacy, the outlines of which first appeared in the foreign policy of King Abdulaziz. Tactically, the Al Saud have relied on discretion, proxies, and covert action. They have avoided confrontations and preferred the role of mediator to that of active participant. When disagreements arose, they have sought to avoid a complete breakdown in relations. Strategically, the Al Saud have relied on three foreign policy tools in order to protect their interests. Cooperation with the regional superpower, which, for the past 50 years, has been the United States; the calculated use of oil production and oil revenue; and a leadership position in the Islamic world.

It would be inaccurate to claim that the Third Saudi State has survived solely because of foreign support, but it would be equally misleading to conclude that this has not been an important factor in its success. As Britain's Minister to Riyadh from 1930 to 1936, Sir Andrew Ryan, wrote, "I believe that Ibn Saud was in his heart hostile to all Western influences including that of Great Britain, but he knew that British friendship was a condition of his survival."[6] Like limiting the expansion of his kingdom's borders, Abdulaziz's willingness to cooperate with foreign Great Powers was a strategic choice that he made differently from his predecessors in the first two Saudi states. It is also a policy that his sons have chosen to maintain.

Abdulaziz began looking for powerful friends early in his career. Before World War I he made repeated requests to Great Britain for the same kind of recognition and protection that London had extended to several smaller Gulf states. When Britain declined because the Amir of Riyadh was not a coastal ruler and London had no interest in the affairs of Central Arabia, Abdulaziz turned to the Turks.[7] Despite opposition from the *ulama*, who considered them to be infidels, and his own long-term objective of gaining independence from Ottoman rule, Abdulaziz traveled to Basra in May 1914 and signed a treaty confirming him, his sons, and grandsons as the Ottoman sultan's governors in the Nejd.[8]

Six months later, Britain was at war with Turkey, and British policy changed. In December 1915, Britain's Political Agent for the Gulf, Sir Percy Cox, traveled to Al Qatif, where Abdulaziz signed an Anglo Saudi Friendship Treaty. Britain gained political influence in Central Arabia, trade concessions, and a significant role in shaping Saudi foreign policy. Abdulaziz secured British protection from foreign enemies; recognition for himself and his heirs as rulers of the Nejd, Al Hasa, and Al Qatif; financial assistance; and a supply of weapons.[9]

Abdulaziz's immediate foreign policy goal at Al Qatif was defensive. He wanted the British Navy to prevent a seaborne return of the Ottomans to the eastern region of Al Hasa. More generally however, the agreement anticipated two important themes of Saudi foreign policy for the next 100 years. First, its overriding goal was to secure the independence of the kingdom and the leadership role of the Al Saud family. Second, unlike most of the Arab leaders who came to regard Britain and France as occupying powers limiting their independence, the Al Saud perceived Britain as an ally against the Ottomans who had invaded their realm three times between 1811 and 1871.[10] In contrast with Egypt, Syria, Iraq, or Algeria, Saudi independence was the result not of a struggle against Western powers but, in part, because of an alliance with them.

After the Battle of Jarab in 1915, Abdulaziz essentially sat out World War I waiting to see which side would win. As he had written to Shakespear, "I am a person who desires to remain quiet so that my state may not become impaired."[11] However, once the war ended Abdulaziz was quick to recognize the new regional superpower, and from then on made a point of cooperating with Britain. He acquiesced to British proposals for his borders with Iraq and Jordan, did not support Ikhwan attacks on British-protected Kuwait, and conquered Ha'il and the Hejaz only with tacit British approval. When the RAF bombed the marauding Ikhwan in Saudi territory, he chose not to confront Britain in a war that he could not win and instead confronted his old allies, whom Britain helped him to destroy. Abdulaziz provided order in a vast, inhospitable desert that the British government had neither the desire nor the resources to police but which bordered on important British interests in Egypt, Transjordan, and Iraq. He protected the flanks of those interests by preventing Central Arabia from becoming an unstable, dangerous place like today's Somalia or Afghanistan, and the British government paid him for his efforts.

During World War II, when most of the Middle East was still under colonial rule, Saudi Arabia was one of the few independent Arab states. As he had done twenty-five years earlier, King Abdulaziz sought to balance relations with both sides while waiting to see who would win. He wrote to Adolf Hitler that "it is our foremost aim to see friendly and intimate relations with the German Reich developed to the utmost limits."[12] He ordered arms from both Germany and Italy, and even discussed plans for a German arms factory to be built near Riyadh. Yet at the same time he maintained close ties to Britain.

In 1941, when an anti-British military coup threatened the Hashemite monarchy in Baghdad, King Abdulaziz received a personal request for support from Hitler, promising to make him "King of all Arabs" if the coup succeeded.[13] Despite his long-standing enmity towards the Hashemites, King Abdulaziz did not turn against Britain to support the coup, and for good reason. Saudi Arabia could not feed itself and, although Britain faced serious shipping shortages, it supplied the country with much-needed Indian rice and Canadian wheat throughout the war, in addition to renewed financial assistance.[14]

It was American, not German, involvement that slowly began to displace British assistance to Saudi Arabia. Rather than send food to Arabia, the United States sent an agricultural mission to help the Saudis grow their own vegetables at a new model farm near Al Kharj, seventy miles south of Riyadh. It was his vehement opposition to the arrival of these infidel agricultural experts that first brought to prominence an obscure young Nejdi cleric who would go on to become the Grand Mufti Abdulaziz bin Baz. Meanwhile, the American-owned Aramco advanced payments on future oil royalties to cover falling pilgrim revenue, and at the same time persuaded Washington to extend Lend Lease aid to Saudi Arabia.[15] Between 1940 and 1947, the United States provided Saudi Arabia with $100 million in foreign aid, only 25 percent of which was meant to be repaid.[16]

Beginning in 1942, Saudi Arabia provided Allied air forces with military access to air routes across the Arabian Peninsula, which remain important to this day.[17] In 1943, the future King Faisal travelled to Washington to meet President Franklin Roosevelt, who declared, "I hereby find that the defense of Saudi Arabia is vital to the defense of the United States."[18] The following year, negotiations began for an American airbase at Dhahran. In the closing months of the war, when the outcome was no longer in doubt, Saudi Arabia declared war on Germany and Japan and thus won a place at the founding conference for the United Nations.[19] Finally, in February 1945, King Abdulaziz sailed from Jeddah on the destroyer *USS Murphy* for an historic summit meeting with President Roosevelt.

Their conversation took place onboard the cruiser *USS Quincy* in the Great Bitter Lake of the Suez Canal. It is regarded as the beginning of the close Saudi American relationship, and today the American ambassador's residence in Riyadh is named Quincy House. The occasion revealed many characteristics of an association that would be filled with cooperation, disagreement, and misunderstanding for the next seventy-five years. A war against the Soviet Union in Afghanistan and the massive effort to liberate Kuwait from Saddam Hussein would be high points of this relationship, while the 1973 oil embargo and the 9/11 attacks would nearly destroy it—and the unresolved Arab–Israeli conflict would remain a persistent source of tension.

Saudi oil was fundamental to the relationship long before the United States became an oil importer. Between 1944 and 1950, Saudi oil production rose from 20,000 to 550,000 barrels a day, with most of this increase going to fuel Western Europe's postwar recovery.[20] Inexpensive energy, much of it from Saudi oil, was as important to post-war

European reconstruction and the containment of communism as the Marshall Plan package of financial aid.

The Saudi American relationship has, however, always been "thicker than oil." The kingdom's strategic location, anti-communist stance during the Cold War, moderating role in the Arab–Israeli dispute, willingness to fund numerous American financial requests, and its unique position in the Islamic world have all made it an important partner. President Roosevelt's successor, Harry S. Truman, restated Washington's commitment when he wrote that the United States was "interested in the preservation of the independence, and territorial integrity of Saudi Arabia [...] And that no threat to your Kingdom could occur which would not be of immediate concern to the United States."[21] These assurances have been repeated many times since.

After World War II the United States continued to support Saudi Arabia, economically and militarily. In 1949 Aramco paid more taxes to the US Treasury than it did royalties to the Saudi government. By 1950, no less than 95 percent of these payments were going to Riyadh through a foreign tax credit arrangement, sometimes called the "Golden Gimmick," which allowed Aramco to deduct foreign payments from its American tax bill. Thus, the oil company's payments to Saudi Arabia increased from $39 million in 1949 to $110 million in 1951.[22] The US Treasury was the loser, but a way had been found to send foreign aid to the still-poor Saudi Arabia.

In 1952 the Eisenhower Administration sold eighteen M-41 tanks to Saudi Arabia and faced the first of many Congressional inquiries on the subject of Saudi arms sales. Senator Wayne Morse (Democrat—Oregon) asked Secretary of State John Foster Dulles how the administration could arm a "totalitarian regime," while Senator Hubert Humphrey (Democrat—Minnesota) expressed concerns that the weapons would be used against Israel.[23] Many of the same concerns were still being raised in 2019 when Congress voted to block precision munitions sales to Saudi Arabia.

Located at the crossroads between Europe and Asia, Saudi Arabia has long been important geographically. For many decades it was on the eastern flank of a Eurocentric world order. In 2020 it has become the western flank of a newly important Indo-Pacific theater. One of the first American requests for Saudi overflight clearance was made in 1943 when President Roosevelt traveled to Tehran to meet Joseph Stalin. Since then, countless US military flights and more than a few cruise missiles have received permission to transit Saudi airspace. The American airbase at Dhahran was originally built as a transit and refueling stop as the Allies began shifting forces from Europe to the Pacific, and evolved into a strategic bomber base on the southern flank of the Soviet Union.[24] The Dhahran base was closed in 1962, but was eventually replaced by Prince Sultan Air Base (PSAB), south of Riyadh. PSAB played a central role in Operation Desert Storm (the 1991 combat phase of the First Gulf War) and in enforcing the Iraqi no-fly zone afterwards. PSAB operated from 1990 to 2003, when its operations were shifted to Al Udeid Air Base in Qatar. Yet in 2019, as tensions mounted with Iran, thousands of American troops again returned to PSAB.

No Arab state supported the United States more consistently during the Cold War than Saudi Arabia. Egypt, Iraq, Syria, and Yemen were all, at one time or another, Soviet clients and harsh critics of the Al Saud's pro-American leanings. Saudi Arabia, on the other hand, supported American anti-communist efforts in Africa, South Asia, Central

America, and Europe. Riyadh did not try to play Moscow off against Washington, never bought Soviet military equipment, and never sent officers for training in the "Godless" Soviet Union. As Lt. Col Oliver North, a National Security Council staff member during the Reagan-era Iran–Contra affair, wrote, the Saudis "knew what it meant to have a communist presence in the neighborhood which is why they were ready to support the Afghan resistance to the tune of hundreds of millions of dollars." North continued, "The Contras would never have survived without the help of Saudi Arabia's King Fahd. President Reagan wanted the Contras kept alive body and soul. The King made it happen with his secret assistance."[25]

Saudi support for the *mujahideen* in Afghanistan and the Contras in Nicaragua form well-known examples of Riyadh's anti-Soviet cooperation with the United States. Less well publicized was their financial support for anti-communist political parties in Europe and alliances with anti-communist intelligence services to fight the Cold War in Africa. Know as the "Safari Club," this particular alliance included Egypt, France, Iran, Morocco, and Saudi Arabia. In 1977 the Safari Club organized an effective response to an invasion of Zaire by Marxist rebels based in Angola. The following year it provided arms to Somalia for its conflict with Soviet-backed Ethiopia. Saudi Arabia's clandestine contributions to the Cold War remained well hidden from public view, but Henry Kissinger noted about this period, "Often I found through other channels a helpful Saudi footprint placed so unobtrusively that one gust of wind could erase its traces."[26]

Egypt's president, Gamal Abdul Nasser, had made no secret of his intention to topple the Saudi monarchy and when the Yemen Civil War broke out in 1962 he saw an opportunity to do so.[27] Nasser sent Egyptian troops to support Yemen's republican forces while Egyptian-backed Yemeni expatriates began a sabotage campaign inside Saudi Arabia—targeting the Ministry of Defense, the US military training mission, oil pipelines, royal palaces, and the Saudi Air Base at Khamis Mushait near the Yemeni border.[28] When Egyptian planes began bombing Saudi border towns, King Faisal made Saudi Arabia's first request for direct American military assistance.

Support arrived quickly. In what became known as Operation Hard Surface, President John F. Kennedy dispatched eight F-100 aircraft, six KB-50 aerial refueling tankers, and more than 500 military personnel to protect Saudi airspace.[29] Based in Jeddah and Dhahran, these forces had orders to "attack and destroy" Egyptian intruders. Fortunately, they never met any and left after six months.[30] Meanwhile, Britain, France, Iran, and Israel all joined Saudi Arabia in supplying clandestine support to the Yemen's royalist forces in another under-reported chapter of the Cold War between the Soviet Union, the United States, and their respective allies.[31]

Saudi Arabia has often sought to compensate its allies for the costs of protecting the kingdom by purchasing weapons—particularly, from Britain and the United States. In the early 1960s, American Secretary of Defense Robert McNamara and his British counterpart Roy Jenkins crafted an arrangement in which Britain sold Saudi Arabia Lightning aircraft and deployed the profit to purchase American F-111s—effectively using the Saudis to fund the RAF.[32] On other occasions, Saudi defense purchases were explicitly based on building security relationships—as when Defense Minister Prince Sultan opted to buy the large, complex, two-engine American F-15 aircraft over the recommendation of senior Saudi Air Force officers that a simpler, more easily maintained aircraft would suffice.

Intended to recycle petrodollars and strengthen defense cooperation, arms sales often placed great strain on Riyadh's relations with Washington and, to a lesser extent, with London. The Saudis obtained their first sixty F-15 aircraft in 1978, only after a bruising battle between Congress and the Carter Administration.[33] When they sought to purchase additional F-15s during the Reagan Administration, the US Congress blocked the sale. In response the Saudis turned to Britain, and created the massive, multi-billion-dollar Al Yamamah program to supply Tornado fighter bombers, Hawk trainers, and the bases to go with them. The F-15s that Congress had blocked would have been sold as purely defensive interceptors. By contrast, two thirds of the Tornados that Britain provided were advanced strike fighters, and all of the Hawks had offensive capabilities.[34]

During the Iraq Iran War (1980–1988), Saudi Arabia felt threatened by the Iranian ballistic missiles that were being fired at Baghdad. Riyadh sought to purchase the conventional Lance missile system from the United States, which had a range of eighty miles. When Congress blocked this sale, too, the Saudis purchased CSS-2 East Wind missiles from China, which were capable of carrying nuclear warheads and had a range of 1,600 miles.[35] They did not inform the United States of this purchase and the missiles were shipped secretly. When they were eventually discovered in 1987 by an American embassy officer and a British nurse on a desert camping trip, the United States was both angry with the purchase and alarmed that Israel might conduct a pre-emptive strike on the missiles.

The East Wind episode tested Saudi American relations. The American ambassador to Riyadh at the time, Hume Horan, believed in retrospect that it was the worst crisis in the relationship before 9/11.[36] On the other hand, the Saudi Ambassador to Washington, Prince Bandar bin Sultan, believed that it could have been worse. He had initially proposed turning to Moscow rather than Beijing because Russia produced better missiles than China. It was King Fahd who had insisted that Prince Bandar arrange a purchase from China rather than Russia so as not to do something that "would have really alienated our American friends."[37]

When Prince Bandar was reprimanded in Washington, he took it as part of his job. However, when Ambassador Horan in Riyadh tried to read King Fahd the riot act he was given twenty-four hours to leave the country.[38] The incident provides a good example of how the never-colonized, financially independent and religiously important Saudis possess a self-confidence that sets them apart from many states that would have been less inclined to dismiss an American ambassador.

Controversial or not, arms sales to Saudi Arabia have created thousands of well-paid manufacturing jobs in Britain and the United States, kept open defense production lines that were otherwise unprofitable, and reduced the unit cost paid by NATO allies for everything from tanks to fighter aircraft. Had the Saudis not purchased Tornados and F-15s, life would have been very different for many working families in Lancashire and St. Louis.

Saudi leaders have sometimes found American gestures of support inadequate. They certainly thought that sending unarmed aircraft to demonstrate support for the kingdom during the Shah of Iran's final, turbulent days a rather feeble gesture by the Carter Administration.[39] However, on August 2, 1990, when 120,000 Iraqi troops seized Kuwait in an overnight blitzkrieg and advanced to within a mile of the Saudi border, they could not complain. Within four days of the invasion, Secretary of Defense, Richard Cheney, and US Central Command's General Norman Schwarzkopf were in Jeddah offering King Fahd American military assistance. The resulting Operation Desert Shield, to protect Saudi Arabia, and Desert Storm, to liberate Kuwait, demonstrated to the Al Saud both the smoldering animosity of their neighbors and the value of powerful friends.

It is certainly true that President George H. W. Bush hoped to establish a new, post-Cold War world order and was determined that unprovoked aggression against Kuwait "would not stand." Yet, in the final analysis, the First Gulf War was about oil. After seizing Kuwait, Iraqi president, Saddam Hussein, controlled 20 percent of the planet's oil reserves. Had he moved south to Dhahran, he would have controlled 40 percent. That was what really could not have been allowed to stand. If Kuwait had instead grown carrots, it is unlikely that the United States would have sent half a million men there (as many as it had deployed at the height of the Vietnam War) in order to restore an Arab monarch.

Despite the exodus of some 300,000 Kuwaiti "refugees" speeding south in their BMWs and calling ahead for reservations at the Riyadh Hilton, King Fahd and his brothers were not convinced that they faced a threat until Secretary Cheney and General Schwarzkopf briefed them on the situation. Deputy National Security Advisor, Robert Gates, and Ambassador Chas Freeman were the only other Americans in the room. The author was sitting outside with a group of majors and colonels, so our only detailed accounts of the meeting are to be found in General Norman Schwartzkopf's *It Doesn't Take a Hero* and Bob Woodward's *The Commanders*.

Secretary Cheney reviewed how the United States had come to Saudi Arabia's assistance in 1962 against Yemen and Egypt, and how in 1987, during the Iran–Iraq War, the United States had launched Operation Earnest Will, the largest naval convoy operation since World War II, to protect Kuwaiti-owned oil tankers from Iranian attacks. He told the king: If you ask, we will come. We will seek no permanent bases. And when you ask us to go home, we will leave. General Schwarzkopf presented satellite imagery revealing that the Iraqis could attack Saudi Arabia in as little as forty-eight hours. King Fahd was convinced that "Kuwait had waited too long and now there was no Kuwait." When his brother, Abdullah, insisted that there still was a Kuwait, the king replied, "and all the Kuwaitis are living in our hotel rooms." After a brief discussion, the king said, "OK."[40] He later told Ambassador Freeman that this was the only time in his decades of government service that he had made a major decision on his own without obtaining a ruling family consensus.[41]

Once agreement had been reached to deploy US forces, the king provided some insight into his reasoning. "We have taken this country from stones and bushes and built it into a nation that has few equals in the world." "We have done this through joint ventures with foreigners." "We are not afraid to learn from people who are better than we are. The Saudi people have no complexes. We want to cooperate with others."[42] And so they did. More than 750,000 foreign troops from thirty-six countries came together to liberate Kuwait—including 45,000 Britons, 36,000 Egyptians, 19,000 Syrians, and 14,000 Frenchmen. The Germans and Japanese wrote billion-dollar checks. The Algerians, Jordanians, Libyans, and Yemenis were no-shows, while the Palestinians cheered for Saddam.

Operation Desert Storm was a massive undertaking. In preparation for the assault, millions of gallons of fuel were stored in inflatable bladders and offshore tankers. The Hajj terminal became a B-52 base, so the war had to be over before the pilgrimage began. Jeddah's civilian airport hosted the largest aerial-refueling operation in history. The dozens of refueling tankers were from Air National Guard units and flown by peacetime commercial airline pilots. For an American the flight line was a moving sight, as each tanker had its home state's name painted proudly on the tail.

After a thirty-four-day air campaign, the ground war lasted barely 100 hours. It too made history, with one of the largest air calvary assaults on record when 300 helicopters carried the 101st Airborne Division into western Iraq. Undetected, another 100,000 troops and 20,000 vehicles had already swung west around the Iraqi flank on a little-used road paralleling the Trans Arabian Oil Pipeline, another Bechtel Project which runs from Saudi Arabia to Lebanon, but is no longer in use. Unfortunately, the war's aftermath also made records when retreating Iraqi forces set fire to 600 Kuwait oil wells and then deliberately created a 4-million-barrel oil spill in the Persian Gulf intended to shut down Saudi Arabia's desalinization plants. For comparison's sake, the Exxon Valdez spilled 250,000 barrels of oil into Alaska's Prince William Sound.

The war cost the United States and Britain very little. Saudi Arabia covered nearly all of their direct in-country costs for food, fuel, and housing—and made a substantial, additional cash contribution. The Saudis not only supported the boots on the ground, they often paid to get them there. Many nations that sent troops had been longtime recipients of Saudi aid, and there can be little doubt that $2.5 billion in direct Saudi aid and credits to Moscow helped to secure a favorable Russian vote on the UN Security Council resolution authorizing the war.[43] Having gone to great lengths to respect Saudi culture while they were in the kingdom, all but 10,000 American forces were gone in a matter of months after Operation Desert Storm ended.

Saudi Arabia has never played a significant military role in the Arabs' wars with Israeli. In 1948 the Saudis contributed two companies, who fought under Egyptian command;[44] in 1967 a Saudi brigade moved slowly towards Jordan, arriving only after the Jordanians had been driven back to the East Bank of the Jordan River and a ceasefire was about to be declared.[45] During the 1973 Arab Israel War, Saudi Arabia sent one brigade to join the Arab Foreign Legion with strict instructions not to become involved in the fighting. By the time they reached Damascus, the hostilities were over.[46] King Abdulaziz was always much more worried about his Hashemite rivals in Jordan and Iraq than about Israel. King Faisal often made it clear to American officials that he regarded Nasser's Egypt, not the Jewish state, as the greatest threat to his kingdom.[47]. In 2020, Saudis are far more concerned with Iran than with Israel.

Nevertheless, nothing has strained the Saudi American relationship more severely or more consistently than the Arab–Israeli conflict. On board the *USS Quincy*, President Roosevelt told King Abdulaziz that the Jews of Central Europe had suffered indescribable horrors at the hands of the Nazis. Eviction from their homes, destruction of their property, torture of their families, and mass murder. Could the king help with their resettlement? The king replied that the Jews and their descendants should be given the best lands of the Germans who had oppressed them. When the president countered that the Jews did not trust the Germans and had no desire to stay in Germany, Abdulaziz replied that if the Allies did not think they could control the Germans then why had they fought such a costly war? Certainly, he would never leave a defeated enemy in a position to strike back in the future. When the president tried again, he received a similar reply. "Amends should be made by the criminal, not by the innocent bystander. What injury have the Arabs done to the Jews of Europe? It is the Christian Germans who stole their homes and lives. Let the Germans pay." Had Roosevelt not dropped the issue, the meeting might well have ended acrimoniously.[48] Although King Abdulaziz was subsequently disappointed with the United Nations' partition of Palestine and the American recognition of Israel, he did not allow this to distract him from his overriding economic and security interests. Despite his genuine anger over Palestine, Abdulaziz never threatened either the Dhahran Air Base lease or the Aramco oil concession.

Saudi Arabia has a very clear interest in maintaining regional stability because like the United States, it has a lot to lose. Anything that disrupts the political and economic status quo—be it Nasser's brand of Arab Nationalism, Bin Laden's brand of Islam, or an unresolved Arab–Israeli conflict—is a threat to Saudi security and prosperity. Widespread Arab sympathy for the Palestinians and a belief that the United States is not an honest broker in the peace process is a domestic issue that complicates Riyadh's relationship with its most important ally. Likewise, public support for Israel has often made strong ties to Saudi Arabia a political liability for American presidents. From the Al Saud's perspective, it would be much better to get the Arab–Israeli problem resolved—and they have tried to do so more than once.

After the 1967 Arab–Israeli War, Arab leaders met in the Sudanese capital, Khartoum, where they adopted their reflexive "Three Noes" with regard to Israel. "No recognition, No negotiation and No reconciliation." With the exception of Egypt, this remained their unanimous, unwavering and unhelpful position for the next fifteen years. Only in 1981 did then Crown Prince Fahd put forth a fresh proposal by which "all states in the region should be able to live in peace." Fahd's plan did not explicitly recognize Israel but it implied that this was now a possibility after the acceptance of a Palestinian state and a return to 1967 borders. Forty years ago, that was a radical policy change. A watered-down version of Fahd's plan was adopted in 1982 by Arab leaders meeting in Morocco. Known as the Fez Plan, it remains Arab League policy to this day. President Ronald Reagan called the Fez Plan "the single largest step towards peace on which the Arab World has been able to agree."[49]

In 2002, then Crown Prince Abdullah repeated the Saudi offer to *The New York Times* columnist Thomas L. Friedman: "full withdrawal from all the occupied territories, in accord with UN resolutions, including Jerusalem, for full normalization of relations" with Israel.[50] The following month Abdullah presented his plan to the Arab Summit in Beirut, where a modified version was unanimously adopted. The Beirut Plan differed from the Fez Plan in that it explicitly stated the conditions under which the Arabs would "consider the Arab–Israeli conflict over and sign a peace agreement with Israel."

At both Fez and Beirut, the Saudi proposal was more forward-leaning than the eventual Arab consensus position adopted. Like Prince Fahd, Prince Abdullah took political risks and spent both political and financial capital convincing other Arab leaders to go as far as they did. Neither proposal was fully acceptable to Israel or the United States as they did not adequately address important issues such as the status of Jerusalem or the return of Palestinian refugees. However, these proposals remain the most constructive plans put forward by the Arabs to date. King Abdullah had them reaffirmed at the 2007 Arab League Summit in Riyadh, and King Salman reiterated them to President Donald Trump during his 2017 visit to Riyadh.

Saudi Arabia has almost always supported independent American efforts to negotiate an end to the Arab–Israeli conflict. The most notable exception was breaking off diplomatic relations with Cairo after Egyptian President Anwar Sadat signed the 1979 Camp David Accords between Egypt and Israel. Yet even when Riyadh followed the Arab consensus in breaking ties with Cairo, the Saudis did not completely cut off financial aid to the Egyptian military or expel thousands of Egyptian guest workers who continued to send home economically vital hard currency remittances.[51]

The eruption of the Second Palestinian Intifada in 2000 put severe new strain on Saudi American relations. Graphically vivid Saudi press coverage of Palestinian suffering resulted in an unofficial, but widespread, Saudi boycott of American consumer products. Crown Prince Abdullah, one of whose wives was Palestinian, felt strongly about the issue. The king also believed that rising public anger over Palestine threatened the position of the United States' moderate Arab allies. In May 2001, King Abdullah declined an invitation to visit the White House, and later canceled high-level military meetings.

Like many of his predecessors, King Abdullah wrote to the President of the United States complaining about the plight of the Palestinians. Unlike his predecessors, in August 2001 Abdullah's missive threatened to re-examine the entire Saudi American relationship if something was not done about it. His Ambassador to Washington, Prince Bandar bin Sultan, read the letter to President George W. Bush, noting that from now on Saudi Arabia would pursue its own defense, security, and economic interests without taking into account the interests of the United States.[52] The president's written reply was intended to avoid such a breach. For the first time, the United States formally agreed to support the creation of a Palestinian state and promised to announce this publicly at the upcoming UN General Assembly meeting.[53] It seemed that another crisis in Saudi American relations had passed.

Only days later, al-Qaeda terrorists carried out a series of coordinated attacks on the United States that killed 2,996 people, injured another 6,000, and caused more than $10 billion-worth of damage. On September 11, 2001 more than 100 people leapt to their deaths from the burning World Trade Center towers; 343 firemen and 72 police officers lost their lives. Civilian airspace was closed for two days and the New York Stock Market remained closed for a week. Hundreds of thousands of tons of toxic debris was scattered over Lower Manhattan. In Washington, DC, Army Deputy Chief of Staff Lt. General Timothy Maude was the highest-ranking officer of the 123 people killed at the Pentagon.

Citing Washington's support for Israel and the presence of American troops in Saudi Arabia as justifications, al-Qaeda claimed responsibility for these atrocities, which killed more people than the Japanese attack on Pearl Harbor almost sixty years previously. When the FBI confirmed that fifteen of the nineteen terrorists were Saudis, Saudi American bilateral relations went into free fall. Draconian restrictions were placed on Saudis traveling to the United States; some in Washington argued that the US consulates in Jeddah and Dhahran should simply be closed. Saudi tourists previously accustomed to receiving American visas in forty-eight hours now faced waiting for six months. Saudi trade missions to the United States disappeared; far fewer military officers from the kingdom went to the US for training; and the number of Saudi students studying in the United States collapsed. For many Americans, it did not matter if another Saudi ever came to the United States. These were precisely the reactions that al-Qaeda's leadership had been hoping for.

Despite allegations that it had supported the attacks, the Saudi government had in fact been at war with Osama bin Laden since 1996 when he issued his "Declaration of Jihad on the Americans Occupying the Two Sacred Places." In this document, Bin Laden had spelled out in great detail how the Al Saud were "agents of imperialist Christians and Jews." Long before 9/11, the Saudi government had stripped Bin Laden of his Saudi citizenship and seized what funds and assets he had in the kingdom. Saudi intelligence chief, Prince Turki al-Faisal, had actually gone to Afghanistan in 1998 in an unsuccessful attempt to persuade the Taliban to expel or even extradite Osama Bin Laden to Saudi Arabia.[54] The Final Report of the National Commission on Terrorist Attacks Upon the United States, informally known as the 9/11 Commission Report, was released in 2004. It identified Saudi individuals as the primary source of al-Qaeda funding but, as noted in Chapter 12, found no evidence of Saudi government involvement. The release of the classified portion of the report a decade later did not change that conclusion.[55]

Still, the view that Saudi proselytizing efforts contributed to the 9/11 attacks continues to color Saudi American relations. Some Saudi apologists argue that this is unfair. They point out that al-Qaeda's world view has more in common with Muslim Brotherhood doctrines and the teachings of militants like Brotherhood theorist, Sayyid Qutb, than with Mohammed Abd al-Wahhab and the views of the Saudi *ulama*. That is partially true. While a young engineering student at Jeddah's King Abdulaziz University, Osama bin Laden had indeed been taught Islamic Studies by Sayyid Qutb's brother, Mohammed Qutb, and by the Muslim Brotherhood ideologue, Abdullah Azzam. These two professors preached a hybrid philosophy of Muslim Brotherhood and Wahhabi doctrines that Bin Laden adopted.[56] On the other hand, it is equally true that for three decades Saudi Arabia exported its fundamentalist brand of Islam by building schools and mosques across the Muslim world. In Europe, South Asia, Africa, and North America the message delivered by these Saudi-funded institutions made global Islam less tolerant, more conservative, and more prone to violence than it had previously been.[57]

The 9/11 Commission Report further noted that "[a] number of FBI and CIA officers complained to the Joint Inquiry (i.e. the 9/11 Commission) about a lack of

Saudi cooperation in terrorism investigations both before and after the September 11 attacks." This state of affairs changed quickly in May 2003 after the al-Qaeda attacks on Riyadh residential compounds. Counterterrorism cooperation became the most significant factor in reviving Saudi American relations. It quietly replaced Soviet containment as the "third leg" of a relationship long based on global energy supplies and Saudi security. More counterterrorism training was provided and more terrorists were apprehended. By 2004, the State Department's Counterterrorism Coordinator, Ambassador J. Cofer Black, was telling Congress that the Saudis were a key ally in the global war on terror.[58]

After 9/11, the Saudi government supported the American campaign to topple the Taliban regime in Afghanistan but opposed the 2003 Anglo-American invasion of Iraq. Crown Prince Abdullah stated clearly that invading Iraq "would not serve American interests or the interests of the world."[59] Many Saudis doubted that the invasion could succeed, and worried that if things went badly the Americans would simply go home leaving them with a large problem on their doorstep. That proved an accurate prediction. Yet despite their misgivings, the Saudis quietly supported Operation Iraqi Freedom—allowing limited air operations from Saudi bases, overflights by US military aircraft and missiles, and the staging of Special Forces operations. They prepared camps to receive anticipated Iraqi refugees and, as they had done many times before, increased oil production to cover anticipated disruptions.[60]

The invasion of Iraq was only one of the policy differences that emerged between the Saudi government and the administration of George W. Bush. Riyadh regarded the new Shia-dominated government in Iraq as an Iranian pawn and refused to give it political or financial support. The Saudis believed that Israel's 2006 war with the Iranian-backed militia, Hezbollah, had only increased Tehran's influence in Lebanon. There was little appetite in the Royal Diwan for President Bush's "Freedom Agenda" to democratize the Arab world.

Yet, there were always tangible signs that both sides still saw value in maintaining a cooperative relationship. Counterterrorism efforts were foremost. In 2008, Saudi Arabia and the United States established The Joint Commission for Critical Infrastructure Protection (JCCIP), which coordinates the activities of numerous US government agencies working with the Saudi Ministry of the Interior to protect oil wells, pipelines, loading terminals, and other critical infrastructure facilities.

Security-related delays in processing Saudi visas post-9/11 had led many Saudis to avoid travel to the United States; a few people seeking medical treatment in the US had, in fact, literally died waiting for a visa. The solution lay not in reducing the scrutiny of security checks but in imposing them less frequently by extending visa validity from two years to five. Extending visa reciprocity for the country many blamed for the 9/11 attacks was not easy. Saudi Arabia's strong counterterrorism cooperation helped—so did the effective lobbying efforts of the American Ambassador in Riyadh, Ford Fraker.

One unclassified cable prepared by the embassy was particularly persuasive. It showed that 50 percent of the members of Saudi Arabia's fledgling parliament, the *Majlis al-Shura*, had American university degrees—as did 70 percent of the Council of Ministers and 100 percent of Saudi Aramco's Board of Directors. It seemed doubtful that outside of Washington there was another country in the world in which half the parliament was American-educated. As the cable pointed out, this source of influence was not likely to continue if Washington kept turning Saudi students away. In 2008, the two nations agreed to grant each other's citizens five-year multiple-entry visas and Americans remain the only Westerners able to obtain such visas to Saudi Arabia.

Incoming President Barak Obama quickly grasped the importance of Saudi Arabia to global economic growth, counterterrorism efforts, and resolving the Arab–Israeli conflict. While he made his 2009 address to the Muslim world from Cairo, he stopped to confer with King Abdullah on his way to Egypt and visited Saudi Arabia more often than any other president. The Obama Administration authorized a dramatic increase in arms sales to Saudi Arabia, which approached $100 billion in total.[61] In 2015, the president broke off important meetings in India to attend King Abdullah's funeral.

Yet President Obama never warmed to the Saudis, and bilateral relations remained strained over many issues. In 2011, the two sides differed over civil unrest in Bahrain, where Riyadh focused on clandestine Iranian involvement and Washington on human-rights violations. In 2012, the two sides differed sharply over Egypt's Arab Spring. Riyadh saw the Muslim Brotherhood as dangerous extremists and Washington cheered their democratic election. Ironically, having advised against the 2003 invasion of Iraq, in 2013 the Saudis strongly argued that a premature American withdrawal would leave a dangerous vacuum to be filled by the likes of ISIS.

When the Arab Spring in Syria deteriorated into a sectarian civil war, King Abdullah became deeply and very emotionally distressed by military attacks on the country's Sunni civilians. During an August 2012 press conference, President Obama gave an answer that the king found reassuring when he said, "We have been very clear to the Assad regime that a red line for us is if we start seeing a whole bunch of chemical weapons moving around or being utilized. That would change my calculus."[62] A year later, when UN inspectors confirmed that the Syrian government had used the nerve gas sarin to kill hundreds of civilians in the Damascus suburb of Ghouta, the king expected the president to act. Although the Obama Administration had credible reasons, both domestic and international, for not attacking Syria, King Abdullah never forgave the president for failing to enforce his own red line, protect Syrian civilians, or seize an opportunity to remove the Assad regime.

Riyadh's confidence in Washington's value as an ally eroded further in 2015, when the five permanent members of the UN Security Council along with Germany and Iran signed the Joint Comprehensive Plan of Action, better known as the Iran Nuclear Deal. The Saudis reluctantly supported the agreement but feared that while it lifted sanctions on Iran, it did not completely dismantle the Islamic Republic's enrichment capacity or close the door on future nuclear weapons development. Moreover, when President Obama then advised the Saudis to "find an effective way to share the neighborhood"[63] with Iran, it sounded to their ears rather like British Prime Minister Neville Chamberlain advising the Czechs to find a way to share the Sudetenland with Germany in 1938.

Since World War II, the United States' relationship with Saudi Arabia has often been more volatile and more strained than with other major allies. This ambivalent partnership between a conservative, theocratic monarchy and a liberal, secular republic has always been about common interests rather than shared values. It has never been popular with the public in either country. It has remained one of the few American alliances implemented through high-level personal relationships rather than formal institutions. In a legal sense there is no treaty of alliance between Saudi Arabia and the United States, only a time-honored understanding between the White House and the Saudi king's Al Yamamah Palace. In both countries the aforementioned factors have made a relationship that benefited both parties vulnerable to leadership changes, shifting public opinion and domestic political agendas. On numerous occasions the partnership has nearly collapsed, and by the end of the Obama Administration wide cracks were appearing in this pillar of Saudi security policy.

Foreign Policy: Deploying Oil and Islam

If the first principle of Saudi foreign policy has been maintaining cooperative relations with powerful friends (see the previous chapter), the second has been an astute use of oil production and oil revenue. Few factors affect the global economy more than energy costs and no nation exerts greater influence over oil prices than Saudi Arabia. Moreover, because of its large oil income and absolute monarchy, the kingdom can distribute large amounts of cash more quickly and quietly than most other nations. Oil has allowed a militarily weak state to maintain a global network of dependent customers and appreciative clients. Although cooperation with powerful nations underpins Saudi defense policy, it is Riyadh's ability to influence oil prices and finance its allies that make it worth defending.

Saudi Arabia remains the world's largest oil exporter and, by any estimation, one of the great reservoirs of global energy. It boasts the world's largest onshore oilfield at Ghawar, in the Eastern Province, and the largest offshore oilfield at Safaniya, in the Persian Gulf. Total Saudi reserves are estimated at 267 bbls (billion barrels) compared with 26.5 bbls for the United States and 20.4 bbls for China.[1] Every day Saudi Aramco produces roughly 11 percent of the world's oil, while the world's largest private oil company, Exxon Mobil, provides approximately 4 percent. At current production levels, Saudi Arabia can supply global energy markets with oil for at least another fifty years.

Not only is there plenty of oil in Saudi Arabia, it is inexpensive to produce and easy to ship. Saudi Aramco is not drilling in the Arctic or North Sea. Most Saudi wells are not especially deep, and many flow from natural pressure. There are no mountains to cross or long pipelines to build in order to get Saudi crude into tankers. Instead, there are well-established production and processing networks that have taken sixty years to build. Most estimates put average Saudi production costs between US$ 3 and 5 a barrel. The average cost of production in the United States, including shale and offshore wells, is roughly $40 a barrel; and Canadian oil-sands production costs are closer to $100 a barrel. In short, Saudi Aramco is not simply the world's largest oil company—it also produces some of the world's least expensive and easiest-to-ship oil.

Yet, it is not total reserves or production costs that make Saudi Arabia the "Central Bank of Global Energy." Nor is the kingdom always the world's largest oil producer; its production sometimes falls behind that of Russia or the United States. What gives Saudi Arabia its importance in global energy markets—and thus, a geopolitical influence disproportionate to its economic size or military strength—is the role that it plays as the producer of last resort. For decades, Saudi Aramco has maintained significant spare production capacity that allows it to rapidly put large volumes of oil into the market. Many countries could reduce their oil production if they chose to; only Saudi Arabia can significantly increase production very quickly by government fiat.

No private oil company or other OPEC member state has anything approaching Saudi Arabia's spare capacity—that is, production that can be brought online in thirty days and maintained for at least ninety days. Saudi Aramco usually keeps at least 2 million barrels a day of spare capacity on hand. This is neither easy nor inexpensive; it is a calculated political policy, not an efficiency-driven commercial choice. If the President of Exxon Mobil spent $50 billion drilling new oil wells and then simply shut them for a rainy day, he would probably be looking for new employment. Yet that is precisely what Saudi Aramco did in the early 1990s, and again between 2005 and 2010 when it increased production capacity from 10 to 12.5 mbd (million barrels a day). To put that into perspective, by doing so Saudi Arabia increased its spare capacity by more than what Kuwait, Iraq, Venezuela, Nigeria, or Iran were producing at the time. In 2018, Saudi Aramco announced that by 2023 it would increase its production capacity by another million barrels a day.

Unlike many OPEC members with much smaller reserves, the Saudis are not trying to make as much money as fast as they can. Instead, their oil policy is usually based on the kingdom's long-term economic interests. Saudi Aramco wants the world to keep buying its oil for as long as possible and recognizes that high prices or extreme price volatility make oil an unattractive energy source. Aramco's management fully understands that high oil prices reduce global economic growth, and thus demand for oil; promote energy conservation and the development of alternative energy sources; and promote oil exploration in new areas. Long-time Petroleum Minister, Ali al-Naimi (Minister 1995–2016), frequently reaffirmed Saudi Arabia's self-interest in preserving reliable oil supplies, stable oil prices, global economic growth, and the "continued importance of oil in the global energy mix."[2]

Consequently, Saudi Aramco has often acted to stabilize or even reduce oil prices. At the start of the Iran–Iraq War in 1980, Saudi Arabia increased oil production from 9.8 mbd to 10.2 mbd in order to prevent a price spike. In 1990, when Iraq's invasion took Kuwaiti crude off the market, Saudi engineers reopened shut-in fields and increased production by 2 mbd, more oil than Texas was producing at the time.[3] In 2003, Saudi Aramco made up most of the 1.2 mbd lost during Operation Iraqi Freedom (the Anglo-American invasion of Iraq). After Hurricane Katrina shut down Gulf of Mexico production in 2005, and again when Arab Spring violence disrupted Middle East production in 2011, the Saudis pumped extra oil in order to keep markets stable.[4] In the thirty-five years between 1980 and 2015, the kingdom raised its annual oil production eighteen times and cut it seventeen times.[5] As Saudi petroleum ministers frequently noted, the country's long-term prosperity was tied more closely to global economic growth than to short-term windfall profits.[6]

When the United States placed economic sanctions on Iran in 2012, it was largely increased Saudi production that prevented a painful spike in global energy prices. In May 2018, as the Trump Administration prepared to re-impose sanctions on the

Islamic Republic, Treasury Secretary, Steven Mnuchin, stated that the US had negotiated ahead of time with oil-producing allies such as Saudi Arabia to boost output and keep prices in check.[7] In reality, the United States' ability to sanction oil producers such as Venezuela or Iran without experiencing a sharp increase in its own energy costs and inflation rate has usually depended on Saudi cooperation—cooperation that other major oil exporters are usually unwilling or unable to provide. The "shale revolution" has not eliminated this basic fact of global geopolitics. Frackers can drill new wells quickly but they maintain very little spare production capacity. More importantly, they respond individually to market forces, not promptly and collectively to government instructions.

The Saudi government began using oil as a foreign policy tool long before it took over full ownership of Aramco in 1980. During the 1956 Suez Crisis, Riyadh instructed Aramco to cut off oil supplies to Britain and France. Aramco's American owners complied, and Texas made up the shortfall. During the 1967 Arab–Israeli War, Saudi Arabia joined Iraq, Kuwait, and Algeria in again stopping oil shipments to countries viewed as supportive of Israel—notably, Britain and the United States. That embargo also proved ineffective because in 1967 the United States imported barely 1 mbd of oil a day and was able to increase its own production, while global markets were cushioned by increased production from non-Arab sources such as Venezuela and Iran. By 1973 the situation had changed. The United States was importing 6 mbd and, for the first time, the US economy was dependent on foreign oil supplies.[8]

October 1973 marked the first occasion on which Saudi Arabia noticeably affected the daily lives of many in the West. When Egypt and Syria launched a surprise attack against Israel, Arab forces quickly crossed the Suez Canal, seized the Golan Heights, and destroyed a quarter of Israel's air force. The Soviet Union organized a massive air and sealift to resupply the Arab armies. The United States responded with Operation Nickel Grass, airlifting 1,000 tons of equipment and munitions a day to the hard-pressed Israelis.[9] On October 16, Soviet Prime Minister Alexei Kosygin arrived in Cairo to inform the Egyptians that, despite Moscow's support, they were losing the war and should seek a ceasefire.[10]

Arab oil producers met the following day in Kuwait to discuss how they might support the faltering Arab offensive. Iraq's representatives demanded an immediate oil embargo against Israel's allies, the nationalization of all American oil assets in the Middle East, and the withdrawal of Arab assets from American banks. Saudi Petroleum Minister, Ahmed Zaki Yamani, had clear instructions from King Faisal not to let events spiral into economic warfare with the United States. He opposed the Iraqi plan and won support for an alternative 5 percent reduction in Arab oil production every month until Israel evacuated the Arab territory that it had occupied in 1967.[11] A day after the Kuwait meeting, Saudi Minister of State for Foreign Affairs, Omar Saggaf, led the delegation of Arab foreign ministers to discuss ending hostilities with President Richard Nixon and Secretary of State Henry Kissinger. In the White House Rose Garden, Saggaf told the waiting press that the Arabs had been promised a "peaceful, just and honorable" settlement.[12]

When the United States then proceeded to provide Israel with an additional $2.2 billion in military aid, the Arabs felt that they had been misled. Abu Dhabi and Libya immediately declared oil embargoes on the United States.[13] King Faisal had previously agreed with Egypt's President Anwar Sadat to support the Arab war effort, though it is not clear whether he ever specified exactly how.[14] Now he halted all oil shipments to the United States as well as Britain, Canada, and Japan.[15]

On October 25, 1973, with the embargo barely five days old, the United States and Soviet Union brokered a ceasefire to defuse their own crisis. A month later Secretary Kissinger warned, "It is clear that if pressure continues unreasonably and indefinitely, then the United States will have to consider what counter measures it may have to take."[16] Ultimately, it was Kissinger's own efforts to negotiate an Arab–Israeli disengagement that led to the ending of the embargo on March 18, 1974.

King Faisal had never been eager to antagonize Saudi Arabia's most important allies and biggest customers; on this occasion, he was reluctantly dragged into the confrontation by more radical Arab leaders. Long before the war began, he had faced mounting criticism for his close relationship with the West. The Libyan press called him a servant of colonialism, while *Al Thawra*, the mouthpiece of Iraq's Ba'ath Party, charged Saudi Arabia with supporting "the U.S.-Zionist scheme to dominate the Red Sea."[17] A month before the war broke out, Faisal had told NBC News, "America's complete support for Zionism puts us in an untenable position in the Arab world and makes it extremely difficult for us to continue supplying the United States with oil."[18]

Ultimately, King Faisal felt compelled to act. The US Congress was offering additional billions of dollars in military support to Israel while American military aircraft were flying equipment directly from Germany to Israeli front-line troops fighting in the Sinai. With the tide of battle turning against the Arabs, Iraq, Libya, Algeria, and Jordan were sending troops to the front. The Arab media was calling him an American stooge, and within Saudi Arabia public anger was rising over his failure to support the Arab cause.[19] Faisal saw no way to remain safely on the sidelines.

Although Israel certainly had many friends in Washington, they were not the sole, or even primary cause for American support. In fact, Washington never sought an overwhelming Israeli victory that would have left it unwilling to negotiate.[20] Like King Faisal, the Nixon Administration needed to balance numerous global, regional, and domestic issues. The Cold War was raging, and Israel was a US ally; the United States was no more prepared to watch its clients defeated by Soviet proxies in the Middle East than it was in South East Asia. The Watergate scandal concerning the 1972 break-in at the Democratic National Committee headquarters in Washington, DC was also raging. On the same day that the Saudis began their oil embargo, President Nixon ordered Attorney General Elliot Richardson to fire Watergate Special Prosecutor Archibald Cox and end his investigation. The attorney general resigned rather than comply. When his deputy, William Ruckelshaus, also refused to dismiss Cox, Nixon fired him too. The number-three Justice Department official, Solicitor General Robert Bork, finally sacked Cox. With such disorder in Washington, it was important that the US still appeared decisive in the arena of foreign policy. Finally, there were credible reports that when the situation looked dire, Israel had loaded nuclear weapons onto its aircraft and was prepared to use them before allowing Egyptian troops to march into Tel Aviv.[21] Preventing a nuclear war was not an insignificant concern.

The Saudi oil embargo initially included US naval forces patrolling the Persian Gulf and fighting in Vietnam. Aramco President, Frank Jungers, was summoned to Washington to meet with Secretary of Defense James Schlesinger and Deputy Secretary William Clements, who asked him to speak to the Saudis about this increasingly severe problem. Once Jungers had assured King Faisal that the request had come from the US government and not the oil companies, the king agreed, so long as this breaking of the Arab oil embargo remained strictly secret. Twenty-five years after the event, Jungers would still not disclose to Anthony Cave Brown for his history of Aramco, *Oil, God and Gold*, just how he had managed this, saying only, "We did figure out a way to do it."[22] Fifteen years after that in his own book, *The Caravan Goes On*, Jungers explained that:

[w]e decided to load a Caltex tanker out of Bahrain, not one of our own terminals, and then secretly offload the cargo to another Caltex ship. To complete the clandestine operation, the second ship was not to be aware of its destination until after it was underway.[23]

Other shipments followed to American naval bases around the globe—many through Singapore—and, although rumors circulated, the transfers did not become public knowledge. Both King Faisal and Aramco had taken substantial political risks to keep the US military sufficiently fueled.

More economic pain was caused by the Saudi production cuts from 8 to 6 million barrels a day than the actual embargo, because the production cuts reduced total global supply. Saudi Arabia had been supplying the United States with less than 4 percent of its daily oil consumption, but the impact of the production cuts was both immediate and long-lasting. Gasoline lines became a new feature of American life as oil prices exploded from $3 to $12 a barrel. US oil imports soared from $4 billion in 1972 to $24 billion in 1974, and for many years remained the single largest element in the United States' trade deficit. Between 1972 and 1974, the US inflation rate soared from 3 to 11 percent, unemployment rose from 5 to 7 percent and GDP growth collapsed from +5 to -0.5 percent. In Britain, Prime Minister Edward Heath announced a three-day working week for most industries, cutting the salaries of 13 million British workers nearly in half as GDP growth fell from +6.5 to -2.5 percent in one year.[24] The 1973 Saudi-led oil embargo initiated years of economic "stagflation," lost jobs, and lost economic growth that could never be recovered. It left deep, permanent scars on the American and British economies as well as on the attitude of their peoples towards Saudi Arabia.

In Washington, the State Department's official history noted:

[t]he embargo laid bare one of the foremost challenges confronting U.S. policy in the Middle East, that of balancing the contradictory demands of unflinching support for Israel and the preservation of close ties to the Arab oil-producing monarchies. The strains on U.S. bilateral relations with Saudi Arabia revealed the difficulties of reconciling those demands.

If the Saudi Foreign Ministry were ever to publish its own official history, it too would surely note the difficulty in "balancing the contradictory demands of unflinching

support for Arab and Islamic causes with the preservation of close ties to Riyadh's most important allies."

Surprisingly, the Arab oil embargo, which caused so much economic havoc, did not permanently damage Saudi American relations. On the contrary, Secretary of State Henry Kissinger and Secretary of the Treasury William Simon recognized the need to recycle petrodollars back into the US economy and avoid this sort of economic disruption in the future. They set out very deliberately to create incentives for Saudi Arabia to become a stakeholder in US economic prosperity, and they succeeded. A treasury attaché arrived in Riyadh, and Saudi Arabia became one of the largest purchasers of US Treasury bonds; the kingdom still is. Military sales, which had totaled only $300 million in 1972, climbed to $5 billion in 1975 as Saudi Arabia became one of the largest purchasers of American weapons; it still is.[25] The US Army Corp of Engineers began a massive $14 billion military infrastructure building program for the Saudis—a program that would prove invaluable during Operation Desert Storm (the combat phase of the First Gulf War: 1990–1991) when arriving US forces found everything from showers to aircraft maintenance facilities, all built to familiar American specifications.[26]

The Joint Economic Commission Office Riyadh (JECOR) opened, and over the next quarter of a century it funneled billions of petrodollars back to the US government through a wide variety of development programs funded by the Saudis. Although managed by the United States Treasury Department, JECOR was effectively a massive foreign aid development program funded by the developing country itself. Although the Saudi government itself did not invest heavily in US equity markets, many private Saudi citizens, including members of the royal family, did so and were made welcome. Some large American refineries were built to operate specifically on heavy Saudi crude, and in 2020 Saudi Aramco still owns the largest refinery in the United States at Port Arthur, Texas. Paradoxically, after the 1973 oil embargo, economic ties between Saudi Arabia and the superpower of the day became closer than ever.

The 1973 oil embargo transformed Saudi Arabia from a hesitant, marginal player on the world stage into a financial superpower. Slowly but inexorably, the Arab world's center of gravity shifted from the banks of the Nile to the sands of the Nejd, as Saudi Arabia became the International Monetary Fund's only Arab board member as well as the only Arab member of the G20 international forum for the governments of the world's largest economies. Unprecedented oil revenue gave Riyadh another foreign policy tool—one that it would subsequently use to promote its own interests in regional stability by funding pro-Western Arab governments.

A prime example of the latter approach is furnished by the Arab Republic of Egypt, whose political stability has long been important to the United States, Europe, Israel, and Saudi Arabia. In 2014, King Salman provided the new Egyptian government of Abdul Fattah Sisi with loans, fuel, and grants worth more than US$20 billion—twice what the IMF eventually provided, and more than ten times the annual value of US aid to Egypt. This Saudi support for its neighbor had implications well beyond the Middle East. Had Egypt, with its 70 million-strong population, collapsed into Arab Spring chaos like Libya, Syria, or Yemen, the wave of refugees flooding into Europe would have swelled into a tsunami of unknown consequences for European unity and security.

In 2015, Saudi aid to Egypt was nearly as important to Europe's political future as Saudi oil had been to rebuilding post-war European economies in 1945.

Other recipients of Western foreign aid—such as Jordan, Lebanon, Morocco, Pakistan, Bahrain, and Oman—are also substantial beneficiaries of Saudi financial support. Riyadh provides them with economic development funding, military assistance, and subsidized fuel. For example, Saudi Arabia has purchased hundreds of millions of dollars-worth of American weapons for Pakistan, and long provided that country with fuel on concessionary terms. For many years, the kingdom sought to stabilize Yemen with payments handled informally by Crown Prince Sultan through tribal networks. At the height of the 2011 Arab Spring, Saudi Arabia contributed to the $20 billion Gulf Cooperation Council aid package designed to stabilize Bahrain and Oman. In 2018, it joined Kuwait and the UAE in placing a billion dollars in Jordan's central bank to help stabilize that country's tottering economy. The list of Saudi foreign aid recipients is long, and this financial contribution to regional stability has been significant.

The third instrument in Saudi Arabia's foreign policy toolbox is Islam, which is often nearly as useful as the support of powerful friends or the judicious use of oil production and revenue. The Al Saud make no claim to being the spiritual or political leader of the world's 1.5 billion Muslims, nor are they trying to re-establish an Islamic caliphate; a transnational political entity intended to unite all Muslims. Their foreign policy is firmly based on the self-interest of an independent nation state, not religion. Nevertheless, they have frequently used religion to provide ideological legitimacy for policies that advanced their national or dynastic interests. Although there are many ministries in Riyadh with striking architecture, only the Ministry of Foreign Affairs has an explicitly Islamic style. This is no coincidence, it reflects the calculated use of Islam to promote Saudi Arabia's international influence.

King Abdulaziz denounced the Ottomans and Hashemites primarily in religious terms.[27] It was the Hashemite Sharif Hussein's claim to be the new Caliph that gave Abdulaziz a pretext to invade the Hejaz. Seeking recognition of his conquest, the new king of the Hejaz sent messages to many Muslim rulers and organizations pledging that he would protect the holy places and improve conditions for the pilgrims—which, as we have seen, he did. Even the Soviet Consulate was impressed, writing to Moscow, "Hajj guides are unhappy as they are no longer able to fleece the pilgrims."[28]

In June 1926, King Abdulaziz convened a Muslim Congress in Mecca, ostensibly to discuss how the holy places should now be governed. Sixty-nine delegates attended from, among other places, Afghanistan, Egypt, India, the Soviet Union, Syria, and Yemen. Once the Congress opened, Abdulaziz made it clear that there would be no debate over how the Hejaz was to be governed. He alone ruled in Mecca and he would tolerate no political activity during the Hajj—especially, no calls for Islamic unity against colonial powers.[29] Deliberations were limited to improving the Hajj services. Some delegates walked out, but most formally recognized that the Al Saud dynasty was now responsible for Islam's holy places.

Since then, the Al Saud's control of Mecca and Medina has given them both prestige and a special role in Islamic affairs. The Great Mosque in Mecca and the Prophet's Mosque in Medina are collectively known as *Al Haramain*—literally, the two sanctuaries, but more often translated as "The Two Holy Mosques." Like the Ottoman sultans before them, Saudi kings have adopted the title Custodian of the Two Holy Mosques, and these two most important sites in the Islamic world come under the supervision of a specialized Saudi government agency known as the Presidency of the Two Holy Mosques. The president is appointed by the king and supervises all matters relating to the mosques—including preaching, teaching, and publishing.

Sermons preached in Mecca are broadcast daily to tens of millions of devout, attentive Muslims around the world. This is Islam's "bully pulpit," a highly effective platform from which to advocate an agenda globally. In 1914, the possibility that sermons from Mecca would call for *jihad* against Christian, colonial powers was a major British concern. In recent years these sermons have been decidedly moderate—regularly condemning terrorism, suicide bombing, unsupervised charitable giving, and going abroad for *jihad*. However, with a less-stable or less Western-aligned government in Riyadh, the threat of radical sermons emanating from Mecca could rapidly re-emerge.

Each year more than two million pilgrims arrive in Mecca for the Hajj. They gather on an empty desert plain to pray; listen to sermons; and sacrifice thousands of goats, sheep, and camels. This immense, multinational congregation provides Saudi leaders with a rich opportunity to promote their views, demonstrate their special role in Islam, and host important Muslim leaders at a spiritually significant moment in their lives. The Saudi government attempts to keep the Hajj apolitical by allocating visas through a national quota system and banning political demonstrations. Providing security, shelter, sanitation, transportation, water, and healthcare for this throng is a major logistical undertaking that preoccupies much of the Saudi government for months beforehand. What was once a major source of income for King Abdulaziz has become a huge government expenditure for his sons, but managing a successful Hajj remains a source of pride for the Saudi people and of legitimacy for the House of Saud.

The Al Saud's most explicit use of Islam as a foreign policy tool occurred in the 1960s, when Arab nationalists sought to free the Middle East from decades of Western domination and centuries of economic backwardness. With his secular, Pan-Arab, anti-imperialist broadcasts on Cairo's Voice of the Arabs radio service, President Nasser captured the imagination of people across the Middle East. He preached vehemently against the twin evils of imperialism and reactionary rulers; overthrowing conservative Arab monarchs aligned with the West was an essential part of his program.

The Egyptian king had been deposed in 1952. In 1956, President Nasser seized the Suez Canal from Great Britain while King Saud renewed the lease on the American air base at Dhahran. That made Saudi Arabia a prime target for Arab nationalists. The Iraqi king and most of his family were brutally murdered during a Ba'ath Party coup in 1958. The few survivors found refuge in the Saudi Embassy, and King Saud sent an aircraft to take them to their cousins in Jordan. By 1960, Saudi pilots and Saudi princes were defecting to Cairo. Saudi Arabia was surrounded by secular, republican regimes in Egypt, Iraq, Syria, and Yemen—all eager to see the monarchy's downfall. Clearly on the defensive at home and abroad, the Al Saud's response was Islam.

In 1962, Crown Prince Faisal unequivocally deployed religion against the secular, socialist ideology of Nasser and the Ba'ath Party of Iraq and Syria. In his capacity as prime minister, Faisal sponsored a conference in Mecca for Islamic religious figures from around the world. Rebuking the Arab nationalists, the conference issued a statement that "[t]hose who disavow Islam and distort its call under the guise of nationalism are actually the most bitter enemies of the Arabs, whose glories are entwined with the glories of Islam."[30] The conference led to the creation of the Muslim World League (*Rabitat al-Alam al-Islami*), which would become one of the world's largest non-governmental organizations.[31] Headquartered in Mecca and funded largely by Saudi Arabia, the Muslim World League encompasses the International Islamic Relief Organization, the World Council of Mosques, and a jurisprudence committee that issues Islamic legal rulings with international standing. Its secretary general has always been a senior Saudi technocrat who consistently functions as a non-governmental Saudi spokesman.

In 1966, King Faisal made official visits to Guinea, Iran, Jordan, Mali, Morocco, Pakistan, Sudan, Tunisia, and Turkey. At each stop, he encouraged the use of Islam against dangerous ideologies, including not only secular Arab nationalism but also imperialism, communism, and Zionism.[32] He found considerable support for his pan-Islamic agenda, and when Israel took control of Jerusalem the following year he found a highly emotive, pan-Islamic cause to promote.

Muslims originally prayed facing Jerusalem, where the Al Aqsa Mosque remains Islam's third most sacred site. When a fire destroyed part of the mosque in August 1969, King Faisal called for the world's first Islamic Summit to address the issue. Because the Al Aqsa Mosque has such great religious importance across the Muslim world, twenty-five heads of state felt compelled to attend the conference held in the Moroccan capital, Rabat. Six months later, Muslim foreign ministers met in Jeddah where they agreed to establish the Organization of Islamic Cooperation (*Munazammat al-Taawun al-Islami*) as an Islamic rival to Cairo's secular Arab League. In 1971 they met again, in Pakistan, and agreed to create an Islamic News Agency, Islamic Broadcasting Agency, and Islamic Development Bank—all of which were eventually located in Saudi Arabia.

King Faisal's Islamic initiative was a success. He transformed the status of Jerusalem from an Arab–Israeli issue into a pan-Islamic one. More importantly, he created an Islamic Block, based in Saudi Arabia and operating through the Organization of Islamic Cooperation (OIC) and the Muslim World League, which consistently supported Saudi foreign policy objectives. In 2005, the OIC adopted King Abdullah's proposal for peace with Israel as the policy of fifty-six Muslim countries. In 2020, the organization has permanent delegations to the United Nations and the European Union. Although Jerusalem remained the OIC's primary focus, Muslim foreign ministers have presented unified positions at the UN on issues ranging from Iraq's invasion of Kuwait to the *Charlie Hebdo* cartoons and the Syrian Civil War.[33]

As the eminent historian of Islam, Princeton Professor Bernard Lewis, noted:

"For many years now there has been an active international grouping at the United Nations and elsewhere, consisting of more that forty Muslims governments, which together constitute the so-called Islamic Bloc. This includes monarchies and republics, conservatives and radicals, exponents of capitalism and socialism, supporters of the Western Bloc, Eastern Bloc and a whole spectrum of shades of neutrality. They have built up an elaborate apparatus of international consultation and, on many issues, cooperation; they hold regular high level conferences, and despite differences of structure, ideology, and policy, they have achieved a significant measure of agreement and common action. In this respect the Islamic peoples are in sharp contrast with those who profess other religions."[34]

The creation of this bloc was largely due to the efforts of King Faisal, and subsequent Saudi kings have continued to play important roles in its leadership.

In 1979, when Sunni jihadists seized the Great Mosque in Mecca and revolutionary Shia clerics overthrew Iran's monarchy, the Al Saud found themselves challenged again, this time not by secular Arab nationalists but by competing Islamic voices. Sunni militants, including the Muslim Brotherhood, condemned the Al Saud for enabling the spread of Western values. Iran's new Shia leadership challenged their fitness to manage the Hajj. With their religious credentials being questioned, Al Saud rule became more conservative and overtly pious. Saudi-funded Wahhabi evangelism became more focused, organized, and energetic.

King Fahd declared there should be "no limit on expenditures for the propagation of Islam," and claimed to have established 200 Islamic centers, 1,500 mosques, and 2,000 schools for Muslim children in non-Islamic countries.[35] He established the Ministry for Islamic Affairs, Proselytizing, and Religious Guidance to support Wahhabi Islam abroad.[36] He also created the Supreme Council for Islamic Affairs, which had explicitly political as well as religious dimensions. The secretary general of the Muslim World League and the ministers of justice, higher education and Islamic affairs were all members of the council, but so were the ministers for defense, foreign affairs, and the presidency of general intelligence. These organizations sent thousands of Saudi missionaries abroad, distributed millions of Qurans, and spent billions of dollars spreading the Wahhabi gospel and Saudi influence—especially in Indonesia, Pakistan, and sub-Saharan Africa.[37]

Another major deployment of Islam as an instrument of Saudi foreign policy emerged in December 1979 when the Fortieth Red Army invaded Afghanistan, thus firmly aligning Saudi and US interests against Moscow. The Saudis had already been worried about the Soviet military presence in South Yemen and Ethiopia. Now Moscow appeared to be threatening their strategic ally, Pakistan, which had kept thousands of troops stationed in Saudi Arabia for many years. In his last State of the Union Address, President Jimmy Carter stated that the Soviet occupation of Afghanistan posed a direct threat to "more than two-thirds of the world's exportable oil." The president was clear that "[a]ny attempt by an outside force to gain control of the Persian Gulf region will be regarded as an assault on the vital interests of the United States of America and will be repelled by all means necessary, including military force."[38]

President Carter's National Security Advisor, Zbigniew Brzezinski, was a determined cold warrior who saw an opportunity to land a blow against the Soviet Union. Hoping that the Soviets could be bogged down and bloodied in Afghanistan as the Americans had been in Vietnam, he wrote a memo to the President stating, "It is essential that Afghanistan's resistance continues [...] This means more money as well as arms shipments to the rebels and some technical advice [...] We should work with Islamic countries, both in propaganda and in a covert action campaign to help the rebels."[39]

In February 1980, Brzezinski traveled to Islamabad and Riyadh to propose that Pakistan's Inter-Service Intelligence Directorate (ISI) covertly distribute Saudi and American funds to those Afghans fighting the *jihad*—or, in Arabic, the *mujahideen*.[40] He found King Khalid and his foreign intelligence chief, Prince Turki al-Faisal, ready to cooperate.[41] They agreed to match US funding in what became a multibillion-dollar partnership, but they deployed very different justifications. As the Saudi Ambassador to Washington, Prince Bandar bin Sultan, explained:

> We did not use East-West arguments or America's anti-communism rhetoric, we used religion. We said "the Communists are atheists, they do not believe in God and we are fighting them for religious reasons. We galvanized the Muslim World behind us, which fitted perfectly with a strategy for fighting the Soviet Union in places the United States could not influence as well as we could."[42]

Defending Islam became the leitmotif of this joint effort. The US government spent $50 million on a "Jihad Literary Project," which printed books for Afghan schoolchildren urging them to attack Soviet troops. These schoolbooks, paid for with American tax dollars, contained lines such as "Doing *jihad* against the infidel is our duty."[43] Mohammed Abd al-Wahhab was certainly smiling somewhere. Meanwhile, Saudi Arabia began to contribute public funds, private money, and thousands of volunteers to the cause. As has been mentioned in Chapter 12, King Khalid appointed the Governor of Riyadh, Prince Salman, to lead private fundraising efforts and the Saudi *ulama* issued a *fatwa* supporting his activities. The Grand Mufti, Sheikh Abdulaziz bin Baz, wrote the forward to an influential book by the Muslim Brotherhood ideologue Abdullah Azzam, entitled *The Defense of Muslim Lands*, which argued that *jihad* against the Soviets was an obligation for all Muslims, not just for those living in Afghanistan.[44]

Soon Saudi men and money were pouring into the Afghan capital, Kabul. The Afghans called them Wahhabis, and were not fully sympathetic to their views. At times the Saudi and Afghan *mujahideen* clashed—especially when, much as the Ikhwan had done in the Hejaz sixty years earlier, the Saudis destroyed graves adorned with markers and flags. Unlike the gentler people of Jeddah, on more than one occasion the bold Afghans killed the Wahhabis for desecrating graves.[45] No one really knows how many Saudis went to fight in the *jihad* against communism, but estimates range between 12,000 and 25,000.[46] The CIA's Islamabad Station estimated that in 1989 there were 4,000 Arab fighters in Afghanistan, heavily supported by the Saudi intelligence service and Gulf charities.[47] Of the 775 al-Qaeda fighters captured in Afghanistan and taken to Guantanamo Bay in Cuba, 17 percent were Saudis.[48] In total, approximately 500 Arab fighters died in Afghanistan compared with tens of thousands of Afghan *mujahideen* and more than a million Afghan civilians.[49]

Ultimately, it was not the number of Saudis who fought in Afghanistan that made a difference but the petrodollars that they brought with them. Again, there are no accurate records of the money sent from Saudi Arabia to the Afghan *mujahideen*. One published account based on CIA contacts estimates that Prince Salman was raising between $20 million and $25 million a month at the height of his campaign.[50] The former director of Pakistan's intelligence service, Mohammed Yousaf, has written:

> It was largely Arab money that saved the system. By this I mean cash from rich individuals or private organizations in the Arab world, not Saudi government funds. Without these extra millions the flow of arms actually getting to the *mujahideen* would have been cut to a trickle before 1983.[51]

Thus, at least by the ISI director's account, private Saudi charitable donations to the *mujahideen* played a decisive role in defeating Soviet forces in Afghanistan and, by extension, helped to end the Cold War.

King Faisal's third son, Saud al-Faisal, managed his country's foreign policy from 1975 until 2015. The world's longest-serving foreign minister, he personified Saudi Arabia's cautious, quiet, private diplomacy. He used personal relationships with seven American presidents, the kingdom's wealth, and importance in global energy markets, as well as its unique place in the Islamic world to enhance Saudi influence. By and large he obtained "peace and lasting security" for his people, but Saud al-Faisal's retirement in 2015 marked the end of an era in Saudi diplomacy.

As we shall see, his successors have been less successful in sparing the Saudi people the cost and pain of foreign conflicts. Whether the Al Saud's foreign policy tools have grown dull or the craftsmen using them are now simply less skillful is a moot point. What matters for Saudi stability is that for the first time since 1934, the Al Saud have been unable to keep their kingdom out of a destructive, prolonged and costly war.

Leading Gradual Social Change

The Al Saud recognized long ago that they must either lead social change or eventually be overtaken by it. Thus, for more than half a century they have supported gradual social evolution—not to satisfy their many Western critics but because it was in their self-interest to do so. They hoped that by cautiously delivering social change they could satisfy demands for liberalization from an increasingly urban, young and well-educated population, while avoiding a backlash from the religious conservatives who remained numerous, well-organized, firmly committed to the past, and potentially violent.

There was no public outcry in Saudi Arabia demanding the end of slavery, the introduction of television, or the opening of girls' schools. In fact, there was considerable opposition to all three. Some of the first cars in the Nejd were burned as tools of the devil,[1] and a protesting prince was killed by the police during violent demonstrations against Riyadh's first television station.[2] When King Abdullah opened his University for Science and Technology in 2009, senior clerics denounced it as a godless institution. For most of Saudi history, reforms that increased religious tolerance or gender equality have been controversial—and in some quarters, they still are. To an extent seldom recognized outside of Saudi Arabia, the monarchy has not been a reactionary force holding back a progressive society. Quite the opposite. The Al Saud have usually moved in step with the majority of public opinion, and at times dictated liberalization in the face of outright opposition.

Nothing illustrates this tendency more clearly than the Al Saud's efforts to modernize the kingdom's legal and educational systems. As discussed in Chapter 8, courts and schools across the Arab world had been firmly in the hands of religious scholars for centuries. In most Arab countries, it was European colonial administrators who dismantled this ancient system by placing courts and schools under civil administration. But again, as we have seen, Saudi Arabia never experienced a colonial administration. In Riyadh, as in much of medieval Europe, it was left to the crown to wrest legal and educational institutions from ecclesiastic control.

King Abdulaziz created a Directorate of Education to begin offering modern, secular education to the Hejaz in 1926. In Riyadh, the religious establishment continued to control the only public education that was available until 1953.[3] King Saud is usually remembered for bankrupting an oil state, but he also created a Ministry of Education and appointed the future King Fahd as its first minister. When Fahd began there was one modern secondary school in the entire kingdom, and that was in Mecca.[4] He had to deal with a population that was 95 percent illiterate, studying a curriculum that was almost entirely religious.[5] Fahd made progress introducing art, English, modern science, and sports for the first time to the Nejd. Stop and think about that. Sixty-five years ago, no school in Riyadh taught chemistry, physics, or French. Chemistry was regarded as magic, physics as heresy, and French as the language of Crusaders.[6] Prince Fahd began to change that, but in doing so made many enemies among the conservative *ulama* and was eventually fired for his efforts.

The *ulama* fought back because they viewed modern, secular education as a Western ploy to invade the Islamic world, spread ideas antagonistic to religion, and, ultimately, undermine their own status. To maintain their influence, the *ulama* persistently worked to slow the development of modern education and to increase the amount of religion, Arabic, and Islamic history taught in Saudi schools. They argued that the only way to protect Islam from foreign influences was to leave education firmly in their hands.

When King Abdulaziz died in 1953, there were no universities in Saudi Arabia. As had happened with secondary education, a two-track university system was created. New secular universities were opened in Riyadh, Jeddah, and Dhahran, but they were balanced with religious universities in Riyadh, Mecca, and Medina. These religious universities were less selective, they offered good stipends and attracted less prosperous and more conservative students. Since most religious studies graduates could not find jobs in the modern economy, they became mosque workers, religious policemen, and, above all, teachers of religion in primary and secondary schools.

Before 1960, Saudi girls were almost completely excluded from education. Many Saudi parents believed then, and some still do, that female education promotes immorality and weakens the family structure. King Faisal's decision to introduce girls' education provoked a firestorm similar to desegregation in the American South. Angry fathers came from Al Qasim by the hundreds and camped outside the city for several weeks, waiting to see the king. Faisal listened, then he made it clear that no one would be required to send their daughter to school, but he would use force against anyone who tried to disrupt those wishing to attend. In 1963 he actually deployed the National Guard to keep girls' schools open in Buraydah, the capital of Al Qasim Province north of Riyadh.[7] This is not ancient history, there are many people still alive who vividly recall these events.

On the issue of girls' education, the Al Saud, as usual, compromised with the *ulama*. Rather than place it under the semi-secular Ministry of Education, Faisal created a separate Presidency of Girls' Education headed by none other than the grand mufti, who now had even more jobs to distribute to the graduates of his religious universities. However, King Faisal was clear about his ultimate goal. He told those whom he sent to negotiate with the *ulama*, let them insist that the girls focus on cooking and sewing; you insist that they also learn to read and write. Once they can read and write the rest will follow.[8] And so it has. Today, there are more women in Saudi universities than men. As Saudi women became more educated, they gained social and economic independence—just as they had in the West, and just as social conservatives in the kingdom had feared.

The Al Saud's compromises with the clergy on modern education was a Faustian bargain. The great benefit was the *ulama's* continued support for the monarchy, and the resulting political stability. That stability, so sorely lacking in most Arab countries, is precious and was by no means inevitable in Saudi Arabia. However, the price has been high. Unlike the Western syllabus, rooted in the values of the Enlightenment, the religiously based Saudi school curriculum has not tolerated the questioning of received wisdom. It has relied heavily on rote memorization and seldom encouraged curiosity or innovation. The result of this has been a labor force poorly prepared for a modern economy.

The Ministry of Education was for decades dominated by conservative, Nejdi clerics and Muslim Brotherhood schoolteachers who fled Egypt during Nasser's rule. Even today, Saudi Arabia's 500,000 government schoolteachers are often poorly trained and come from the bottom of their university class.[9] Thousands teach religious studies, which still constitute a third of the curriculum in some grades. Needless to say, these teachers and their allies, the *ulama*, continue to resist curriculum changes that reduce the importance of religious studies.

King Abdullah recognized these problems and made a major effort to improve primary and secondary education, expand universities, and send Saudis abroad to study. Both as crown prince and during his reign from 2005–2015, he provided substantial funding for education and placed capable people in charge. In 2008, he appointed his son-in-law, Prince Faisal bin Abdullah, minister of education and subsequently appointed Saudi Arabia's first female deputy minister to head girls' education. His gradual transformation of girls' education from an independent organization managed by the religious establishment into a part of the Ministry of Education managed by women was noteworthy.

In 2014, King Abdullah asked Prince Khalid al-Faisal to become minister of education. A son of King Faisal, the governor of Mecca, brother of the foreign minister, a graduate of Oxford, and former governor of the southern Asir Province with more than 40 years of government service, Khalid al-Faisal was one of the most accomplished and respected people in the kingdom. His transfer from governing Mecca, which was traditionally one of the kingdom's most important posts, to become minister of education, which traditionally lacked such status, clearly demonstrated the importance that King Abdullah placed on improving education.

The reforms initiated by the two princes whom King Abdullah put in charge of education went well beyond removing intolerant passages from old textbooks. International consultants ranging from Columbia University to McKinsey were contracted to develop a long-term strategic plan for educational reform. New curricula were designed to encourage "critical thinking skills for a globally competitive, knowledge-based economy."[10] Many of the new textbooks adopted were direct Arabic translations of Western counterparts. More math, science, and English language teaching was gradually introduced at the expense of religious studies.

Although King Abdullah had no university degree, he did not neglect higher education. More than a dozen new universities were opened during his reign, mostly in provincial cities. More private universities were founded, and they won some independence from government oversight of their curricula and social policies. Abdullah's most ambitious projects were Princess Noura University for Women in Riyadh and King Abdullah University for Science and Technology (KAUST) located on the Red Sea coast north of Jeddah.

Naming a university after Princess Noura bint Abd al-Rahman was not an accident. Noura was King Abdulaziz's favorite full sister. His personal battle cry was, "I am Noura's brother," and she is often credited with helping her brother to found the Third Saudi State. Dedicating a university to her was intended to emphasize the role that a woman had played in creating Saudi Arabia. At a cost of more than $2 billion dollars, King Abdullah pushed for the rapid construction of what is now the largest women's university in the world. It has more than 40,000 female students, 12,000 employees, a 700-bed teaching hospital, and its own monorail. Some Saudi feminists condemn Princess Noura University as a "gilded cage." Why, they ask, should there be a purely women's university? They have a point, but it was a step in the right direction in a country where in 1960 girls could not go to elementary school, yet in 2020 they comprise 60 percent of university graduates.

KAUST was an even more ambitious and expensive project than Princess Noura University. Built by Saudi Aramco, which the king rightly judged to be the most competent engineering organization in the country, it is an entirely postgraduate, research university with no tuition fees. The university uses one of the world's most powerful supercomputers for research into areas directly relevant to Saudi Arabia such as solar energy, water desalinization, clean combustion, and desert agriculture. Unlike other Saudi universities, KAUST operates under its own authority, fully independent of the Ministry of Education. English is the language of instruction for the university's nearly 1,000 male and female students who come from more than sixty countries. With some 6,500 applicants each year for 250 places, its masters and doctoral programs have one of the lowest acceptance rates of any university in the world. The KAUST endowment of more than $20 billion is the world's third largest, falling behind only Harvard and Yale.

King Abdullah wanted KAUST to rekindle the once-celebrated, but long dormant, legacy of science in the Islamic world. Once he had decided to fund such a large enterprise, he created a committee of Saudi scientists and scholars to develop the project. The committee was sharply divided; some argued for a university based in Mecca that admitted only Muslims and would revive "Islamic Science;" others believed that there was only "International Science" and that the university should be made as attractive as possible to foreign scholars. King Abdullah sided with the internationalists.

The king's religious critics did not remain silent as he pushed Saudi Arabia towards global norms. Clerics claimed that KAUST was a source of secularism and disbelief in Saudi Arabia and that its program were not based on religious principles. Its president and most faculty members were infidels, its finances were based on interest from an endowment fund, no courses were offered on the Quran, and, worst of all, it totally lacked any surveillance by the religious police. All of these observations were completely accurate. Most of the KAUST faculty are non-Muslim, collecting interest is prohibited in Islamic law, and the religious police do not provide their usual moral supervision to KAUST students. Dismantling KAUST became, and remains, a priority for Saudi Arabia's religious conservatives.[11]

King Abdullah realized that reforming Saudi schools and building world-class universities would be a difficult process taking many years. His short-term solution was to get Saudi students out of Saudi Arabia and into higher education abroad. When he announced his scholarship program in 2005, he stated that it was so that young Saudis "would learn to know the world and the world would learn to know them." The program provided full tuition, room, and board; health insurance; and a stipend to study science and engineering-related subjects at foreign universities.

At the height of the first oil boom in 1980, there were nearly 10,000 Saudi students in the United States. After 9/11, there were barely 2,000. In 2017, because of King Abdullah's scholarship program, there were 68,000 Saudis studying at hundreds of American universities and thousands of others in Britain, Canada, and Australia. Only China, India, and South Korea had more students in American universities, and when measured as a percentage of its total population, Saudi Arabia was far ahead of the others. The scholarships are competitive, and 20 percent of them are reserved for women. In what must be a truly unique benefit, a woman on the program can, if her father requests, be accompanied by a male relative to serve as chaperone; his expenses are also covered by the government. By 2020, more than 200,000 young Saudis have used the program to study in some forty countries.

In 2015, the new King Salman returned Prince Khalid al-Faisal to his former post as governor of Mecca and appointed a new minister of education who was, if anything, even more liberal. The new minister, Dr Ahmed al-Issa, came from a family of religious scholars but also held a PhD in Education from Penn State University. As president of the private Al Yamamah University, he had allowed the performance of a mixed-gender play, that focused on the problem of intolerance and was accompanied by music. Outraged conservatives had physically attacked the actors and stopped the performance. As a result of the incident, Dr. al-Issa had lost his job.

King Salman's removal of the long-serving, conservative Khalid al-Angari from the Ministry of Higher Education and appointment of the progressive Ahmed al-Issa to lead a newly combined Ministry of Education and Higher Education was a bold move. Within Saudi Arabia it caused as much comment in early 2016 as the international reaction to the proposed Aramco IPO. After three years, King Salman appointed Dr Hamad bin Mohammed Al al-Sheikh to become minister of education. Dr Al al-Sheikh was not only a member of Saudi Arabia's most prominent religious family, he also held a Doctorate in Economics from Stanford.

In 2020, none of King Salman's ministers faces a more daunting task than his minister of education. Saudi Arabia spends nearly 25 percent of its published budget on education, which makes its per-student spending nearly equal to that of the United States. Illiteracy is now less than 2 percent for those aged ten to fifteen. Yet despite investing a great deal of money and effort, progress has been slow and uneven. Results on standardized tests remain poor, even when compared with nations benefiting from much less funding or political stability. Although Saudi Arabia has produced highly competent doctors, engineers, and corporate managers, it must still produce more people who can live comfortably with the outside world if it is to succeed in the globalized twenty-first century.

As with education policy, there is a long-running struggle between the Saudi crown and the Wahhabi religious establishment for control of the kingdom's legal system. This struggle between ecclesiastic and lay judicial prerogatives has often strained the Al Saud's venerable alliance with the *ulama*, just as a similar twelfth-century struggle strained relations between England's King Henry II and his Archbishop of Canterbury, Thomas Becket. Although many Westerners have heard of the blind Grand Mufti Abdulaziz bin Baz (1910–1999), who at one point thought that the Earth was flat, few are familiar with his more prominent predecessor, Sheikh Mohammed Ibrahim Al al-Sheikh (1893–1969), who, like Becket, vigorously defended canon law and ecclesiastic courts.

Mohammed Ibrahim was a giant among the *ulama*, a man holding great moral and political authority second only to the king. In his day, there was no minister of justice or minister of Islamic affairs whom the monarch could appoint or dismiss. As grand mufti, Mohammed Ibrahim firmly controlled all Saudi courts and mosques as well as much of the education system. He ran the religious police and was even in charge of the rudimentary welfare system, including orphanages and charitable religious endowments. One of his sons was deputy grand mufti and another was in charge of religious schools. One brother ran the religious police in the Hejaz while another brother did the same for the religious police in the Nejd. This was a very formidable political machine. Such a concentration of power in the hands of a cleric had not been seen in Western Europe for centuries and has never existed in the United States.

Unlike Henry II, King Faisal never asked, "Who will rid me of this troublesome Mufti"? He just waited for Mohammed Ibrahim to die of natural causes, and then moved quickly to dismantle the grand mufti's empire. Much as King Abdulaziz had sought to reduce the power of paramount tribal leaders, King Faisal wanted collective leadership of the religious establishment—not one supreme religious figure, or even one dominant family. Accordingly, Faisal never appointed a new grand mufti and the position remained vacant for twenty years. Instead, he created the Committee of Senior Scholars, the majority of whom have never been drawn from the Al al-Sheikh family.

Faisal went on to create a Ministry of Justice to deal with administrative issues such as judges' salaries, promotions, and retirement; and a Supreme Judicial Council to serve as something of a supreme court for the kingdom. A new body was established under Abdulaziz bin Baz to issue *fatwas*, which are religious rulings on broad legal principles as well as specific cases. The two branches of the religious police in the Hejaz and Nejd were unified and, more significantly, began reporting directly to the king. Eventually, a Ministry of Islamic Affairs was created to administer mosques, a Ministry of Social Affairs took over orphanages and charities, and a Ministry of Hajj and Awqaf took over the pilgrimage and religious endowments. All of these new ministers and council chairmen reported to the king, not a grand mufti. These organizational changes brought the Saudi legal system under secular control. They were augmented by a series of civil and criminal procedural reforms that began to clarify issues such as bar exams and attorney client privilege, about which the *sharia* had nothing to say.[12]

In 2007, King Abdullah issued the Law of the Judiciary, which provided nearly $3 billion for legal reform.[13] New courthouses were built; more judges were hired, and their training improved. New regulations called for the creation of specialized labor, commercial, and family courts. Ministerial tribunals would continue to provide administrative settlements made by bureaucrats, but these decisions could now be appealed to and enforced by the new, specialized courts staffed by *sharia* judges—

implying that decisions of the new courts would be based on codified regulations, not general *sharia* principles.

The judicial appeals system was strengthened. The previous so-called Supreme Court had really been just a committee that considered procedural issues. Under the new regulations, a Supreme Court with nine members would consider points of law as well as procedures. This was intended to ensure that *sharia* principles were applied consistently and produced more predictable verdicts than had been the case. The chief justice of the new Supreme Court was given the rank and pay of a cabinet minister. These were major changes to a system that had long supported the final authority of individual judges, and decisions based on personal interpretations of *sharia* rather than legal precedents.

In 2020, implementing these reforms remains a work in progress. A Supreme Court building has been built, a few cases heard, and the judgments published. More cases are being heard in the new commercial court, and in late 2018 the new labor courts began hearing cases. A new bankruptcy law has made business restructuring easier, and a new enforcement court has made collecting final judgments much quicker. Foreign arbitration has been revived after having been defunct since the 1950s, when Saudi Arabia lost an arbitration dispute with Aramco over using Greek shipping magnate Aristotle Onassis's tanker fleet to ship Saudi crude oil.

Nevertheless, according to the kingdom's Basic Law of Governance, the *sharia* remains the only source of law in Saudi Arabia. The Ministry of Justice is still trying to codify laws, and published precedents are not always followed by *sharia* judges. Not all of the new specialized courts are functioning. The dual system of ministerial tribunals and *sharia* courts still exists, and jurisdiction between them often remains unclear— banks still try to collect debts through Ministry of Finance tribunals, for instance, and *sharia* courts continue to reject the concept of interest payments. In short, much more needs to be done to make the Saudi legal system transparent, predictable, and accommodating to foreign investors.

Transforming educational and legal institutions can be pursued by administrative fiat; transforming the social values of an orthodox and insular population is a more complicated task. For many centuries, the people of the Central Arabian Plateau remained cut off from the mainstream of global culture. Even today, geneticists looking for long-isolated human gene pools travel to Iceland, Tierra del Fuego, and the Nejd. The first two islands were isolated by stormy oceans; the Nejd was isolated by a parched sea of sand, which very few Europeans or other Asians had crossed before World War II. Such isolation has caused problems. It has, for example, led to the persistence of social values in the Nejd that, while once common in the West, have long been abandoned in Europe and the United States. The most persistent of these anachronistic social values are religious intolerance and gender inequality.

Explicit nineteenth-century Wahhabi texts stipulate the need to avoid contact with "the others"—meaning Christians, Jews, Shia, and even non-Wahhabi Sunni Muslims.[14] King Abdullah sought to modify these views. Beginning in 2003, he sponsored a series of National Dialogue conferences, at which selected community leaders discussed controversial topics including improving relations with "the others." There were no Christian or Jewish Saudis to attend, but Shias, Sufis, and non-Wahhabi Sunnis were invited. Some well-known Wahhabi divines, such as Safar al-Hawali, refused to attend a meeting with "deviants," while others, like Salman al-Auda, agreed to join the debate.[15] Their conclusion was that all people have the right to "honor, freedom of conscience and property," and that Islam supports "becoming acquainted with other people, communicating with them and cooperating with them in whatever is beneficial to humanity."[16]

In most places, such anodyne statements would be entirely unremarkable. Not in Saudi Arabia. Mohammed Abd al-Wahhab's grandson, Suleiman, had categorically condemned traveling to non-Muslim lands or befriending idolaters. The Ikhwan, who helped unite Saudi Arabia in the early twentieth century, firmly believed in their right to seize the property of nonbelievers, as did ISIS nearly 100 years later when it seized half of Iraq.

King Abdullah was nevertheless encouraged and went a step further. In November 2007, he traveled to the Vatican for the first ever meeting between a Saudi king and a pope. The king and Pope Benedict XVI agreed on the "value of collaboration between Christians, Muslims and Jews," and on the need to find a "just solution" to the Arab–Israeli conflict. For the Custodian of the Two Holy Mosques and *imam* of the Wahhabis to be photographed with the Bishop of Rome in front of a crucifix was a remarkable rebuke to the more xenophobic *ulama*.[17]

The following year in Madrid, King Abdullah went on to co-host a conference on interfaith dialogue with Spain's King Juan Carlos. 300 religious figures from fifty countries gathered to hear the Saudi monarch blame many of the world's problems on the abandonment of religious faith and the decline of the family. Participants included not only Jews, Christians, and Muslims—all of whom Islam considers to be "People of the Book"—but also Buddhists and Hindus, who have often been denounced as pagans, polytheists, and cow-worshipers by the Wahhabis. The king went so far as to state that if God had wanted to, he could have created only one religion, and obviously he chose not to. The *Christian Science Monitor* likened the conference to an Islamic Vatican II (the Second Ecumenical Council of the Vatican), at which, in 1962, the Catholic Church had formally accepted the validity of other religions.[18]

The Vatican II analogy was probably going too far. Neither King Abdullah nor any other Saudi ruler has every claimed to formally speak for all Muslims, and there are still no churches or synagogues in their kingdom. The conference was, however, a serious effort by a man of faith to promote religious tolerance. Abdullah followed up by founding and funding an Interfaith Center in Vienna to carry on the conference's work. King Abdullah was trying to nudge his country into the mainstream of global culture in much the same way as he had sent thousands of Saudi students abroad for Western educations and promoted Saudi entry into the World Trade Organization as a means of tying Saudi Arabia into the global economy.

The Al Saud are far more interested in maintaining political stability than in mediating culture wars; however, there are some social conflicts that they cannot avoid. Nothing has tested their ability to balance competing stakeholders' agendas more than gender equality. The powerful religious establishment firmly holds that God ordained and scripture prescribed fundamentally different roles for males and females. It regards enforcing conservative gender roles as a fundamental duty of any Islamic ruler; has traditionally opposed women's employment outside of the home; and suspects that gender mixing, dancing, and alcohol may be catalysts for lust. Some *ulama* regard the liberalization of gender roles as a direct challenge to their authority, and a cause to question the king's right to rule.

The *ulama* are not alone in opposing women's rights. Some Saudi men see their personal honor tied directly to conservative behavior on the part of their female family members. They agree with Sheikh Abd al-Rahman bin Nasir al-Barrak, who wrote in a religious ruling, "anyone who accepts his daughter, sister or wife working with men or attending mixed-gender schools cares little about his honor as this is simply a form of pimping."[19] Some tribal leaders view women's independence to live and work wherever they please, and ultimately to marry whomever they choose, as a direct threat to tribal cohesion—and thus, their own political futures.

There are also Saudi women who prefer the status quo. To Western ears their views seem extraordinary, so perhaps it is best to let them speak for themselves. Noura Abdulrahman worked for the Saudi Ministry of Education and defended male guardianship in an interview with *The New York Times*, saying, "In Saudi culture, women have their integrity and a special life that is separate from men. As a Saudi woman, I demand to have a guardian."[20] Rowdah al-Yousef began a petition called, "My Guardian Knows What is Best for Me," which obtained 5,400 signatures in two months.[21] The British-educated Saudi journalist, Sabria Jawhar, noted that

Non Saudis presume to know what is best for Saudis, like Saudis should modernize and join the twenty-first century or that Saudi women need to be free of the veil and abaya [their traditional full-length garment] [...] by freeing Saudi women, the West really means they want us to be just like them, running around in short skirts, nightclubbing and abandoning our own religion and culture.[22]

It is also worth noting that the oppression of women is not inherently part of either Islam or the teaching of Mohammed Abd al-Wahhab. 1,400 years ago, the Prophet Mohammed preached a radical gospel of social justice for women. For the first time in Arabia, women became parties to, rather than the objects of, marriage contracts; dowries became payable to brides, not their fathers; female infanticide, which had been common, was outlawed; and daughters were given inheritance rights—not equal to sons, but before Islam they had received nothing at all.[23]

Likewise, considering the importance that Mohammed Abd al-Wahhab placed on scripture, it is not surprising that his commentaries on marriage and divorce always emphasized a woman's God-given rights, even when these were at odds with local custom.[24] For example, Abd al-Wahhab condemned prostitution and, in contrast with some *bedu* practices, insisted on strict enforcement of women's inheritance and property rights.[25]

Thus, the source of widespread gender segregation and inequality in Saudi Arabia today is not strictly Wahhabi theology. Rather, it can be traced to a combination of conservative social values; local traditions; tribal practices; and the complicating fact that the Quran is, for Muslims, immutable. What might have been progressive in 610, such as leaving a daughter half as much as a son, is reactionary today—but the word of God is difficult to amend.

In spite of such widespread traditional attitudes, increasing numbers of Saudi women have been exposed to Western gender norms through travel, education, and the media. Many no longer accept having the religious police tell them how to dress, when to pray, or where to sit in a café; they asked why they could not drive, and why there are still no female judges or religious scholars. Some certainly regard male guardianship as a restriction, not a protection. Until recently they chafed at Saudi dress codes, and still resent the legal or cultural restrictions women face in managing their personal affairs or running a business.

King Abdullah sought to balance these competing social forces by promoting slow, incremental change. In 2000, women could not attend the Jeddah Economic Forum; a year later, they were allowed to sit in a segregated section. Then in 2004, a woman spoke from the podium, and many women have done so since. In 2003, women were permitted for the first time to study geology, architecture, and engineering in Saudi universities. In 2006, against strong clerical opposition, women were issued with national ID cards that allowed them to conduct some official business, such as enrolling a child in school without a male guardian being present. In 2008, the ban on women staying in hotels without a male guardian was lifted. In 2009, the first women were elected to the Board of the Jeddah Chamber of Commerce, and the Ministry of Justice announced that female attorneys would be allowed to represent female clients in family law matters.[26]

When he came to the throne in 2015, King Salman abandoned his brother's gradualist approach, and Crown Prince Mohammed bin Salman has emphatically rejected the religious restrictions imposed after the 1979 Mecca Mosque seizure.[27] For those who did not know Saudi Arabia in 2000 or even 2010, it is difficult to appreciate the scope and pace of social change that this father-and-son team have introduced. The examples are numerous and dramatic. Tolerance is actively promoted, entertainment opportunities have increased exponentially, and overt gender discrimination has noticeably declined. Those who say that these changes were long overdue have a point, those who argue that they are superficial are either ignorant or malicious.

Ten years ago, there was no public entertainment in Riyadh. No theaters, no cinemas, no sporting events that women could attend, no restaurants where unmarried couples could dine together, and no music anywhere. All of this has changed. After having been closed for thirty-five years, cinemas have reopened and ground has been broken for a large theme park outside of Riyadh. Many restaurants now play music and no longer strictly enforce gender segregation. Religious policemen, who once shot holes in satellite television dishes and publicly smashed mobile phones containing cameras, now limit themselves to complaining about tinted contact lenses.

A new Entertainment Commission has introduced hundreds of events, including rock concerts, magic shows, opera performances, auto races, and even female wrestling matches. The commission now pays young musicians to perform in public places, and Saudi parents bring talented children to audition for these places. Each month the Entertainment Commission offers something new, such as free ice skating lessons for any Saudi who wants to learn. A new Culture Commission openly promotes theater, music, figurative art, film, and literature—all but the last of which were once firmly opposed by the *ulama*.

Valentine's Day, which had long been banned as a pagan holiday, is now openly celebrated, with florists no longer arrested for selling red roses and students no longer sent home for wearing red. National Day, which was once banned by the religious authorities as a secular holiday, is now celebrated with public parties. In 2020, the New Year was greeted with large-scale public fireworks in a kingdom that still officially uses only the Hijra calendar.

All of this was intended to give young Saudis something to do other than "drifting," or *tafheet*, the dangerous game of driving a car on two wheels and trying to change the two airborne tires while doing so.[28] By the spring of 2018 there were designated parks where young people of both genders could meet for coffee, but to get in they sometimes needed a key. These keys were hidden around the city, with clues on where to find them given out on social media. While many young Saudis engaged in this state-sponsored treasure hunt, the police cracked down hard on the few remaining drifters.

Unlike some Muslims countries where the United States spent billions of dollars trying to change the status of women, in Saudi Arabia their status has improved far more rapidly than would have been expected only a few years ago. Since 2016, there has been a clear affirmative action program for Saudi women, with the government and multinational firms leading the way. For the first time women have joined the Saudi diplomatic service and public prosecutor's office. The chairman of the Saudi Stock Exchange is a woman—as are the CEOs or chairmen of several major Saudi banks. Women are back as anchors on television news programs and the Ministries of Labor and Education have female deputy ministers. Dhahran has a female deputy mayor, and a woman now sits on the Saudi Aramco Board. In a country where girls could not take gym classes five years ago, the Presidency of Youth and Sport now has a women director for female sports. Female falconers have even begun to participate in this very traditional, very male, Saudi sporting event.

Debate over which jobs are appropriate for women has receded, with women now permitted to hold nearly all positions—including that of taxi driver. In major cities Saudis are no longer surprised to see women working as cashiers, salespeople, or even waitresses, and the stigma against Saudi women working as nurses or in retail sales has begun to fade. Women remain excluded only from occupations deemed dangerous, such as mining.

Although women remain segregated at King Saud University, there is now a Women's Empowerment Program on campus The arrival and departure of female students is no longer monitored by the religious police and, although no formal announcement was ever made, female students are now allowed to wear trousers instead of dresses. At the private and progressive King Faisal University, women used to sit in classroom balconies; this was considered very liberal and was as close to gender mixing as any Saudi university had got. In 2017, women moved out of the balconies and began sitting in a separate section of the men's classroom. Most students expect that even this segregation will fade in time, and students are already mixing on their way in and out of class.

When plans for allowing women to drive were first announced, most Saudis anticipated that change would come gradually—as it usually had in the past. They expected to start with something along the lines of: now, married women over 40 will be allowed to drive during daylight hours. Not this time. Mohammed bin Salman said that women would drive just like men, and when Saudi women began applying for driving licenses in 2018, they were examined and licensed just like male applicants.

More liberating for many women, and much more complicated to enact legally, were the amendments to numerous guardianship regulations made in 2019. These changes removed restrictions on employment, travel, education, and healthcare. No longer was a male guardian's permission needed to obtain a passport or travel abroad; no longer was a woman legally required to live with her guardian or face police detention for the crime of absenteeism. The World Bank's annual report, "Women, Business and the Law," uses a series of indicators to evaluate female economic opportunity and the relationship between legal gender equality and women's entrepreneurship and employment. In 2020 it ranked Saudi Arabia as the most improved of the 190 economies reviewed.[29]

Nonetheless, Saudi women still face significant discrimination. Just because a woman has a driving license does not mean that her husband will pay for a car, gasoline, or insurance; a female lawyer may appear in court, but she will still cover her face before a *sharia* judge; and a woman can win custody of her child in a divorce settlement, but her father will accompany her to court. The long-established practice of child marriage was nominally abolished in 2020, but it remains to be seen whether or not *sharia* judges will enforce the new regulations. Women still need their guardian's permission to marry or to study abroad.

In 2020, social change in Saudi Arabia is meeting opposition from both the left and right. Some find the reforms inadequate, and demand even faster change; others think it has already gone too far and want gender and entertainment liberalization stopped or even rolled back. Complaints from conservatives over a women's wrestling match led to the sacking of the Entertainment Commission's first director. The director of the Small Business Administration was fired, in part for allowing a less-than-modest fashion show. A prominent local dairy withdrew an advertisement celebrating liberated female behavior when it provoked a firestorm of angry comments on social media. Even in liberal Jeddah, the White "nightclub"—part of an upscale chain of playgrounds for the chic, based in Beirut—was closed after its first night, remodeled, and rebranded as a high-end café.

The Crown Prince is betting on demographics. Most Saudis are under 30 and half of them are female. His strategy has been to win the support of the majority in the center by giving them what they want, while firmly suppressing outliers at both ends of the political spectrum. In September 2017, dozens of activists, ranging from hardline clerics opposing any social change to liberal intellectuals demanding much more, were arrested. The Crown Prince made it clear that he sets the pace of social change, and that loud, public dissent—either for or against it—will not be tolerated.[30]

For the time being, this carrot-and-stick approach appears to be succeeding. Saudi women and young people have overwhelmingly embraced both the Crown Prince and his liberalizing social program. But public support can be fickle, and those cheering the prince today are less organized and less committed than his critics. Should the economy falter, the war in Yemen drag on, or succession come into dispute, Mohammed bin Salman's efforts to reshape Wahhabi Islam and liberalize Saudi society will have given his enemies a clear avenue to attack his legitimacy. In difficult circumstances, the entertainment-based support of the crowd may prove less reliable than the interest-based support of the old, established elites.

Part V

Meeting New Challenges

For decades, the Saudi throne stood firmly on four legs: respect for the Al Saud's historic role in creating the nation; their ability to repeatedly manage peaceful successions; their preservation of elite cohesion, through a careful balancing of specific stakeholders' interests; and their provision of competent government, including extensive welfare benefits, to the citizenry at large. In all of these areas the monarchy has a noteworthy record of success. The future of the House of Saud, however, will depend not on past achievements but on the ability of King Salman and his son to maintain stability under a changing set of circumstances.

As they seek to navigate a new course, one that will inevitably alienate those Saudis whose values and institutions are threatened by change, a relatively short list of interwoven issues will most likely determine their success. This section identifies these most pressing economic, political, and security problems facing Saudi Arabia, examines how its leaders are responding, and suggests a set of indicators to measure their success in preserving the foundations of Saudi stability.

Economic Challenges: Kicking the Oil Habit

Oil has been both a blessing and a curse for the Saudi people. It has funded the dramatic economic development that long underpinned the kingdom's political stability, but at the same time, it has sorely aggravated many of the kingdom's most pressing problems. Saudi Arabia remains excessively dependent on oil to sustain current economic activity, fund investment and drive future growth. At the same time, oil prices, and thus government revenue, remain notoriously unstable and unpredictable. Between 2007 and 2011 oil prices doubled, halved, and doubled again in four years, greatly complicating any rational economic planning. Moreover, the highly capital-intensive oil industry creates few Saudi jobs even as it distorts the non-oil labor market.

On average, oil revenue has accounted for 80 percent of the government's budget and 40 percent of Saudi Arabia's GDP, though in fact much of non-oil GDP remains dependent on the government's spending of oil revenue. During the early years of the twenty-first century, oil producers around the world, including Saudi Arabia, enjoyed a boom that had more to do with growing Chinese demand than OPEC production quotas. Prices rose from $12 a barrel in 2000 to $140 a barrel in 2008. Then, the market turned, sending oil tumbling to $32 a barrel by early 2016.[1] The Saudi budget became a gusher of red ink as falling oil revenue could barely cover the country's civil service salaries.[2] Many long recognized concerns now became pressing, bringing immediate problems for the new King Salman.

The underlying, long-term cause of falling oil prices has been the rapid growth of non-OPEC energy production. Largely due to the development of shale oil, US production doubled between 2008 and 2018 as new fracking techniques, horizontal drilling, and 3D seismic technology added billions of barrels to global reserves.[3] These game-changing developments herald a new world order for energy. Shale oil is not going to put Saudi Aramco out of business or reduce Saudi Arabia's importance as the only producer with abundant spare capacity, but it will most likely cap oil prices at well below $140 a barrel for many years to come.

Commentary about the "break even" oil price for the Saudi budget is largely meaningless because revenue depends on the volume of oil exported as much as its price. Here, too, Saudi Arabia's oil habit has left the country dangerously exposed. OPEC quotas do sometimes reduce Saudi oil export volumes, but it is growing domestic consumption encouraged by very low local prices that has emerged as the principal long-term constraint to Saudi export volume. Take for example refined diesel fuel, a mainstay of Saudi truckers and farmers. At times, it has been sold domestically at well below the global price of unrefined crude oil. Saudi Arabia also uses crude oil to produce electricity. For many years the average Saudi household paid just 2.5 cents per kilowatt hour for electricity—as compared with 13 cents per kilowatt hour in the United States and 45 cents in Germany. At these prices, it was no surprise that Saudi electricity consumption increased by a third between 2010 and 2015.[4]

This has been the pattern for water prices as well. It costs roughly $1 to produce a cubic meter of desalinated water in Saudi Arabia and another $2–3 to transport it from the coast to the interior, a move that requires pumping it across hundreds of miles and up hundreds of feet in elevation. Most households have been able to buy this $3 water for 3 cents a cubic meter.[5] Desalinated water, most of it produced by boiling sea water with crude oil, has been so cheap that desert-dwelling Saudis have become used to consuming nearly as much water per capita as water-rich Canadians and three times as much as the average European.

Saudi industry has also been an inefficient consumer of energy. In 2016 non-oil industry provided roughly 15 percent of Saudi GDP but consumed 40 percent of the country's heavily subsidized energy production. As a result, Saudi Arabia used more energy to produce a dollar's worth of GDP than any other G20 nation—and that ratio had been moving in the wrong direction. Between 1986 and 2010 the energy needed to produce a dollar of Saudi GDP rose by 3.8 percent per year, while in the United States it was falling by 0.8 percent annually.[6]

As a result of generous subsidies for gasoline, as well as oil-generated water and electricity, by 2015, Saudis consumed roughly 20 percent of their own oil production. That had not been the case in 1973—or even in 2000. Petroleum Minister, Khalid al-Falih (in office 2016–2019), warned that if current growth rates for domestic oil consumption continued, Saudi Arabia would be burning three quarters of its production by 2030. Then, domestic oil consumption could crowd out oil exports—just as happened in the United States during the 1960s.

Falling oil prices and rising domestic demand are not the only threats to Saudi oil revenue. The country does possess vast reserves of crude oil and can maintain current production levels for many years to come, but there is nothing magical about Aramco's oil wells. Oilfields deplete in Saudi Arabia just like they do in West Texas or the North Sea. Developments on the horizon—such as renewable energy, electric cars, and environmental legislation—could all reduce demand for Saudi oil long before its wells run dry. As former Petroleum Minister, Ahmed Zaki Yamani (in office 1962–1986), famously observed, "The Stone Age did not end because we ran out of stones."

In addition to making Saudi Arabia dangerously overdependent on a single, unstable source of income, oil revenue has badly distorted its labor market. Before the 1973 oil boom, the kingdom's labor market operated much like that in a normal country. Most work was done by Saudis working in the private sector, and many women were gainfully employed. Oil revenue changed all that and transformed most Saudis from productive citizens into government employees. By 2015, that arrangement was becoming untenable.

King Abdulaziz oversaw the administration of his kingdom with only a handful of permanent civil servants. Since government employment was not an option at that time, Saudis worked in the private sector and were willing to travel long distances— often on foot—to find work. People in Taif, just south-east of Mecca, remember how their grandfathers walked to Palestine to pick oranges during the British Mandate. Those from Wadi Dawasir in the southern Nejd left their villages each year to work as divers and pullers in the pearling fleets of Kuwait and Bahrain. In the provinces of Al Qasim and Asir, many people have relatives who walked to find work in the new oilfields on the other side of the country.

All of this changed in the 1970s when King Faisal sought to distribute swelling oil revenue through new social services and state-sponsored employment. By and large, his plan worked. Today, Saudi Arabia has modern infrastructure, many social services, and a great many unproductive bureaucrats. Saudis have come to prefer government employment—which offers better pay, shorter hours, and greater job security—than working in the private sector. Some also consider working for the state more respectable because this is simply claiming your fair share of the oil patrimony, while working for a fellow citizen can be demeaning.

As Saudis left private employment for public-sector jobs, the government encouraged the arrival of foreign workers on a scale that very few other countries would tolerate. To reduce the wages needed to attract foreign workers, the government subsidized their utilities, food, and fuel. It initially paid for their healthcare as well. Like Saudis, these foreign workers paid no income tax. Trade unions were prohibited and a new sponsorship system legally bound foreign workers to a sponsor, thus further reducing job mobility and wage demands. Expatriates, who accounted for only 10 percent of the kingdom's population in 1970, comprised nearly a third of the total population by 2015 and filled more than 85 percent of non-oil private sector jobs.[7]

Oil-funded economic growth led to social change as well. In 1960, 70 percent of Saudis lived in rural areas where women participated in traditional economic activities such as farming and animal husbandry. Performing economically essential activities provided them with social status. In contrast, by 2015, 85 percent of Saudis were living in cities.[8] Farming had largely disappeared, and what little remained was done by foreign workers—even the *Bedu* hired immigrant Sudanese to herd their camels. Housekeeping was left almost entirely to Indonesian and Filipino maids. Meanwhile, as their traditional sources of employment dried up, Saudi women were barred by law and custom from pursuing new opportunities in the modern economy.

Many important stakeholders in the Al Saud's historic coalition have strong vested interests in maintaining these labor market distortions. Government employment has played a vital role in the kingdom's welfare system and is regarded by many Saudis as an entitlement. Saudi businessmen have become dependent on a steady flow of semi-skilled, well-disciplined, and inexpensive foreign workers. The religious and judicial communities consider it their moral duty to keep women out of the workplace. Yet by 2015, the direction of necessary change was clear. Saudi workers needed to return to the private sector, many foreign workers needed to go home, and Saudi women needed to rejoin the workforce.

King Salman inherited an economy crippled by its dependence on oil revenue, but as oil prices fell his government had few policy tools with which to respond. For example, since 1986 the Saudi riyal had remained pegged to the US dollar at an exchange rate of 3.75 riyals to the dollar. As a result, Saudi interest rates are effectively set by the Federal Reserve in Washington. The use of tax policy in a country with no personal taxes was also problematic. The expansionary impact of government spending was limited in an economy where most goods were imported rather than manufactured locally, and most private sector wages were sent home by foreign workers. At the same time, the government was providing a level of economic benefits and subsidies unique in the G20 group of large, developed economies. Because of large entitlement programs and falling oil revenue, the Saudi economy was listing badly. As Crown Prince Mohammed bin Salman stated, it was time for Saudi Arabia to kick its "addiction to oil"[9]—but this will certainly prove easier said than done.

King Salman's solution was Vision 2030, an ambitious plan to turn Saudi Arabia into the world's fifteenth-largest economy by the end of the decade. The Vision lays out dozens of social, cultural, and economic objectives for everything from increasing life expectancy and personal-savings rates to entertainment and the amount of time that Saudis spend in the gym. The core of the project, however, focuses sharply on balancing the government's budget, diversifying the economy, and creating jobs.

Of these priorities, balancing the budget has been the most pressing. In 2015, the government's budget deficit reached an unsustainable 16 percent of GDP.[10] For comparison, the US and UK budget deficits in the same year were 4.5 percent and 4 percent of GDP, respectively.[11] Balancing the Saudi budget required cutting expenditure, increasing revenue, and diversifying the sources of that revenue. Progress was made in all three areas.

Capital spending was slashed and payments to government contractors delayed. Subsidies for gasoline, electricity, and water—which, together, consumed 10 percent of GDP in 2015—were cut back.[12] In 2016, gasoline prices rose by 50 percent and electricity bills doubled. The price of ethane—a petrochemical feedstock—increased by 133 percent, and industrial firms were warned that they would have to pay the full cost of fuel and water within a few years. As a result of these changes, years of growth in domestic energy use began to slow. In addition, fees for visas and fines for traffic infringements rose sharply; an airport tax was introduced, as were levies on vacant land, luxury goods, soft drinks, and tobacco. In January 2018, a 5 percent VAT tax went into effect. These significant fiscal reforms were commended by both the International Monetary Fund and the World Bank. Together with rising oil prices, these reforms helped the Saudi budget deficit to fall to 7 percent of GDP in 2018 and, until the 2020 oil price collapse, a budget surplus was projected for 2023.

Similar subsidy cuts or tax hikes have triggered violent protests in many other Arab countries. Saudi Arabia avoided that outcome in part because a significant effort was made to shield the most vulnerable members of its society. A specialized agency known as the Citizens Fund was established to compensate the poor. Starting in June 2018, families of six making less than $5,500 a month received a substantial offset payment to help cushion the blow from the fiscal reforms. For those earning less than $2,000 a month, the compensation payment actually exceeded the expected loss from gasoline and electricity price hikes. Thus, wealthy and middle-class households bore the full cost of rising energy prices, while the poor were largely shielded from them.

As part of its strategy for diversifying government revenue away from oil, Vision 2030 calls for a significant expansion in public borrowing. From a negligible 1.6 percent of GDP in 2014, national debt is projected to rise to a still very healthy 50 percent by 2030. In October 2016, Saudi Arabia floated its first international bond issue in many years, quickly raising $17.5 billion. Oversubscribed by 300 percent, this was the largest emerging-market bond issue in history.[13] In a break from past practice, Saudi Arabia sold bonds in American, European, and Asian markets rather than to local banks and government institutions. Following on from the successful October 2016 round, international bond issues have continued and have all been heavily oversubscribed. By 2019, the kingdom was approaching a debt-to-GDP ratio of 30 percent. Again, for comparison, the gross government debt ratios for the US and UK in 2018 were 107 and 86 percent respectively.[14]

Perhaps most importantly, Saudi Arabia's Public Investment Fund (PIF) has been restructured as a sovereign wealth fund intended to produce a substantial and predictable source of government revenue unrelated to oil exports. Founded in 1971, the PIF functioned for many years as a domestic development bank, investing almost exclusively in local projects. Although it held the government's share in many largely state-owned, enterprises such as SABIC, it had very few income-generating foreign assets. Barely two months after King Salman came to power, oversight of the PIF was transferred from the Ministry of Finance to the new Council of Economic and Development Affairs (CEDA) chaired by Mohammed bin Salman. The Vision 2030 goal is to grow the PIF into a $2 trillion vehicle that, like Norway's sovereign wealth fund, would invest abroad, generate income, and supplement the budget when oil prices fall.

Funding for these foreign investments is expected to come from selling a small percentage of the state-owned Saudi Aramco to the public, and selling the PIF's share of SABIC to Saudi Aramco. In 2019, the Saudi government raised $25.6 billion by selling 1.5 percent of Saudi Aramco to local investors. This was the largest IPO in history. It was oversubscribed in part because the government encouraged wealthy Saudis and government pension funds to purchase shares. The SABIC sale is expected to be finalized in 2020..

Improved government finances are only one part of Vision 2030. The program also seeks to diversify the economy beyond crude oil production, refining and petrochemicals into other sectors where Saudi Arabia has some form of comparative advantage—specifically, mining and tourism. The country holds 7 percent of the world's phosphate reserves as well as significant deposits of bauxite, gold, copper, and zinc. Yet in 2015, mining contributed only 4 percent of Saudi GDP. Vision 2030 sets out to double that number and create 90,000 jobs in the mining sector.

A new industrial city at Ras al-Khair has become for mining what Jubail and Yanbu are for petrochemicals (see Chapter 13). Located north of Jubail on Saudi Arabia's East Coast, Ras al-Khair is the terminus of a new 850-mile rail line that carries bauxite and phosphate rock from recently opened mines in the north. Bauxite is now mined, refined, smelted, and rolled into aluminum sheets locally, exploiting

Saudi Arabia's still relatively low energy costs. With phosphate rock from its mines, ammonia from its petrochemical plants, and sulphur from its refineries, Saudi Arabia is also well on the way to becoming one of the world's largest—and lowest-cost—producers of fertilizer. By 2019, Saudi Arabia's Ma'aden was the tenth-largest mining company in the world. A mining institute has opened to train Saudi miners, who provide 80 percent of Ma'aden's labor force—including 50 percent of those working underground—albeit at twice the wages of their foreign co-workers.[15] In a 2019 government restructuring indicative of mining's new importance, mineral-resource development was spun off from the old Ministry of Petroleum to form the new Ministry of Minerals and Industry.

Religious tourism—once a mainstay of the Saudi economy—is another area of comparative advantage for the kingdom. There is only one Mecca, and the world's 1.5 billion Muslims are all enjoined to visit the city at least once in their lifetime. Vision 2030 sets out to capitalize on this natural monopoly, and to restore religious tourism to the economic prominence that it enjoyed before the discovery of oil. The goal is to increase the total number of pilgrims coming to Mecca each year from 8.5 million in 2015 to 30 million by 2030. By 2018, the number of pilgrims visiting Mecca had already risen to more than 15 million, benefiting local hotels, restaurants, retailers, and even charter airlines.[16] Unfortunately, Mecca has historically been a major transmission point for pandemics and this promising start was badly disrupted when King Salman shut down the pilgrimage in response to the 2020 COVID-19 outbreak.

One of Vision 2030's most ambitious goals is to reduce overall unemployment from 12 percent in 2015 to 10.5 percent by 2022 by creating private-sector jobs. One partial solution lies in the country's retail sector. In 2015 the Saudi retail sector provided 1.5 million jobs, but 80 percent of them were filled by expatriates. The government intends to close small corner shops—which are dominated by foreign workers—and encourage the growth of large retail stores, to be staffed primarily by Saudis. In a number of areas, such as women's clothing and electronics, this change is well under way. In more than a dozen other retail sectors—from car dealerships to opticians—new regulations requiring 70 percent of staff to be Saudi are being enforced.

Saudi Arabia's female labor force participation rate, which is one of the lowest in the world, is scheduled to increase to 30 percent by 2030. This is a realistic goal considering that the female labor force participation rate in both the United States and the United Kingdom is nearly 60 percent. To meet this objective, the Saudi civil service and many large corporations have established aggressive affirmative action programs for women. Specific programs have also been initiated to provide childcare and transportation for women who wish to go out to work.

As a result of these initiatives, the Saudi labor market is slowly changing. Male unemployment—which was 7.5 percent in 2018—had fallen to 6 percent a year later; youth unemployment declined from 42 percent to 30 percent in the same period; a rate lower than Greece or Spain and not much higher than Sweden. Female unemployment remains at 32 percent, but as more women have moved into the workforce, the female labor force participation rate has edged up from 20 to 23 percent. Meanwhile, the proportion of private-sector jobs taken by Saudis rose from 20 to 22 percent[17]—in part, because more than 2 million expatriate workers left Saudi Arabia between 2017 and

2020. More will certainly follow as steep annual increases in the levies placed on foreign workers and their dependents continues to rise.[18]

Nevertheless, as the rating agency, Fitch, warned in March 2017, Vision 2030 faces a number of significant problems. Specifically, the report noted:

> The scale of the reform agenda risks overwhelming the government's administrative capacity. In addition, the economy may not be able to absorb rises in administered energy prices, which could severely affect energy-intensive industries and the planned expat levies could undermine large parts of the domestic private sector.[19]

These predictions have, in fact, all materialized to some extent, and in 2020 the list of obstacles facing Vision 2030 remains long. The departure of so many foreign workers has decreased aggregate demand, raised labor costs, and forced many small businesses into bankruptcy. Real estate prices and retail sales have fallen sharply. Few truly new jobs have been created and Saudis are not willing to fill every job vacated by a departing expatriate. Most new jobs are in the service sector where much needed productivity growth is historically low. Despite government support for small businesses, the kingdom's banking and legal systems still do not encourage Saudis to become entrepreneurs. In addition, rising energy costs have reduced Saudi competitiveness and the privatization of government assets, which is essential to creating more private-sector jobs, has barely begun. Most worrying is the lack of local human capital.

At the professional level, Saudi Arabia has made noteworthy progress in developing its human resources. In 1980, most international law firms in Riyadh had one Saudi partner who was used for marketing, while the legal work was done by Western expatriates. Forty years later, those ratios have completely reversed. In many firms you will find one Western partner, who helps with marketing, while the core work is now done by Western-trained Saudis. In 1980, 10 percent of Aramco engineers were Saudis. By 2018, that number has risen to 90 percent.[20] Likewise, most Saudi banks are now managed by local employees.

Below the professional level, however, the story is very different. Technical and vocational training has been weak and unpopular. English language skills are often poor. Even Saudis who have the skills needed to work in the private sector often lack the motivation or discipline to do so. Employment regulations have added to the problem. Only in 2015 was Article 77 of the Saudi Labor Code changed to make firing a Saudi employee a realistic option. As a result, even in the kingdom's modern petrochemical sector productivity is low by global standards.

Still, there are some reasons to believe that Vision 2030 will prove more successful than previous Saudi development plans. Unlike many of his predecessors, King Salman inherited a realm with a modern infrastructure, full treasury, and well-regulated banking system. Moreover, the Saudi population was no longer exploding. The annual population growth rate, which had reached 6.4 percent in 1982, had fallen to 2.0 percent by 2017.[21]

At the same time, the political structure of the Saudi government has changed. The consensus-based oligarchy that had existed for decades was by its nature less capable of making difficult policy choices than the more centralized and autocratic rule of King Salman. The old partnership of brothers is gone, and with it the once-independent, uncoordinated ministerial fiefdoms controlled by powerful sons of King Abdulaziz. King Salman is now the only second-generation prince left in the Council of Ministers and since 2015 he has replaced every other member. All of the new ministers, including the royals, are the king's men. No Saudi ruler since King Faisal has exercised such firm authority in the Council of Ministers.

Unlike the previous nine National Five Year Plans, Vision 2030 has high-level political support. It was not, as has happened previously, issued by the Minister of Planning and then promptly forgotten by everyone else. Instead, it is the personal project of an undeniably powerful crown prince who has linked his political future to its success. Mohammed bin Salman has told all of the consultants and advisors around him, "Time is our enemy. We can no longer wait to reform our country."[22]

A good example of this new-found vigor was the introduction of the 5 percent value added tax in January 2018. Many consultants, international financial institutions, and local accountants advised that such a tax had never before been implemented so quickly in a large economy. They all counseled that a 2018 launch was unrealistic. When the Crown Prince made it clear that delay was not an option, the consultants worked overtime to create a system of attractive incentives and stiff penalties to encourage compliance. Accountants were offered training programs, small businesses were given an extension, and for most of the economy the tax was successfully implemented as scheduled.

Understanding that the Crown Prince is watching and expects meaningful results quickly, his advisors' proposals have become more focused and measurable. Now, all large civilian projects must be approved by CEDA, which Mohammed bin Salman chairs. Projects are approved only if they explicitly support Vision 2030 objectives, and they will receive additional funding only when set targets are met.

The administrative structure of the Saudi government has also changed. As we saw in Chapter 10, a dozen Vision Realization Programs (VRPs) have been created to promote intra-ministerial coordination. A powerful Strategic Management Committee, composed of ministers and chaired by the Crown Prince, overseas the VRPs. A Strategic Management Office reviews the strategies of various ministries, supervises the VRPs, and acts as the executive arm of the Strategic Management Committee. Nothing like this degree of structured, ongoing, high-level political engagement in the planning process has ever happened before in Saudi Arabia.

Vision 2030 is also much more comprehensive than previous plans. It has been expanded to include a National Transformation Plan, as well as specific programs for fiscal balancing, government restructuring, project management, regulation review and performance measurement. These programs were created and staffed by legions of consultants from firms such as Bain, Boston Consulting Group, Booze Allen, Deloitte, McKinsey, Oliver Wyman, Price Waterhouse, and many others. All of these programs now have detailed objectives and dozens of key performance indicators at the national, ministerial, and sub-ministerial levels—all of which are now closely monitored on a quarterly basis by a central office. Clearly the consultants have been busy designing organization charts, but their efforts have begun to correct the uncoordinated, undisciplined, chaos that long characterized the Saudi bureaucracy. At least at the top, the Saudi government has become more focused and coordinated.

Mohammed bin Salman's plan should probably have been called Vision 2050. That would have represented a more realistic, if a less-motivating, time frame for transforming the Saudi economy from one based almost entirely on oil-funded transfer payments into one based partially on productive enterprise. Even by 2050, the existing plan will not turn Saudi Arabia into South Korea. Then again, it does not need to. Like Canada and Australia, Saudi Arabia will rely heavily on natural resource exports for many years to come, but its economy can become more diversified. Nor will Vision 2030 dismantle the Saudi welfare system, but it can make it more sustainable. Finally, like everything else in Saudi Arabia, Vision 2030 will be affected by oil prices. With $100-a-barrel oil, the entire plan will look inspired; at $40 a barrel parts of it will appear ill-conceived and at $20 a barrel it will largely need to be rewritten.

In 2020, the jury is still out. Most Saudis want Mohammed bin Salman and his reform plan to succeed, but they worry about more subsidy cuts, fewer public-sector jobs, and destabilizing social change. There is uncertainty. Should I buy land, or will real estate prices fall further? Will my son find a good government job like I did when he returns with a degree from California? Will I need to get him private health insurance? Will my brother's contracting business pick up? No one knows the answer to such questions. Hope that Vision 2030 will succeed and fear that it may not are among the most prevalent emotions expressed by Saudis today.

Those wanting to understand Saudi Arabia will need to assess its economic reforms long before 2030, and they will need a set of publicly available benchmarks. Here are a few to watch.

Oil exports. Will growing domestic oil consumption crowd out oil exports as former Petroleum Minister, Khalid al-Falih, warned—or will additional subsidy cuts continue to reduce domestic gasoline, water, and electricity consumption?

Government revenue. Will the domestically oriented PIF become the world's largest sovereign wealth fund, help balance the budget by 2023, and increase non-oil government revenue to 30 percent of the budget by 2030?

Regulatory reform. Will the Saudi legal system become more transparent and predictable, moving towards global standards for accounting and bankruptcy law while consistently enforcing foreign judgments and arbitration claims?

Private-sector expansion. Will the privatizations of state-owned enterprises—including airlines, railways, grain silos, and banks—materialize, thus helping to expand the private sector from 40 to 65 percent of GDP?

Human resources. Will foreign workers continue to leave? Can Saudi labor force participation and productivity improve along with student scores on standardized international examinations? Will public-sector employment and youth unemployment continue to decline?

Not all these indicators need to be satisfied for success but, looked at collectively, they will provide a sound understanding of whether or not the Al Saud can continue to provide the politically-stabilizing prosperity that their people have come to expect.

Figure 27. Downtown Riyadh in 2019, a city of over seven million people and the largest in the Arabian Peninsula.

Figure 28. Muslim pilgrims circle the Kaaba in Mecca's Great Mosque, the spiritual focal point for a quarter of the world's population.

Figure 29. Traditional architecture in old Jeddah which was much more sophisticated than anything found in Riyadh in 1924. See Figure 3.

Figure 30. Ancient tombs at Mada'in Saleh, a 2,000 year old Nabatean city near Medina and Saudi Arabia's first UNESCO World Heritage Site.

Figure 31. The Ottoman Hejaz Railway ran from Damascus to Medina and was the primary target of T.E Lawrence's exploits. Its stations (seen here) were stone blockhouses. The author is on the right in this 1984 photograph.

Figure 32. A rural village near the great oasis of Najran in southern Saudi Arabia. While most Saudis have moved to large cities, twenty percent still live in villages.

Figure 33. A modern-day camel race outside of Riyadh. No longer a necessity, camel breeding is now a hobby for many urban Saudis, but the prizes to be won are large.

Figure 34. The Arabian Oryx was hunted nearly to extinction, but is again thriving, having been reintroduced and protected by the Saudi Wildlife Commission.

Security Challenges: Containing Iran and Managing Yemen

No Saudi king ever came to power facing greater regional instability than Salman bin Abdulaziz. In January 2015, the very existence of Iraq, Lebanon, Libya, Syria, and Yemen was in question. The Islamic State, or ISIS, had become the first terrorist organization with its own capital city and oil production. Iran was supporting the Houthi insurgents in Yemen, who had just taken the capital, Sanaa, and were on the verge of capturing the entire country. Not since the Mongol invasions of the thirteenth century had the Arab world seen such widespread chaos—and all of it threatened Saudi security.

The instability confronting King Salman was rooted in the 2003 Anglo-American invasion of Iraq and the 2011 Arab Spring. As many Saudis had feared, the invasion destroyed Iraq's Sunni-dominated political structures. Almost immediately, resentful Sunnis and disempowered Ba'ath Party members began to resist the new Shia government and its Western allies. Their movement's early leader, Abu Musab al-Zarqawi, was killed by a US airstrike in 2006. The following year, a surge of 20,000 American troops nearly eliminated what remained of Sunni armed resistance. Regrettably, the persistent anti-Sunni policies of Nouri al-Maliki's government in Baghdad (2006–2014) ultimately reignited Sunni violence, just as the departure of American forces in 2013 weakened Iraq's central government.

The Arab Spring was a series of anti-government protests and armed uprisings that began in Tunisia and rapidly spread to Bahrain, Egypt, Jordan, Libya, Morocco, Syria, and Yemen. Following the overthrow and death of its head of state, Muammar Qaddafi, Libya ceased to exist as a coherent state. In Cairo, Saudi Arabia's long-standing ally, Egyptian President Hosni Mubarak was replaced by a decidedly anti-monarchist Muslim Brotherhood government. The collapse of central authority in Syria allowed ISIS to establish sanctuaries from which its self-proclaimed Caliph Ibrahim led an assault on Iraq. In 2014, ISIS captured Mosul and nearly reached Baghdad. In the kingdom's most populous neighbor, Yemen, the Saudi-supported thirty-year rule of Ali Abdullah Saleh ended in civil war.

The Sunni Arab Saudis viewed these security problems through a lens of deep, persistent mistrust of Shia, Persian Iran. To the Saudis, the Iranians were an arrogant people, overly boastful of their civilization, whose religious practices were a heresy. Long before revolution overthrew its monarchy in 1979, Tehran had frequently challenged Riyadh's preeminent role in OPEC. After the revolution, Iranian broadcasts to the Arab world regularly challenged Saudi Arabia's control of Mecca and the legitimacy of the Saudi monarchy.

These tensions stemmed in part from very different histories. Iran has often felt politically beleaguered and economically abused by Western powers. Its people remain deeply resentful of nineteenth- and twentieth-century British, Russian, and American involvement in their domestic affairs—such as the bitterly contested nationalization of the Anglo Persian Oil Company in 1951, which is still celebrated as a national holiday in Iran, and Operation Ajax, the 1953 Anglo-American covert operation supporting a military coup against Iran's nationalist prime minister, Mohammed Mosaddegh. For Iranians, the Iran–Iraq War (1980–1989) was a bloody Iraqi invasion funded by the Gulf Arabs and supported by the United States. Consequently, from Iran's historical perspective, driving Western influence from the Middle East and confronting America's regional allies, Saudi Arabia and Israel, are necessary, defensive measures. More recently, exporting its revolutionary values also became a legitimizing exercise for the Islamic Republic.

The Saudis, on the other hand, had a largely cooperative relationship with American oilmen and have been closely aligned with the United States for seventy years. Staunch supporters of the status quo, the Al Saud have always opposed Iran's efforts to export its Islamic revolution. From Riyadh's perspective, the 2003 invasion of Iraq unleashed Iranian expansion and the 2011 Arab Spring accelerated the process. In Iraq, Lebanon, Syria, and Yemen, the Saudis nervously watched Tehran's influence grow as it funded pro-Iranian governments and armed Shia militia. Riyadh felt increasingly beleaguered, and by 2015 most Saudis regarded it as self-evident that Iran's long-term ambitions were to acquire nuclear weapons; become the region's dominant military power; take Saudi market share in global oil markets; challenge Saudi Arabia's role in the Muslim world; and, if possible, topple the House of Saud. These were not unfounded fears.

In early 2011, most of the world was focused on the fall of President Hosni Mubarak in Egypt rather than the more historic events unfolding elsewhere. Mubarak was one of dozens of pharaohs who have come and gone on the banks of the Nile. His successor, Mohammed Morsi, lasted little more than a year. The truly historic event that spring was the rise to power of Prime Minister Najib Mikati in Beirut, with the backing of the Iranian-supported Lebanese Shia political party and militia, Hezbollah.[1] In 2011, Iran became a major player in Lebanese politics and, following the lead of their patrons in Tehran, Hezbollah's media outlets became relentless in their attacks on the House of Saud. It was not lost on the Saudis—or, for that matter, the Israelis—that this was the first time in 1,000 years that a Persian-based regime had allies and significant influence on the shores of the Mediterranean.

Neighboring Syria is a predominantly Sunni nation ruled by a regime drawn largely from the semi-Shia Alawite sect. Iran and its Hezbollah allies have long supported the Assad regime in Damascus for both religious and geopolitical reasons. Syria physically connects Iran's allies in Baghdad and Beirut while also providing another front from which to challenge Israel. During the Syrian Civil War, Hezbollah sent thousands of fighters to support the Assad government, and often occupied areas in which the latter had lost control. On the Syrian side of the Golan Heights, Israeli drone strikes killed not only Hezbollah fighters but also Iranian Revolutionary Guard General

Mohammed Ali Allahdadi who was there assisting them fully five years before a similar American strike in Iraq eliminated the Revolutionary Guard's commander Qassim Suleimani. Having watched Iraq leave the region's Sunni camp, Riyadh hoped to help Syria rejoin it.

In 2014, Iranian trained and armed Shia militia known collectively as the Popular Mobilization Forces *(al-hashd as-sha'abi)* helped to stop the ISIS assault on Baghdad. These same militia helped to drive ISIS from Iraq's Sunni cities of Tikrit in 2015, Fallujah in 2016, and Mosul in 2017. Although American-enabled Iraqi armed forces and US Special Forces units fought the decisive battles against ISIS, Iran's Revolutionary Guard commander, General Qassim Suleimani, nevertheless became a more prominent figure in Baghdad than any Western general at the time. Referring to Saddam Hussein's hometown, Saudi Foreign Minister, Saud al-Faisal, complained in 2015, "The situation in Tikrit is a prime example of what we are worried about. Iran is taking over the country."[2] A year later, the new Saudi Foreign Minister, Adel al-Jubeir, noted, "Sectarian division in Iraq grew from Iran's divisive policies. How can there be peace in Iraq with Qassim Suleimani now leading a campaign against the Sunnis?"[3]

In addition to this insecurity on their northern border, the Saudis were facing instability to the south in Yemen, an impoverished country with a local population larger than Saudi Arabia's. Every day, Saudi border police detain and immediately deport hundreds of economic migrants attempting to cross the kingdom's long, porous border with Yemen. For the United States, it would be as if Mexico had a population of 400 million with an annual per-capita income of $1,500 instead of 125 million people with annual incomes of nearly $10,000.[4] This large, poor, often ungovernable country is Saudi Arabia's soft underbelly. Egypt's President Nasser tried to destabilize Saudi Arabia through Yemen in 1962, when revolution overthrew Imam Mohammed al-Badr. Iran attempted to do the same after the 2011 Arab Spring overthrew President Ali Abdullah Saleh. Well aware of the security problems that Iranian-supported Hezbollah militia created for Israel on its northern border, the Saudis were determined to prevent Iranian-supported Houthi militia from creating similar problems on their southern border.

As the new King Salman faced this worrying array of security threats, he found all of Saudi Arabia's traditional foreign policy tools in poor repair. Falling oil prices and large budget deficits placed limits on Saudi checkbook diplomacy. Iran was challenging Saudi Arabia's role in the Muslim world, including its right to govern Mecca and administer the Hajj. Washington seemed more interested in improving ties with Tehran than in maintaining them with Riyadh. Thus, King Salman and Mohammed bin Salman set out to first strengthen the instruments of Saudi foreign policy and then deploy them more assertively. The results have been mixed and controversial.

Concentrating authority was King Salman's first priority in all areas, including security matters. The largely ineffective Saudi National Security Council, consisting of coequal ministers who met infrequently, was replaced by the Committee for Political and Security Affairs, which, under Mohammad bin Salman, met weekly. Foreign Minister, Saud al-Faisal—who, as the son of a king with many years of foreign policy experience, had exercised considerable authority—was replaced by a technocrat with much less autonomy. The semi-independence of the National Guard, Ministry of

Defense, and Ministry of the Interior leadership was ended. Decision-making became faster and more decisive, but also less consensual.

King Salman's next order of business was securing the financial resources needed to maintain domestic stability and finance Saudi foreign policy. King Abdullah and his oil minister, Ali al-Naimi (in office 1995–2016), had focused on maintaining market share rather than high prices. In February 2016, with the benchmark crude West Texas Intermediate (WTI) falling to $27 a barrel, King Salman and his new oil minister, Khalid al-Faleh (in office 2016–2019), adopted a different approach. By forging a strategic energy alliance with Russia, al-Faleh engineered a 2 percent reduction in global oil production. These were the first OPEC production cuts in eight years, and they were successful. Global inventories returned to normal levels and WTI had risen to $70 a barrel by September 2018.

Coordinating effective OPEC cooperation, enlisting Russian support, and maintaining historically high Saudi production levels—all while North American shale production surged—was a noteworthy Saudi achievement. The increased oil revenue that it produced paid for an expensive war in Yemen; billions of dollars-worth of Western weapons; and substantial financial assistance to moderate, Western-aligned regimes in Bahrain, Egypt, Jordan, Morocco and Pakistan. In July 2019, when OPEC and Russia formalized two and a half years of combined effort with a Charter of Cooperation, it was largely because of the work of Russian President Vladimir Putin and Saudi Crown Prince Mohammed bin Salman.

Oil market cooperation between Saudi Arabia and Russia collapsed in the spring of 2020. Although demand was falling sharply because of the COVID-19 virus pandemic, Russia would not cooperate with OPEC plans to further restrain production. Saudi Arabia responded with a significant production increase that sent oil prices tumbling. This response was not primarily an effort to slow the growth of North American shale oil production or take market share from Russia, although these were both welcome secondary outcomes. The principal motivation behind the Saudi production increase was the very painful lesson learned between 1982 and 1986.

Then, in an effort to maintain prices in the face of rising North Sea production, Saudi Arabia had cut its production by more than six million barrels per day. Although most OPEC members continued to produce above their quotas, Aramco's monthly revenues fell from $8 billion in 1982 to $700 million in July 1986. Adjusted for inflation, Saudi GDP fell by 25 percent. The Saudis determined never again to cut production alone and this has been their guiding principle ever since. Former oil minister Ali al-Naimi significantly increased production in 2014 when OPEC would not support Saudi production cuts. Not surprisingly, the current oil minister Prince Abdulaziz bin Salman followed the same course when Russia would not cut production in early March 2020.

In a nutshell, as a low-cost producer in an oversupplied market, Saudi Arabia normally suffers less from production increases than it would from unilateral production cuts. However, the COVID-19 pandemic created an unprecedented situation in which global oil demand collapsed by 20 percent in one month. By mid April 2020, Saudi Arabia, OPEC, Russia and even the United States were all implementing production cuts to preserved their energy sectors from long-term damage.

Not since King Faisal deployed Islam against Arab nationalism in the 1970s had a Saudi monarch used religion as a foreign policy tool so explicitly as King Salman. In 2015, he formally founded the Islamic Counterterrorism Coalition. Headquartered in Riyadh, this association of thirty-nine Sunni Muslim states began to share intelligence and conduct joint military training exercises. When Saudi Arabia hosted Northern Thunder, a large-scale military exercise near the Iraqi border in 2017, more than twenty Muslim heads of state attended the closing ceremony. Likewise, King Salman strongly supported the pan-Islamic organizations that King Faisal had created—particularly, the Organization of Islamic Cooperation and the Muslim World League, where he installed his own new leadership teams.

Saudi Arabia could not easily replace its security relationship with the United States. In 2015, the US had two fleets and 54,000 troops deployed in the Near East and South Asia, far more than any other outside power. Almost all of Saudi Arabia's major defense platforms—from F-15 fighter aircraft and Apache attack helicopters to Abrams tanks and Patriot air-defense systems—are American. Any shift away from US equipment would take decades and cost billions of dollars. When Donald Trump became president, improving relations with Washington became a Saudi priority. As Saudi Foreign Minister, Adel al-Jubeir, (in office 2015–2018) noted during the Munich Security Conference in February 2017, "Trump believes in destroying ISIS. So do we. He believes in containing Iran. So do we. He believes in working with traditional allies. So do we. We look forward to working with this administration very very closely."[5] President Trump shared the Saudis' desire for improved relations and made Riyadh his first official foreign destination.

In 2008, President Barak Obama had chosen to address the Islamic world from Cairo before an almost entirely Egyptian audience. During the next eight years, the Egyptian government was overthrown twice while the country's economy collapsed before becoming heavily dependent on financial support from Saudi Arabia and the International Monetary Fund. Over that same period, Saudi Arabia, with a GDP nearly three times that of Egypt, remained stable, saw the peaceful transfer of power from one king to another, and initiated significant economic reforms. President Trump's decision to address Muslim heads of state assembled in Riyadh acknowledged that difference. It marked a watershed in the leadership of the Arab, and at least Sunni, Islamic worlds that had been in the making for decades.

The Saudis welcomed the new president with unprecedented pageantry. During his three days in Riyadh, President Trump held three summits—one with King Salman, one with the six members of the Gulf Cooperation Council, and one with leaders from fifty-five Muslim nations who gathered to meet him. By emphasizing important policy alignments with Washington, demonstrating Saudi Arabia's massive purchasing power, and confirming its special place among Muslim nations, King Salman began to repair relations with his kingdom's most powerful friend.

The king and president formally codified Saudi-American relations in a Joint Strategic Vision. Confronting Iran and its proxies across the Middle East was at the top of their agenda. President Trump made his views every clear, stating, "From Lebanon to Iraq to Yemen, Iran funds, arms and trains terrorists, militia and other extremist groups that spread destruction and chaos across the region." He specifically praised

Saudi Arabia for listing Israel's sworn enemy, Hezbollah, as a terrorist organization, and placing economic sanctions on senior Hezbollah leaders. Likewise, he praised Saudi actions against the Houthis in Yemen and committed the United States to work with Saudi Arabia on border security, cybersecurity, and air defense.[6]

Economic cooperation was a second major theme, with Saudi Arabia agreeing to make large investments in US industry and jobs. Billions of dollars-worth of contracts and letters of intent were signed with major industrial firms such as General Electric, Exxon Mobil, Dow Chemical, and Boeing. Aramco signed contracts with American oilfield service firms. Defense companies—including Raytheon, Lockheed Martin, and General Dynamics—initiated further military sales. The Saudi Public Investment Fund committed $20 billion to an infrastructure development fund being launched by the American private equity giant, Blackstone. One observer noted that the value of all the agreements signed was twice the GDP of Lebanon, but much of this was public relations. Some deals had been in the works for years, others were non-binding letters of intent, and still others dealt with potential sales years in the future. Nevertheless, the bottom line was clear. Saudi Arabia had the capacity to make a significant contribution to American economic growth and employment.[7]

King Salman used Saudi Arabia's special role in the Islamic world to court Washington by assembling an unprecedented number of Muslim kings, *amirs*, presidents, and prime ministers to greet the new American president. This too was a noteworthy transition. It had been fifty years since Egypt's President Nasser led the Arabs into a disastrous war with Israel; the ideas that he had used to ignite the Arab world—Arab nationalism; Arab socialism; and, to a large extent, secular liberalism— had all been discredited in the Middle East. Now, it was the Custodian of the Two Holy Mosques—the leader of a conservative, market-oriented, religiously based monarchy, firmly aligned with the West—assembling Muslim leaders to meet with an American president who agreed to work with their new military coalition directed not against Israel, but against ISIS—and, indirectly, Iran.

A year later in April 2018, Crown Prince Mohammed bin Salman made an extended visit to the United States where he met four American presidents, business leaders, media moguls, and entertainment celebrities. He did more than any other Saudi leader had ever done to explain that Saudi Arabia was more than a weird place where mosques promoted intolerance and women were second-class citizens. He emphasized the need for change and the opening up of Saudi society. By and large, the prince and his message were well received.

These efforts to restore Saudi American relations and improve Saudi Arabia's global image were largely undone in October 2018 by the brutal murder of *Washington Post* contributing columnist Jamal Khashoggi. Khashoggi had once been the prominent editor of Saudi Arabia's *Al Watan* newspaper, but after becoming increasingly critical of his government he had gone into self-imposed exile in the United States. When he was killed by Saudi government agents inside the Saudi Consulate in Istanbul, the Saudi government produced one implausible explanation after another while the Turkish media leaked lurid details fueling speculation. Western ambassadors in Riyadh were given ever-changing explanations, that made the Saudis look either incompetent or dissembling.

Most Western press reports focused on the depraved circumstances of Khashoggi's death rather than its political motivation. Consequently, it is worth noting that Jamal Khashoggi was not killed because he was a dissident journalist criticizing the Saudi government. Nor was he killed because of his contacts with the Muslim Brotherhood—avowed ideological opponents of the Saudi monarchy. Many Saudi dissidents have been far more provocative, disruptive, or threatening than Jamal Khashoggi ever was. Whether Saudi operatives intended to kill or kidnap him is not entirely clear, but in either case the mission's justification was accepted by those who carried it out.

Jamal Khashoggi's grandfather, Mohammed Khashoggi, had been the private physician of King Abdulaziz, and as a result of this privileged position the Khashoggi family had grown wealthy. Jamal's uncle, Adnan Khashoggi, used his proximity to the royal court to amass a very large fortune acting as a door opener who took commissions from many companies seeking lucrative contracts—especially in the defense sector. In the eyes of the Al Saud, everything that the Khashoggis had in the world came from them; hence, every member of the Khashoggi family owed the Al Saud their personal loyalty. In Arabic, the expression "I am between your hands" (*Ana bayna yadaikum*) sums up Jamal's position vis-à-vis the Al Saud. In the final analysis, Khashoggi was a trusted retainer who stabbed the family in the back. At a time when his country was fighting a costly war, he went to a foreign capital and denounced its government. Writing critical articles and endorsing the Muslim Brotherhood were elements of his infidelity but not the cause of his death. In the eyes of the Al Saud, Jamal Khashoggi's egregious crime was betrayal or treason.

Such reasoning may justify kidnapping or killing in Riyadh but, like public beheadings, it flies in the face of acceptable Western norms. Although no serious observer believed that the Saudi government had been complicit in the 9/11 attacks, its official involvement in Istanbul was obvious to all. Though no evidence directly linked the Crown Prince to the murder, many found it very difficult to accept that such an organized operation against such a well-known Saudi citizen could have taken place without his knowledge.

In an effort to calm the situation, King Salman spoke with both US President Donald Trump and Turkish President Recep Erdoğan. A great-granddaughter of King Abdulaziz, Princess Reema bint Bandar bin Sultan, was sent to Washington as Saudi Arabia's first female ambassador. Efforts to reform the Saudi intelligence services were announced. Saudi prosecutors charged that the killing was a premeditated murder, not a botched rendition. Seventeen Saudi security officials were arrested, and the death penalty sought and obtained for five of them. The Crown Prince denied any foreknowledge of the operation, but accepted responsibility for an event "that happened on my watch."[8] In spite of these efforts, the murder of Jamal Khashoggi proved to be a foreign policy disaster for Saudi Arabia that has had lasting negative consequences for its relationships with the United States and Britain.

The incident did not have nearly the same impact on Saudi relations with some of its other friends. At the November 2018 G20 Summit in Buenos Aires, Indian Prime Minister, Narendra Modi, made a specific effort to meet privately with Mohammed bin Salman. That was not surprising considering that Saudi Arabia had become India's fourth-largest trading partner and was home to 2.5 million Indian workers who sent home $11 billion in remittances each year. The two had much to discuss, including upcoming petrochemical and refinery investments as well as joint naval exercises. The following year, Prime Minister Modi was the keynote speaker at Saudi Arabia's gala Future Investment Initiative—an event that in 2019 was attended by twelve heads of state, compared with only two the year of Khashoggi's death.

China and Saudi Arabia had opened diplomatic relations in 1990 and raised the Sino–Saudi relationship to the level of "comprehensive strategic partnership" in 2016. Both King Salman and Mohammed bin Salman made official visits to Beijing, where they initiated billions of dollars-worth of joint venture refining and petrochemicals projects. During President Xi Jinping's 2017 visit to Riyadh, the Saudis agreed to participate in China's New Silk Road initiative, specifically agreeing to serve as the Western hub for Chinese exports to the Middle East and Africa.

As China's share of global oil consumption rose from 4 percent in 1993 to 12 percent in 2016, Saudi Arabia sought to make it clear that Saudi Aramco, not the Iranian National Oil Company, was the most dependable supplier for China's energy needs. These Saudi efforts to gain market share were successful, and by 2018 the People's Republic had become the kingdom's largest trading partner as well as the largest customer for both Saudi Aramco and SABIC. By August 2019, Saudi Arabia was China's most important source of imported oil, while Iran was supplying it with less oil than Oman.[9]

As Saudi ties with the West frayed because of a lack of shared values, its ties to Israel grew stronger as a result of common interests. Both nations fear Iran's growing regional influence. Both support Egypt's President, Abdul Fattah Sisi, oppose Syria's Assad regime, and want to find a solution to the Arab–Israeli conflict. When Egypt agreed to transfer the Red Sea islands of Tiran and Sanafir to Saudi Arabia in April 2016, the Saudis agreed to honor the terms of the Israeli–Egyptian Peace Treaty—a document that, in 1979, had caused them to break off diplomatic relations with Cairo.

In June 2016, the Saudi news network, Al Arabiya, strongly condemned the killing of four Israelis by Palestinians in Tel Aviv.[10] Saudi Arabia joined the United States and the European Union in listing parts of the Palestinian resistance movement, Hamas, as a terrorist organization. In 2017, when Saudi Arabia initiated an economic boycott against Qatar, in part because of the Qatari media attacks on the Saudi monarchy, Israel closed the local office of the Qatar-based Al Jazeera television station.

Though not officially sanctioned, Saudi businessmen and academics have begun to openly visit Israel.[11] Israeli businessmen traveling on second passports have started going to Riyadh. Saudi and Israeli speakers have appeared together publicly in Washington to discuss the threat of Iranian expansionism.[12] Reports of counterterrorism intelligence sharing and discreet sales of sophisticated Israeli counterterrorism equipment to Saudi Arabia through third parties became public knowledge. "The message is clear. Israel is willing to sell to these countries security related technology as they forge closer ties with it in its strategic battle against Iran."[13]

In October 2019, the Saudi national football team, which had always played the Palestinian national team in a third country, traveled to Israel to play them in a much publicized game in Ramallah. After the match, the team went to Jerusalem to pray at the Al Aqsa Mosque. In Ramadan 2020 Saudi television broadcast the mini-series Mother of

Aaron which portrayed a Jewish family living in Kuwait in 1948 and had some dialogue in Hebrew. As with the operation against Jamal Khashoggi, that broadcast seems unlikely to have happened without at least the tacit approval of Mohammed bin Salman. If not yet thawed, the ice between Saudi Arabia and Israel is clearly getting thinner.

King Salman recognized that Iranian influence in Baghdad was best constrained not by Saudi-supported Sunni politicians, but by Shia Iraqi nationalists. Saudi Arabia began reaching out to Iraqi Shia leaders—hosting visits to Riyadh by Prime Minister Haider al-Abadi (in office 2014–2018) and even militia-leader-turned-politician, Muqtada al-Sadr. Saudi ministers, business delegations, and the path-breaking Saudi national football team all visited Iraq. Air links and border crossings were reopened as bilateral trade and Saudi foreign aid increased. The Saudis hoped that Iraqis' Arab identity would eventually override their affinity for fellow Shia Muslims in Tehran. Repeated attacks on the Iranian consulates in the Iraqi cities of Basra, Karbalah, and Najaf appear to support that view.

Finally, Saudi Arabia established a robust alliance with the United Arab Emirates (UAE). These two neighbors created a very powerful bloc, producing between them nearly half the Arab world's GDP and 40 percent of OPEC's oil. Their crown princes, Mohammed bin Salman and Mohammed bin Zayed, were close personally and professionally. Although their interests were not completely aligned, from 2015 onward the two neighbors fought together against the Iranian-supported Houthis in Yemen. In June 2018, their de-facto alliance was given a formal structure through a new Saudi–Emirati Coordination Council. Led by the two crown princes, the new body issued a "Strategy for Resolve" listing forty-four joint economic and military projects that the two nations planned to carry out over the following five years.

The dominant theme in Saudi efforts to improve relations with China, Iraq, Israel, Russia, the UAE, and the United States has been containing Iran. All of these efforts have been partially successful; other Saudi foreign policy initiatives involving Qatar, Lebanon, and Yemen have proven far less productive. These have left an impression of impulsive and reckless policy-making, which has not been helped by Saudi Arabia's habitual lack of transparency and poor public diplomacy.

Tensions between Riyadh and Doha are long-standing and complex.[14] Had it not been for British protection, King Abdulaziz would have very quickly incorporated Qatar into his expanding realm. More recently, Saudi Arabia has accused Qatar of terrorism because of its support for the sometime-terrorist, always anti-monarchist Muslim Brotherhood and of interfering in Saudi domestic politics through sponsorship of the routinely anti-Saudi Al Jazeera television news channel. With the two countries supporting different factions in Egypt, Syria, and Libya, the possibility of boycotting Qatar had been openly discussed in Riyadh during King Abdullah's reign.

In June 2017, these simmering tensions boiled over when Saudi Arabia—along with Bahrain, Egypt, and the UAE—imposed an economic and transportation boycott on Qatar until it met a list of thirteen demands, including closing Al Jazeera. Qatar refused to meet the Saudi demands. In 2020 the boycott continues, costing Riyadh much less than Doha but appearing to many observers as an unreasonable overreaction—one that has needlessly undermined the Gulf Cooperation Council and pushed Qatar closer to Iran and Turkey.

Saudi Arabia had long supported Lebanon's Sunni Prime Minister, Saad Hariri, politically and financially with the expectation that he would limit Iran's influence in Lebanon. In Saudi eyes, Hariri had tried to walk both sides of the street, taking their money while remaining much too friendly with Iran and Hezbollah. In November 2017, the prime minister of Lebanon was summoned to Riyadh, detained at the Ritz Carlton, and pressured to announce his resignation in a televised address. Intervention from Washington and Paris led to his release and return to Beirut, where he immediately rescinded his resignation. A year later at the 2018 Future Investment Initiative conference, ironically also held in Riyadh's Ritz Carlton Hotel, a smiling Hariri appeared on stage with Mohammed bin Salman, who joked about "kidnapping" him, capping off what many considered a bizarre and troubling episode.[15]

Even more controversial has been the Saudi air campaign against the Houthis in northern Yemen. Although this conflict developed international implications, its roots lay in long-standing local, tribal, and regional issues. The Treaty of Taif (1934; see Chapter 3) and the more-recent Treaty of Jeddah (2000) had demarcated a political border between Saudi Arabia and Yemen with very little regard for either demographics or geography. Border tribes were divided but given documents allowing them to cross back and forth freely. These border tribes came to regard smuggling food, fuel, medicine, electronics, and livestock as a legitimate right. Smuggling liquor, drugs, weapons, explosives, and people was less legitimate, but equally important to the local economy. For many years the Saudi government maintained border security by turning a blind eye to smuggling and providing extensive patronage to local tribal leaders, but that stable situation gradually eroded.

The people of Sa'dah Province along the Saudi border are largely Zaydi Shia who in the 1960s supported the royalist Imam during Yemen's revolution and civil war. The victorious republican government repaid them with decades of economic neglect and political marginalization. Moreover, what limited development assistance the Yemeni government did provide in Sa'dah Province was, like financial aid from Riyadh, focused more on co-opting tribal leaders than supporting the local population. It was these economic imbalances on both the national and local levels, combined with changes in Saudi border policy, that prepared the ground for the Houthi revolt.[16]

In 2003, when al-Qaeda launched a bloody campaign of terror across Saudi Arabia, the government suddenly became much more concerned about smuggling weapons through its porous southern border with Yemen. These concerns only intensified when Saudi al-Qaeda terrorists fled to Yemen and joined forces with local militants. As a result, Saudi security forces began to demarcate, fortify, and control their border with Yemen much more forcefully than had been the case in the past. Not surprisingly, these efforts, which limited the traditional movements of the Yemeni border tribes, provoked fierce opposition—just as British efforts to secure Iraq's borders from the Ikhwan had in the 1920s.

The term "Houthi" is a derogatory epithet coined by opponents of the *Ansar Allah* or Supporters of God movement. The movement was founded in the 1990s by Hussein al-Houthi and has been led since his death in 2004 by his brother, Abdul Malik al-Houthi. Hussein al-Houthi was a gifted orator. His message called for a Zaydi religious revival energized by opposition to local tribal elites, the central government in Sanaa, and the United States. The Houthi's signature chant is, "Death to America, death to Israel, curse the Jews, God is great."[17] They also regard the government in Riyadh as an American pawn and resent the presence of Saudi-sponsored Sunni missionaries in Yemen.

In what became known as the Sa'dah Wars, the Houthis revolted against the central government in Sanaa six times between 2004 and 2010. This fighting sometimes spilled over into Saudi Arabia, encouraging even tighter Saudi border control. In 2011, when Arab Spring protests led to the collapse of Yemen's central government, the Houthis quickly seized control of Sa'dah Province and expelled most of the traditional tribal *sheikhs*. Gone in a matter of months were the local leaders whom Saudi Arabia had paid for years and relied upon to maintain border security.

In 2014, the Houthi militia joined forces with military units still loyal to deposed dictator, Ali Abdullah Saleh, to seize power from the internationally recognized transitional government of Abdu Rabbu Mansour Hadi. With Saleh's help, the Houthis advanced from their homeland in the north all the way to the south-western port of Aden in barely two months. They eventually fell out with Saleh and killed him while President Hadi fled to Riyadh and requested military support from the Arab League and Gulf Cooperation Council. The United Nations Security Council passed resolutions firmly condemning the Houthi coup and human rights abuses, placing sanctions on Abdul Malik al-Houthi, and authorizing the inspection of vessels bound for Yemen in order to confiscate weapons.[18]

As should be clear from the above, Iran did not create this confused situation. However, it took every opportunity to vigorously exploit it. Tehran does not control the Houthis, but it funds, arms, trains, and advises them. In terms of history, religion, politics, and even family ties, Iran's recent relationship with the Houthis differs substantially from its deep, long-standing connections with Lebanon's Shia community and Hezbollah—but the Saudis found little solace in that distinction.[19] From Riyadh's perspective, the Houthis have been a vehicle for Iranian expansion into the Arabian Peninsula. That is something that no Saudi king could accept, and planning for intervention in Yemen began while King Abdullah was still on the throne.

When the UN Security Council demanded that the Houthis "withdraw forces from all areas they had seized," Saudi Arabia's two largest trading partners, China and the United States, voted in favor of Resolution 2216. Russia, the kingdom's new oil market ally, abstained. Saudi Arabia assembled a coalition of Arab aircraft from Bahrain, Egypt, Jordan, Kuwait, Morocco, Qatar, and the UAE for air strikes against the Houthis. For three and a half years the United States facilitated this campaign with mid-air refueling support, while Saudi and Egyptian naval forces imposed a blockade on Yemen that, under UN authority, the United States and Pakistan subsequently joined. The UAE sent ground forces into Yemen; Britain, France, and the US provided intelligence and logistics support. In 2015, all of these governments agreed that intervention in Yemen was a reasonable Saudi response to a genuine threat.

It is worth remembering that in March 2015, Mohammed bin Salman was not the king of Saudi Arabia nor the crown prince or even the deputy crown prince. He was the newly appointed minister of defense. The experienced Saud al-Faisal, though ill, was still foreign minister. The popular view that 30-year-old Mohammed bin Salman recklessly took his country to war and that ten sovereign states, including Britain and the United States, blithely followed him, is a misreading of history. King Salman made the decision in order to stop the "Hezbollahization" of Yemen. Major Western powers supported the Saudis in order to prevent the expansion of Iranian influence into the Red Sea, especially in the strategically important Bab al-Mandeb strait, and to maintain Saudi support for then-ongoing nuclear negotiations with Iran.

Whatever its origins or justifications, the Saudi-led intervention to restore the Hadi government resulted in a prolonged stalemate and appalling human tragedy. Although the Houthis were driven from most of Yemen, they retained control of the country's mountainous northern plateau—essentially, the former North Yemen or Yemen Arab Republic, which had existed from the collapse of the imamate in 1962 until the unification of Yemen in 1990. Regions controlled by the Houthis became a vicious police state while the rest of the country fragmented, with every tribe and region going its own way. International human-rights-monitoring groups have accused all sides of war crimes.

The multinational naval blockade led to great suffering in a country that imported most of its food, fuel, and medicine. The destruction of infrastructure—most notably, bridges—as well as rampant corruption at Houthi-controlled ports, made it difficult to distribute any aid that did arrive. In what became a multi-party civil war, thousands of Yemeni civilians died not only from air strikes but also from widespread ground fighting, starvation, and disease. Millions more were put at risk of starvation or made homeless. Despite numerous efforts at negotiation and more than $14 billion of recent Saudi aid to Yemen, the war has dragged on for five years—exposing Saudi Arabia to mounting international criticism.

The war has also exposed the kingdom to damaging physical attacks. The Houthi have fired more than 200 ballistic missiles into Saudi Arabia; oil tankers and oil pipelines have been targeted; and civilian airports in the southern cities of Abha, Jizan, and Najran have been closed. On at least two occasions, Tehran launched highly effective cyber attacks against Saudi Arabia, one of which destroyed thousands of Saudi Aramco computers.[20] In September 2019, Iranian cruise missiles and drones attacked the kingdom's Abqaiq oil processing facility and succeeded in taking nearly half of Saudi oil production offline for several weeks.

Ordinary Saudis now worry that Iranian attacks may shift from economic to civilian targets. Should that happen, Iran's extensive array of short- and medium-range missiles could devastate the desalination plants on which Riyadh depends for drinking water. An even greater fear is that Iran will acquire nuclear weapons. Saudi leaders have always questioned the efficacy of the 2015 Joint Comprehensive Plan of Action, better known as the Iran Nuclear Deal. From their perspective, the agreement allowed unlimited uranium enrichment after twelve years while doing nothing to curtail Iran's ballistic-missile-development program or disruptive involvement in Arab countries.

In some ways, the Saudi response to Iran has followed its long-established security policies; spend billions of dollars on advanced weapons and turn to traditional partners for support. In 2019, Riyadh made the first payments on an estimated $15 billion contract for Lockheed Martin's Terminal High Altitude Area Defense (THAAD) air defense system. That summer, Saudi Arabia reopened the Prince Sultan Air Base for the deployment of US aircraft, air defense missile batteries, and several thousand soldiers and airmen. Yet in other ways the Saudi response under King Salman and Mohammed bin Salman has been unconventional and may become even more so. Launching an independent air campaign in Yemen or investing seriously in a domestic defense industry were new approaches. Most worryingly, as the former head of Israel's National Security Council Yaakov Amidor warned—a nuclear armed Iran would not only surround Israel with a "ring of fire," it would very likely drive Turkey and Saudi Arabia to seek their own nuclear weapons.[21]

Foreign policy currently presents King Salman with some of his most complex challenges—in part, because outcomes depend on decisions made not in Riyadh, but in Tehran, Jerusalem, London, and Washington. Below are some of the principal issues that he will need to address if he is to maintain the international peace and security which has for many years underpinned Saudi domestic stability but is now in question.

Human rights. Will Saudi Arabia punish those responsible for the death of Jamal Khashoggi, take steps to improve its human rights record, and thus strengthen its ties to the West—or will Western nations place economic sanctions on the kingdom and senior Saudi officials?

Yemen. Can Saudi leaders negotiate a political solution to the war in Yemen—one that accommodates the Houthis, their Yemeni political rivals, and South Yemen's aspirations for greater autonomy while also addressing Saudi Arabia's legitimate security concerns?

Iran. Saudi leaders do not expect the Islamic Republic to abandon its quest for regional hegemony—but can they limit Tehran's influence in Arab states without war, or will they be forced to cede regional hegemony to an aggressive, nuclear-armed revolutionary government?

Israel. Will Saudi Arabia make peace with Israel, either because it fears Iran or because it is confident enough to act in its own self-interest?

Russia. Faced with growing shale-oil production, will Saudi Arabia and Russia continue coordinating their own oil exports to maintain prices?

Moderate Arab neighbors. Will Saudi oil revenue remain sufficient to sustain substantial financial assistance to Bahrain, Egypt, Jordan and Oman? Will this be enough to maintain those moderate pro-Western governments or will they be replaced by anti-Saudi regimes?

Map 5. Areas of Significant Iranian Influence.

Political Challenges: Sectarianism, Corruption, and Autocracy

Vision 2030 is a paradox. It introduced disruptive social and economic reform in order to preserve a traditional monarchy. Yet, replacing state-dependent subjects with self-sufficient citizens will eventually alter the relationship between a people and their government. Bringing rock concerts and unmarried, female tourists to the Land of the Two Holy Mosques will anger religious conservatives. Eliminating long-established entitlements will cause widespread resentment. There are now many points of political friction in this rapidly changing society, but three merit particular attention: discrimination against the Shia population, elite corruption, and rising authoritarianism.

The Eastern Province of Saudi Arabia sits on both the world's largest oil field and a cultural fault line between Sunni and Shia Muslims. Very few Shia live in the remainder of Saudi Arabia. In the Eastern Province, nearly half the population is Shia. Before 1913, Shia families such as Al Juma, Al Khunaizi, Al Jishi, and Al Nasrallah had long been prominent landowners, merchants, religious scholars, and judges. They maintained important patronage networks, served as tax collectors, and used their access to Ottoman political authority to maintain their own local influence. As he did elsewhere, King Abdulaziz sought to work with these local notables and most made peace with the Al Saud. One exception was Abd al-Hussein al-Juma who after calling for the return of the Ottomans, became the first Shia dissident publicly executed by the Al Saud.[1]

Over time, more and more restrictions were placed on the Saudi Shia. Limits were placed on the authority of Shia courts, the building of Shia mosques, and the holding of Shia religious processions, although the rigor with which these controls were enforced varied over time. As restrictions grew, the Shia *aiyan* found themselves in an increasingly difficult position; the Al Saud expected them to maintain order, but never made them fully part of the Sunni-dominated stakeholder elite.

Unable to end economic and social discrimination against their clients, the traditional Shia leading families slowly lost influence to more assertive local leaders, who were more often political activists than religious scholars. The most influential among them followed the radical Iranian Grand Ayatollah Mohammed Shirazi (1928–2001) who was exiled to Kuwait until the Iranian Revolution in 1979. Known collectively as the Shirazi Movement, their local Saudi leader was Hasan al-Saffar.

The 1979 Iranian Revolution emboldened the Shirazis and triggered the Muharram Intifada. These comprised the largest street riots ever seen in Saudi Arabia. In the Eastern Province towns of Al Qatif and Safwa, banks and government offices were ransacked by crowds carrying posters of the Ayatollah Khomeini. The revolutionary government in Tehran encouraged the rioting with radio broadcasts, leaflets, and cassette tapes calling for Shia solidarity with Tehran and the overthrow of the corrupt, hypocritical Al Saud.[2] The traditional Shia notables were powerless to stop the violence and some twenty protesters died when the National Guard used force to restore order.[3] Hundreds were arrested and the leaders of the Shirazi Movement fled into exile.

While his brothers were simultaneously busy dealing with the mosque seizure in Mecca, Deputy Minister of the Interior Prince Ahmed bin Abdulaziz traveled to the Eastern Province to meet Shia notables and moderate clerics. He promised infrastructure improvements, and the government soon launched a large-scale program to upgrade electricity connections, roads, schools, hospitals, street lighting, and sewage systems.[4]

In 1993, King Fahd offered a general amnesty to Shia leaders in exile, and lifted travel bans on those who had remained in the kingdom. He met personally with the returned exiles, not as usual in his office at the Royal Diwan, but at his home. Dressed informally, King Fahd spent the first half hour in friendly conversation asking his rather perplexed guests for advice about his upcoming knee operation. Then, before the activists could raise their complaints, the king himself recited a long and very detailed list of Shia grievances, acknowledged the need for change, and asked for time to deal with the Sunni *ulama*.

Many prominent Shia activists including Hasan al-Saffar and Jaffar al-Talib accepted King Fahd's olive branch and sought to improve conditions for the Shia by working with the government. As the Shia notables had done previously, the leaders of the Shirazi Movement reaffirmed their allegiance to the Saudi state while continuing to call for an end to discrimination against its Shia citizens. And like the notables before them, they faced a long, uphill battle against Wahhabi preachers who continued to criticize any efforts to co-opt idolaters with material benefits.

Eighteen years later when the 2011 Arab Spring swept across the Middle East, very little happened in Saudi Arabia. King Abdullah had always been a popular leader, and in early 2011 he spent billions to keep the peace. He gave state employees a two-month salary bonus, hired 60,000 new civil servants, and increased the minimum wage for Saudis. Civil and religious authorities repeatedly warned people not to participate in public protests and most had little interest in doing so. The "Day of Rage" promoted by Saudi dissidents living abroad turned out to be a non-event. In Riyadh's Cairo Square, where the protest was to have taken place, unarmed traffic police kept cars and pedestrians moving along without incident.[5]

The only Arab Spring-related protest in Saudi Arabia occurred in the Eastern Province, where Saudi Shia held large demonstrations in support of ongoing protests by Bahrain's predominantly Shia population. The Shia cleric Nimr al-Nimr emerged as the leader of these Saudi protests. His arrest in July 2012 sparked further demonstrations leading to several deaths. More demonstrations erupted in 2014 when Sheikh Nimr was sentenced to death. He was executed in May 2016, along with more than two dozen convicted Sunni al-Qaeda terrorists and the radical Sunni preacher Faris Zahrani. Although most of those executed were Sunnis, al-Nimr's death set off demonstrations across the Shia world and the burning of the Saudi Embassy in Tehran. In response, the

Saudis broke off diplomatic relations with Iran. They have yet to be restored. Sporadic demonstrations and arrests have continued in the Shia areas of the Eastern Province ever since.

Nimr al-Nimr remains a controversial figure. He had studied in Tehran for more than a decade and belonged to a clerical family that had vehemently opposed Al Saud rule for three generations. Although popular with Shia youth, he had never been considered a senior Shia cleric. His sermons openly denounced the Al Saud government as tyrants and called for secession of the Eastern Province if treatment of the Shia did not improve. He celebrated the death of Interior Minister Prince Naif, welcomed Iranian intervention in the Eastern Province, and supported Iran's quest for nuclear weapons.

Al Nimr was shot during his arrest, allegedly because he fired on the police. Numerous human rights groups have criticized his arrest, trial, and execution on the grounds that his protests amounted to nothing more than an exercise of free speech and that he was denied adequate defense at his trial. The Saudi government's view is that he encouraged violent demonstrations, armed resistance to the police, foreign intervention in Saudi domestic affairs, and physically attacked the security forces.

What is not in dispute is that in the months before al-Nimr's execution, more than a dozen Saudi police officers were killed in the Shia districts of the Eastern Province. Most were from the Bani Khalid and Ajman, and these important Sunni tribes were demanding retribution against the killers of their fellow tribesmen. To the Saudi authorities, this appeared a deliberate, and possibly Iranian-backed, attempt to start a sectarian tribal war in the oil rich Eastern Province.

All this notwithstanding, there have been gradual improvements in the status of Saudi Shia under the last three Saudi kings. In 2020 they have their own mosques and courts, which deal with civil and property matters. They manage their own religious endowments. They are allowed to hold religious processions, which government security forces protect from Sunni extremists. Members of the Shirazi Movement have won seats in the local municipal council elections. Members of the old notable families continue to be appointed to the *Majlis al-Shura*.

The use of derogatory terms describing Shia in the Saudi media has largely ended. The aggressive demonization of Shia ideology in Saudi schools has declined and many Shia have received government scholarships to study abroad. Except for security-related positions, government job opportunities for Eastern Province Shia have increased. As noted earlier, in a move strongly opposed by some of the Wahhabi *ulama*, King Abdullah included Shia in National Dialogue meetings.[6]

Al Nimr's own village of Awamiyah offers a fascinating window into the Al Saud's strategy for dealing with Shia dissent. For years it had been a hotbed of unrest with narrow, crowded alleys the security forces found difficult to patrol. In 2016, a massive urban renewal program delivered large, modern homes, paved roads, better utilities, a community center, and recreation facilities. Some residents of Awamiyah welcomed these substantial, government-financed improvements to their village. Others lamented the loss of historic buildings and still others claimed it was all done just to make policing more effective. All were probably correct to some degree.

Despite these improvements, discrimination against the Shia remains a fact of life in Saudi Arabia where the monarchy legitimizes itself with a form of Sunni nationalism.

The principals at most Shia schools are still Sunni. Shia parents must still comfort their children when they are asked, "are you a Muslim or a Shia?" Infrastructure in Shia areas, particularly roads, still lags behind that in Sunni neighborhoods. Mixed marriages remain extremely rare. Outside of Al Qatif, there are few Shia judges in Saudi Arabia. Although Shia account for at least 10 percent of the Saudi population, they hold barely 4 percent of seats in the *Majlis al-Shura*. There has never been a Shia member of the Council of Senior Scholars, or a Shia Saudi Ambassador to any country but Iran.

The war in Yemen has encouraged a sense of a Sunni, Saudi nationalism while tension with Iran has made Sunni-Shia relations more tense. The people of the large Al Qatif oasis are almost entirely Shia and, like all Shia, make financial contributions to their spiritual leader or *marji*. From Al Qatif, the faithful send their financial contributions largely to Ayatollah Sistani in Iraq; a practice which makes their loyalty suspect in the eyes of Sunni Saudis. On social media, Saudi Shia have increasingly been portrayed as a disloyal fifth column.

Some Saudi Shia have spoken openly of seceding from the kingdom or disrupting oil production in the Eastern Province. That will not happen. Large detachments of the National Guard based in the Eastern Province give the Saudi state the ability to deploy overwhelming force; which it would use if seriously threatened. To date, however, the security forces have preferred to use minimal force, avoiding the use of live ammunition, downplaying police casualties, and generally allowing protests within Shia villages. King Salman has spoken of the need to end division in Saudi society. He has met publicly with Shia religious leaders including Hasan al-Saffar, and appointed a representative from Awamiyah to the *Majlis al-Shura*. There is now a Shia minister of state, president of Saudi Aramco and chairman of the Crown Prince's mega-project NEOM, an ambitious and yet unproven concept for an entirely new futuristic city to be built near the border with Jordan.

Nevertheless, the failure to create a society in which the Shia community feels fully Saudi remains a difficult and potentially destabilizing problem. The author has visited Al Qatif and Hofuf on many occasions and found that although some Saudi Shia are implacably opposed to the government in Riyadh, most wish to remain loyal citizens. They recognize that violent Shia opposition to the Al Saud would almost certainly fail. Even more tellingly, they fear that the most likely replacement for the Al Saud monarchy would be an even more intolerant Sunni Salafi regime. Few would deny that their circumstances have improved over the past forty years, or that more remains to be done to repair decades of systematic discrimination.

A second threat to Al Saud legitimacy, and thus their kingdom's stability, is corruption. Politically sanctioned profiteering has been an endemic problem in Saudi Arabia, at least since the first oil boom when King Fahd tacitly accepted it as part of the oil income redistribution processes. Powerful sons of King Abdulaziz who controlled bureaucratic fiefdoms siphoned off large sums for their own use and provided an umbrella of protection for select subordinates to do the same. Contractors were paid for flood controls that they never built. Firms were paid to treat sewage that they actually dumped into the sea. Public land was given away and then sold back to the government at exorbitant prices. Senior military officers owned companies that imported munitions or repaired aircraft. It could have been worse. Unlike many oil-fueled kleptocracies, Saudi roads, schools, and hospitals actually got built; the water is drinkable; and the electricity works—but, of course, it all could have worked much better with less corruption.

When he came to the throne, King Salman understood that royal entitlements and high-level corruption were the most serious complaints that most Saudis had against the monarchy. In fact, he shared that view. As governor of Riyadh, Salman had often limited the abuses of junior princes, but had been frustrated by his inability to control his elder brothers or their sons. As crown prince, Salman had watched corruption grow worse as King Abdullah grew physically weaker. Royal gifts, known as *minha*, had become excessive—especially regarding public land. Grants were effectively being made by senior court officials, not the ailing king. When Salman became king, he employed both international accounting firms and the Ministry of the Interior to secretly investigate many dubious government contracts and land transactions.

As we have seen, succession—not fighting corruption—was the new king's first order of business. Salman's concentration of power was well planned, gradual, relentless, and successful. He removed most of King Abdullah's sons from positions of authority during his first month on the throne, but waited two years before retiring National Guard Commander Mit'ib bin Abdullah. During those two years, he gradually replaced all ministers and heads of important government agencies, most provincial governors and deputy governors, and many senior military officers. He created a new Presidency of State Security, neutered the Ministry of the Interior, transferred the State Prosecutor's Office from that ministry to the Royal Diwan and significantly expanded the Royal Guard. He removed one crown prince and then another. Mohammed bin Salman became minister of defense and took firm control of Saudi Aramco as well as the kingdom's Public Investment Fund before eventually becoming crown prince.

Only when they controlled all the levers of power and faced little effective resistance, did the king and his son strike out at corruption. The same day that King Salman eliminated the independence of the Saudi Arabian National Guard by removing its commander Mit'ib bin Abdullah, he created an independent Supreme Committee for Combating Corruption (the SCCC) chaired by Mohammed bin Salman. This new body was given extensive legal authority to investigate corruption, block travel, make arrests, and seize assets. Its members included Saudi Arabia's chief prosecutor or attorney general, Sheikh Saud al-Mojeb, and the head of the Presidency for State Security, General Abdulaziz al-Howairini.

The SCCC wasted no time making its presence felt. Within hours of its creation, General al-Howairini efficiently conducted a wide-scale series of detentions unprecedented in Saudi history. Those taken to Riyadh's five-star Ritz Carlton Hotel included eleven princes, four serving ministers, dozens of former ministers, deputy ministers, and prominent businessmen. The first princes detained were three sons of King Abdullah, including former National Guard Commander Mit'ib bin Abdullah and former Riyadh Governor Turki bin Abdullah. Several sons and sons-in-law of the late King Fahd and Crown Prince Sultan were detained—as was the Chairman of Kingdom Holdings Prince Waleed bin Talal.

Many non-royal members of the old guard were rounded up, including King Abdullah's chief of staff and chief of protocol. The serving minster of economy and planning, the former CEOs of the national airline Saudia, the Saudi Telecom Company and the Saudi Arabian General Investment Authority as well as a son of a former petroleum minister were all detained. Among the merchants arrested was the chairman of the Jeddah Chamber of Commerce as well as several half-brothers of Osama Bin Laden, including the chairman of the family construction firm. In total, more than 300 prominent members of the Saudi establishment were taken into custody. One group left untouched was the senior *ulama* who were allowed to keep their own very substantial land holdings and would soon sign off on the issues of women driving, movie theaters, and gender mixing.

King Salman went on television stating, "The law will be upheld and applied firmly to all those entrusted with public funds(. . . .). This is part of the reform agenda against abuses that have hindered our development for decades." The kingdom's powerful Committee of Senior Scholars endorsed the crackdown, declaring, "Islamic law and the national interest demand that corruption be combated (. . .). Fighting corruption is as important as fighting terrorism."[7] The attorney general added that, "All accused will be treated equally regardless of status or position."[8] In a country where corruption had long been widespread and those with the status of prince or position of minister protected from prosecution, such words coming from senior officials were news indeed.

The anti-corruption decree stipulated that assets gained illicitly could be seized by the state. All the assets of the detainees were frozen, along with those of an additional 1,700 of their associates and family members. Private jets were grounded and a no-fly list for "persons of interest" distributed at airports. The assets of several deceased sons of King Abdulaziz were frozen, including some of Princes Sultan's, Mishal's, and Abd al-Rahman's vast holdings. Just over $100 billion were ultimately recovered.[9] Assets seized included cash, stock portfolios, and real estate. One prince lost his collection of 185 antique cars and more than one royal stable was dismantled. Some assets were located abroad; a hotel in Los Angeles, office towers in Istanbul—even a shopping mall in Boulder, Colorado. A new government agency was created to collect and dispose of the confiscated assets.

One often overlooked aspect of the Ritz Carlton detentions was land reform, a topic more often associated with Egypt than Saudi Arabia. Because of high land prices, not construction costs, urban housing in Saudi Arabia is very expensive. In the United States the value of residential real estate is roughly 25 percent land and 75 percent the actual building. In Saudi Arabia, these ratios are reversed, with the lot often being worth much more than the house itself. This anomaly is the result of very large tracks of undeveloped land being held by a few princes and speculators who, expecting land prices to rise over time and paying no real estate taxes, choose to hold raw land as a long-term investment rather than develop it.

As a result of high land prices, most Saudis cannot afford their own home. After the 2011 Arab Spring unrest in neighboring countries, King Abdullah sought to address this longstanding social problem. He established a new Housing Ministry and ordered the construction of 500,000 new government-funded homes. Unfortunately, very little happened because building lots were not available. To free up undeveloped urban real estate, the king implemented a tax on so-called "white land." However, many plutocrats still found ways to circumvent the tax by building shacks or dividing parcels into small plots not covered by the new law. Again, few new houses were being built.

King Salman was more decisive. The overwhelming majority of the 100 billion dollars in assets obtained from the Ritz Carton detainees was not in cash or equities, but in raw land. Well over 50 percent of the undeveloped urban real estate in Riyadh and Jeddah was returned to government ownership. Along with a new mortgage law that finally found a way to deal with *sharia* opposition to foreclosures, this new stock of available building sites has begun to resolve the Saudi housing shortage.

Few Saudis were surprised by the names of those detained for corruption, but many were struck by the handful of prominent brigands who managed to escape the roundup. There was an explanation. Months before the Ritz detentions, King Salman had met personally with numerous princes and businessmen and spoken bluntly—the country was at war, running a deep budget deficit and going through major economic restructuring. For years, they had done very well out of government contracts, subsidies, and concessions—in some cases too well. None of them had ever paid any Saudi income taxes. The king asked for contributions and suggested how much each owed. Some were shown evidence of their corruption. Many paid up. Those who did were allowed to keep roughly one third of what many believed were ill-gotten gains.

Those who ignored the king's request found themselves "guests" of the Ritz Carlton, where a lavish ballroom became the guards' barracks and an entire section was set aside for female guests. Room service was available, but National Geographic was the only television channel permitted. Detainees were isolated, but family visits were allowed in the hotel lobby. Some detainees claimed to have been interrogated for as many as six hours a day, while others said they had hardly been questioned at all. They were all shown detailed evidence of their corruption and a few successfully explained away the charges against them or were only used as witnesses—but most made settlements, after which they were released. Upon release they were asked to pay the bill for their months-long stay at the Ritz, but fortunately they were all allowed to keep the loyalty points they had earned.

Those released were placed under a travel ban and required to wear GPS tracking ankle bracelets. They were instructed not to talk about what had happened to them or the settlements they had made. Most would not discuss their time at the Ritz even with their families. Military doctors who had been called to the hotel to provide detainees with regular checkups, were required to sign agreements not to discuss what had gone on. Even those, like Prince Waleed bin Talal, who gave several press interviews after his release, refused to publicly discuss the terms of any settlement that they had reached. All that can be known for certain is that most detainees lost a considerable amount of weight, one came home deaf in one ear, and one died in the Ritz.

The cause of the lone fatality—Major General Ali al-Qahtani, who was Prince Turki bin Abdullah's office manager—has never been made public. Some claim that he committed suicide or suffered a heart attack. The most widely heard explanation is that he was severely beaten by his guards after he threw an ashtray at one of his interrogators, which gravely injured the man's head. Like the deaths of Jamal Khashoggi in 2018 and the shooting of Royal Guard Commander Major General Abdulaziz al-Faghem in

2019, the lack of transparency surrounding General al-Qahtani's death has damaged Saudi credibility, which would benefit from a credible official explanation.

When the Ritz reopened for normal business in early 2018, not all of the detainees had been released. Nearly sixty of them were eventually transferred to Al Haier maximum security prison in Riyadh to face criminal charges.[10] These included those involved in the long troubled King Abdullah Economic City development project who were charged with defrauding shareholders, as well as those whose alleged corruption had caused deaths. Most notable among these were Ministry of Health officials who had allegedly embezzled funds intended to fight an outbreak of dengue fever in the kingdom.

Unlike the Arab Spring uprisings, Riyadh's anti-corruption campaign was implemented by the highest levels of government in order to preserve, not overthrow, a government. Furthermore, it was not, as is often heard, a power-grab by an ambitious young prince. By November 2017, Mohammed bin Salman and his father had already neutralized any serious opposition. It was, however, a clear display of Royal power and an obvious warning to any would-be opponents. Just as importantly, seizing the property of corrupt princes and business barons was a calculated move to win public support for painful austerity measures and subsidy cuts.

Local public reaction to the detentions was overwhelmingly positive. Statements such as, "They were leeches, sucking the blood of the country to pay for their million dollar cars and villas," and "It should have been done long ago," were typical of comments heard by the author at the time. The local press stoked public indignation and celebrated the detentions. The local paper *Okaz* ran a headline in bold red type demanding, "Where Did You Get the Money"? Something it would never have dared to have asked a prince only the week before. Well-known political science professor Waheed Arab Hashim wrote approvingly, "Since he became governor of Riyadh, King Salman has shown no tolerance for corruption." Women's rights activists and university professor Hatoon al-Fassi wrote in *Al Riyadh* of the "massive public support" for a move that "feels like a new era"—though perhaps not so new as she thought as she was herself detained for her activism the following year.[11]

The majority of Saudi businessmen who had never benefited from sweetheart deals were pleased, and commented on a sharp decline in the level of corruption, though hardly its complete elimination. American and British businessmen in Riyadh who operated under the US Foreign Corrupt Practice Act or the UK Bribery Act had long felt disadvantaged in competition with firms from less scrupulous nations. They saw the Saudi anti-corruption campaign as leveling the local playing field. Among business circles in London and New York, however, the crackdown prompted concerns about a lack of transparency and due process that led some to reconsider future investments in Saudi Arabia.

There was no let up in the anti-corruption campaign that continued throughout 2018, 2019 and 2020. Dozens of new investigations were opened on businessmen and royal family members less prominent than those taken to the Ritz. The bank accounts of civil servants are now being monitored for unexplained deposits. Many new travel bans were imposed. Hundreds of additional bank accounts were frozen as investigators sought to uncover illegal activity. Assets have been seized on charges of money laundering, which became as common as those for embezzlement. Some foreign firms accused of making bribes in the past have been blacklisted.

In March 2018, a new anti-corruption division was created in the Public Prosecutor's Office. Two months later, the king issued regulations protecting and rewarding whistleblowers reporting government corruption. In 2019, the head of the government's anti-corruption agency announced that he would now be going after corrupt low- and mid-level bureaucrats. . In March 2020, more than 300 government functionaries who posed no political threat to the monarchy were arrested on embezzlement and bribery charges. Corruption still exists in Saudi Arabia, but it is no longer openly accepted. A large public relations campaign reminiscent of the one against al-Qaeda has been launched against corruption. All of this appears to be a sustained effort to reduce corruption in the kingdom. Whether it does so or simply makes room for a new crew of pirates remains to be seen.

The third, and most worrying source of potential instability in Saudi Arabia, is increasing autocracy. King Abdulaziz ruled as an absolute monarch; King Faisal brooked no dissent from Arab nationalists, and imprisoned many of them. The Basic Law of Governance issued by King Fahd confirmed that Saudi Arabia was a monarchy, governed by a king and ruled by a dynasty. Yet their kingdom differed from many of the Arab World's post-colonial military dictatorships insofar as it was not a police state governed solely by fear. Comparisons with Saddam's Iraq, Stalin's Russia, or even Sadat's Egypt are misleading. Instead, the Al Saud practiced their own brand of autocratic paternalism combining generous economic cooption with authoritarian political control, and generally placing more emphasis on the former.

This distinctive balance of fear and favor has now shifted. Under King Salman, Saudi Arabia has become more autocratic. Civil liberties, which were never prominent, have become even more restricted. The sophisticated electronic surveillance systems developed to monitor violent terrorists have been used to detect nonviolent political dissent. Saudis who once spoke freely have become hesitant to criticize their government. Opponents of Mohammed bin Salman have been intimidated or eliminated. Accusations of torture have resurfaced. According to the crown prince himself, some 1,500 political activists ranging from conservative Salafi preachers to women's rights activists have been arrested in a process that has become more arbitrary and opaque.[12] Fear has become noticeably more prevalent.

Determined to preempt any hint of opposition, the crown prince established the Center for Media Affairs, which not only coordinates Saudi Arabia's propaganda wars with Qatar and Iran, but also aggressively monitors social media. The Center's director, Saud al-Qahtani, has urged Saudis to report those critical of the government or supportive of Qatar using the hashtag, "Black List."[13] Saudis posting messages critical of the crown prince have been pressured to publish apologies. Al-Qahtani was eventually implicated in the murder of Jamal Khashoggi, fired from his job, sanctioned by the U.S. Treasury Department, and banned from leaving Saudi Arabia. He was, however, ultimately acquitted and the institution he created to control public discourse remains very much in place.

As dissent has been discouraged, the age-old system of individual access to senior leaders for personal matters has unraveled. An ever-growing population and new security concerns make it impossible for everyone to meet with King Salman in the same way that their fathers met his. Senior princes who receive hundreds of written petitions a week find it increasingly difficult to reply promptly. As royal authority becomes more concentrated, the number of princes who can actually help the average citizen has sharply declined. People have begun to feel less personally connected to their leaders.

At the same time, Saudi Arabia's nascent participatory institutions—the appointed national *Majlis al-Shura*, elected municipal councils, and the royal family's Allegiance Council—have all been marginalized. The *Majlis al-Shura* continues to meet, but has never voted on important social or economic issues such as women driving or the new VAT tax. The local municipal councils never gained significant authority and have become irrelevant to most Saudis. Allegiance Council members were consulted individually, but never met to collectively approve the accession of the last two crown princes.

For now, Saudis seem willing to trade social liberalization and anticipated economic opportunity for greater political participation—but for how long? Saudi Arabia's unique framework of Wahhabi religious doctrines, tribal social customs, and oil-funded distributive economics is evolving quite rapidly towards global norms while the traditional Saudi mechanisms for creating consensus are not. Ultimately, the same forces that historically led to change in other monarchies are emerging today in Saudi Arabia. Among these changes are an end to social deference, the decline of religion, and the emergence of the individual.

Travel and satellite TV have exposed Saudis to the outside world. Students returning from study abroad have experienced societies far less respectful of princes and priests than their own. As a result, submission to the male-dominated, hierarchical, tribal, and family structures upon which the Saudi political order rests is slowly fading.

Like the village priest in parts of Europe 100 years ago, members of the Wahhabi religious establishment were once the best educated people in many Saudi communities—often, the only ones who could read. This gave them a social and political influence outside the mosque. Such influence has faded in a society where even the remotest village has a school and tuition-free university education is widely available. Indeed, one subsidy that has not been cut is the stipend paid to students to attend university.

Sixty years ago, Saudis worried more about their immortal souls on judgment day than their utility bills. That is changing. The kingdom will not become a secular state by 2030, but religion is no longer the central feature of many Saudi lives. As the unquestioned authority of the *ulama* declines, so does acceptance of its quietist approach to supporting any government that enforces *sharia* law regardless of how it came to power or how autocratic its rule might otherwise be.

A growing number of Saudis either no longer belong to traditional stakeholder groups, or have developed views clearly at odds with the majority of their communal group. Saudi women and young people, in particular, have interests that have not been promoted by the kingdom's religious, tribal, business, or bureaucratic elites. Individual identities are becoming more important as deference to established hierarchies wanes. Those who find themselves outside the old social system are expressing a desire for new ways in which to participate in the political process.

The Crown Prince deserves credit for leading a difficult, long-overdue social and economic revolution. He has reached out to young Saudis and women—especially poor women; those who could not afford a driver and now have a Hyundai and a job. He has won considerable public support, but also faces determined opposition. Although he has championed some parts of society, he has crushed others. For some Saudis, "He threw a stone into a stagnant stream and started it flowing again." For others, "The iron fist is here to stay. Now we are becoming a police state just like every other Arab country."[14] There is clearly a tension between Mohammad bin Salman's modernizing agenda and his authoritarian means of achieving it. In the short run, that may be unavoidable; in the long run it is likely to be unsustainable.

In the long run, the objective of Vision 2030 is a politically stable and economically prosperous country that is integrated into the global community. The motivation, innovation, and investment needed to produce prosperity will require not an iron fist but a more tolerant society; greater respect for the rule of law; increased popular participation in government; and less corruption. Without these, the sparkle of social liberation will soon fade. Here are a few specific indicators to help judge progress on these fronts.

Religious tolerance. Will a century of prejudice and systematic discrimination against the Saudi Shia continue to fade? Will other religious minorities such as Christians, Jews, Hindus, and non-Wahhabi Muslims be allowed to worship freely?

Corruption. Will awarding tenders, issuing licenses, enforcing compliance, and providing services become less prone to bribery and malfeasance? Most importantly, will anti-corruption efforts themselves be conducted through a transparent process?

Participation. Will the *Majlis al-Shura* become an elected rather than an appointed body with significant budget and oversight authority? Will the Saudi press become more independent? Will labor unions be permitted to organize and operate in the kingdom?

20

Evolving Arabia

Do not try to do too much with your own hands. Better the Arabs do it tolerably than that you do it perfectly. It is their war. You are here to help them, not win it for them. Actually, also, under the very odd conditions of Arabia, your own practical work will perhaps not be as good as you think it is.

T. E. Lawrence, *The Arab Bulletin*, August 20, 1917

Much has gone right in Saudi Arabia, yet at the same time addressing many long-recognized social and economic problems was routinely postponed. That is no longer the case. The country is in the midst of a dramatic social revolution accompanied by attempts at significant economic restructuring. Women are driving and joining the workforce, the religious police are defanged and abayas are coming off, tourists are arriving, music is being heard, and corruption declining. These efforts to transform Saudi Arabia into a more open society with a more productive economy need to be recognized and encouraged. Nevertheless, anything that is changing rapidly is by definition less stable than something standing still, and Saudi Arabia is no exception. Although reform should eventually increase Saudi stability, in the near-term rapid change is disruptive. Reform is creating new winners and losers. It is overturning established relationships, institutions and expectations thus making Saudi Arabia less stable in 2020 than it was in 2015.

Saudi Arabia is not Egypt with the unifying geography of the Nile, a homogeneous population, and 4,000 years of centralized government. It is a recent creation—still a collection of regions and tribes unified by a monarchy, with separatist sentiments lingering in the Hejaz and Eastern Province. Although Saudis over sixty-five years of age can remember an impoverished past of hunger, illiteracy, and high infant mortality, those under thirty now take security for granted—along with roads, schools, Starbucks, and iPhones. In 2020, gratitude for the Al Saud's historic role in national unification is a waning source of regime legitimacy.

Since 2015, Saudi Arabia has witnessed an uncommon degree of turmoil within the House of Saud. The long-running political partnership of King Abdulaziz's sons is gone, and with it the reassuring succession from brother to brother. Two competent and well-liked crown princes have been removed and a young, inexperienced prince elevated ahead of many older, more experienced—and now resentful—rivals. The king's health is uncertain, and no deputy crown prince has been named.

Royal family members have watched their privileges erode and their status decline. Princes have been arrested, had their bank accounts frozen, property seized, and travel restricted. Most no longer receive free airline tickets, free electricity, or large blocks of land that they can profitably sell. Maintaining royal family unity while downsizing the Al Saud is problematic. Although the possibility of a contested succession remains slight, it is nevertheless greater than at any time in the past sixty years.

All of the stakeholders in the Al Saud's traditional coalition, tribal leaders, religious scholars, leading merchants, and senior technocrats have felt the unsettling effects of rapid change. Across Saudi society, the reshaping of social and economic norms has weakened elite cohesion as well as the patronage networks that the Al Saud relied on to monitor and alleviate dissent.

Tribal leaders have watched their authority ebb as their followers become economically more dependent on the state than the tribe, and as rural people move to cities The *sheikhs* understand that ongoing changes can only undermine their social status, local political influence, and, ultimately, their importance to the monarchy. They have already been abandoned by some of their constituents and, over time, may become no more politically significant than Scottish clan chiefs are in modern Britain.[1]

The *ulama's* loss of control over the pace of social liberalization has strained the Al Saud's alliance with religiously conservative segments of society. Make no mistake, Saudi Arabia remains a deeply religious society. The westerner schooled in the Enlightenment and looking for objective truth in his religion may find little of value. The Saudi seeking a sense of community, comforting rituals to relieve life's anxiety, and traditions linking him to his forefathers finds much worth preserving. Conservative Saudis are angry when they see women driving, going to cinemas, checking into hotels alone, and no longer requiring their husband's permission for a caesarean delivery. The senior official religious scholars continue to support the government, but they now risk being outflanked by more conservative, dissident clerics vehemently opposed to recent social reforms. Although progressive activists often garner Western media attention by calling for even faster change, in Saudi Arabia others would like the Crown Prince to slow down before he ignites a reactionary backlash.

Reshaping the economy undeniably threatens some of the business elite who have depended on protected markets, generous subsidies, cheap foreign labor, and corruption. Prominent business leaders have been detained and dispossessed. The Hejazi commercial elite feels that it has been singled out for particularly harsh anti-corruption treatment. Many merchants wonder how they will compete in a less subsidized, more open, globalized economy. Some are trying to leave the kingdom, while others have already left.

Senior Saudi technocrats have seen long-familiar structures dismantled and replaced by new bureaucratic experiments managed by fresh faces. Outside of a few newly empowered agencies, bureaucrats begrudge their loss of authority, sinecures, and opportunities for corruption. They resent Vision 2030's explicit commitment to reduce the size of both the civil service and the public sector. Some technocrats who resisted reform or privatization have lost their jobs.

For the average Saudi citizen, peace and generous welfare programs were the mainstay of Saudi stability. Yet in 2020, the kingdom is entangled in a prolonged and increasingly unpopular war in which costs and casualties mount daily. Important Saudi oil facilities have been targeted and severely damaged. At the same time, economic benefits long taken for granted have been withdrawn. Saudis are unhappy about the imposition of taxes as well as sharp rises in gasoline and electricity prices. Some hope that Mohammed bin Salman will fail because they believe that this will lead to a reinstatement of the old order.

Should we care? Does it really matter who governs in Riyadh or that many of the influences which kept Saudi Arabia stable for decades are now eroding. Certainly there is ample reason to conclude that Saudi oil production is increasingly irrelevant. Largely because of hydraulic fracking and remarkable West Texas production growth, the United States became a net exporter of petroleum products in late 2019.[2] Because of the conservation efforts and a shift away from energy-intensive manufacturing, the amount of oil needed to produce a dollars-worth of GDP in an increasingly service-based American economy continues to fall. At the same time, the share of renewables in the energy mix continues to grow. Whether or not these trends continue, the US economy has already become much less reliant on oil in general, and foreign oil in particular, than it was in 1973.

None of this changes the fact that oil is traded on a global market. It continues to supply a third of the world's energy needs and its price remains a significant factor determining economic growth, especially for less-developed countries. Most years, global demand for oil is still growing. Saudi Arabia remains by far the world's largest oil exporter and the only "swing producer" capable of rapidly ramping up production to meet either natural or politically created shortages. Unless the United States prohibits oil exports or imposes a tariff on imported oil, its own domestic prices will continue to reflect global supply and demand. Even with an export ban, high or unstable oil prices in the rest of world would affect economic growth in America's trading partners and thus the US economy.

Those arguing that any successor to the present Saudi government would still need to sell its oil may wish to consider the historical record. To sell oil, you must first produce it. Iran produced 6 mbd of oil under the Shah but, following the 1979 Iranian Revolution, saw production fall to less than 2 mbd in 1980. The oil-price spike created by Iran's revolution caused far greater economic damage to the West than the Saudi-led 1973 oil embargo. Forty years later, because of poor reservoir management and inadequate investment, Iran can still barely produce 4 mbd. More recently, political instability has caused sharp production declines in Libya, Venezuela, Iraq, and, at times, Nigeria. It is quite possible that Saudi Aramco's commercial orientation and efficient production procedures would not survive the collapse of the Saudi monarchy. Then, a dozen tribes fighting over the world's largest oilfields could affect global energy prices far more than any OPEC meeting.

Whatever else they may be doing, King Salman and his crown prince have publicly embraced a more tolerant form of Islam. The king told the *Majlis al-Shura*, "There is no room for extremists who oppose modernization and exploit our tolerant religion to achieve their goals."[3] Mohammed bin Salman has repeatedly stated that, following the 1979 seizure of the Great Mosque in Mecca, Islam in Saudi Arabia was "hijacked," losing an entire generation to a distorted view of their religion.[4] King Salman's meeting with Vatican officials in Riyadh, Saudi press support for Pope Francis's 2019 visit to Abu Dhabi, and the Crown Prince's meeting with Coptic Pope Tawadros in Cairo were all very public steps towards greater religious tolerance in a society where intolerance has long been a normative value.

Within Saudi Arabia this has meant an end to many long-established social restrictions, particularly those on women. That is certainly welcome news, but it is ultimately a domestic Saudi issue. What matters to the rest of the world is Saudi preaching outside of the kingdom. The United States and its allies can kill terrorists and cut off their funds, but they cannot delegitimize Salafi extremists in Muslim eyes—only respected Islamic scholars can do that. If Saudi clerics are now more willing to help, we should welcome their assistance.[5]

In 2016, King Salman appointed the moderate Abd al-Karim al-Issa to serve as secretary general of the Muslim World League (MWL). A former Saudi justice minister and member of a prominent Nejdi clerical family, al-Issa has condemned Holocaust deniers and told Muslims living in Western countries to obey local laws or leave. In 2020, he attended the seventy-fifth anniversary of the liberation of Auschwitz and announced that Saudi Arabia would no longer be funding foreign mosques.

The US State Department's 2018 Country Report on Terrorism noted, "The current MWL leadership has presented a significantly more inclusive and less discriminatory message than previous MWL leaders and Saudi religious authorities."[6] For example, after the October 2018 shootings at a Pittsburgh synagogue, al-Issa sent a letter of condolence to the Washington Institute for Near East Policy, which read in part:

> Personally, and on behalf of the Muslim World League from Mecca, where hundreds of millions of Muslims around the world follow us and on behalf of the World Organization of Muslim Scholars, we feel the pain of this terrorist incident and extend our deepest condolences to the families of the victims.[7]

That is a very different Saudi message from the one that Israel's Ambassador to the United Nations, Dore Gold, clearly documented twenty years ago in his book *Hatred's Kingdom*.[8] It is also a message unlikely to be repeated should Islamic populists or Salafi jihadists get hold of the microphone in Mecca.

Nor should we undervalue Saudi Arabia's direct contributions to our own counterterrorism efforts. The grand mufti has repeatedly denounced both al-Qaeda and ISIS as enemies of Islam. Again, in the rather dry language of the state department's most recent Country Report on Terrorism, "Saudi Arabia continued to maintain a strong counterterrorism relationship with the United States and supported enhanced bilateral cooperation to ensure the safety of both U.S. and Saudi citizens within Saudi territories and abroad."[9] In the rather clearer language of the Obama Administration's Counterterrorism Coordinator, Daniel Benjamin, "On more than one occasion, (Saudi Minister of the Interior) Mohammed bin Naif's work saved American lives."[10]

Finally, nothing has been more destabilizing for the modern Middle East than the Arab–Israeli conflict, and no solution to that problem is likely without the endorsement of the Custodian of the Two Holy Mosques. As status quo rulers seeking regional stability, the Al Saud have long wanted to see this issue resolved. Both Crown Prince Fahd in 1981 and Crown Prince Abdullah in 2002 put forward proposals that moved the Arab position towards compromise. Now seeking allies in a cold war with Iran, Crown Prince Mohammed has told *The Atlantic's* Jeffrey Goldberg " . . . the Israelis have the right to have their own land."[11]

More than his predecessors, Mohammed bin Salman has emphasized the economic benefits of peace. "Israel is a big economy compared to its size and it's a growing economy, and of course there are a lot of interests we share with Israel — and if there is peace, there would be a lot of interest between Israel and the Gulf Cooperation Council countries."[12] It is probably not a coincidence that the Crown Prince's futuristic city, NEOM, is being built 40 miles south of Eilat with the involvement of Egypt and Jordan, the two Arab states that have already signed peace treaties with Israel.

More than his predecessors, King Salman has also been willing to take unilateral steps to improve relations with Israel, for example by terminating the seventy-year-old ban on commercial airline flights bound for Tel Aviv transiting Saudi airspace. Although the Saudis have been clear that expanding formal relations will require a peace agreement, their attitude towards Israel nevertheless contrasts sharply with that of Iran, Hamas, Hezbollah, the Houthis, al-Qaeda, or for that matter the Muslim Brotherhood government that ruled Egypt between 2012 and 2013.

Undoubtedly, serious social change is underway in Saudi Arabia and meaningful economic reform is a possibility—but what of the kingdom's political system? A decade from now, Saudi Arabia's political landscape will appear different than it does today. Perhaps the Al Saud will be swept away like the kings of Egypt, Iraq, and Libya. In that event, is the new government likely to be a reliable partner of the West?

If a successor government came to power by the ballot, it would almost certainly be an Islamist populist regime. In Algeria's first post-independence election (1991), the Islamic Salvation Front would have won enough seats to change the constitution had the army not intervened. In Saudi Arabia's 2005 municipal elections, well-organized groups ideologically aligned with the Muslim Brotherhood won in many urban areas.[13] Similar candidates won more recent national elections in Egypt, Turkey, and Palestine. On the other hand, if a new government came to power through violence, it would most likely be a jihadist organization such as ISIS or al-Qaeda. Although their methods may differ, all of these groups view Western culture as a threat to traditional Islamic values and institutions. It is exceedingly unlikely that the Saudi monarchy would be replaced by a stable, liberal, parliamentary democracy. Indeed, there are no examples of such an outcome anywhere in the Arab world, and it is wishful thinking to believe the Wahhabis will become the first.

Alternatively, the present Saudi government could evolve into either a more autocratic and repressive regime or a more accountable and limited monarchy. At present it is clearly becoming more autocratic. King Abdullah was a force for gradual political liberalization who initiated municipal elections, expanded the *Majlis al-Shura*, and formalized the royal family's voting procedures with an Allegiance Council. His successor, King Salman, has largely ignored all of these institutions, centralized authority, and punished dissent. Hardline religious conservatives and human-rights activists have found themselves side-by-side in Salman's prisons.

Yet King Salman and his crown prince have also made far more sweeping social and economic changes than King Abdullah, who never opened cinemas, allowed women to drive, or marginalized the religious police. Nor did King Abdullah ever cut energy subsidies or introduce significant new taxes. Had King Salman left any of these controversial issues with the *Majlis al-Shura*, the debate would still be going on. For better or worse, the monarchy remains the only institution in Saudi Arabia capable of moving against the narrow demands of various interest groups. Businessmen who want cheap foreign labor, princes who won't pay their electricity bills, clerics who want women to stay at home, or technocrats trying to derail the Aramco IPO.

The ill-fated Arab Spring caused political turmoil and human misery in Egypt, Libya, Syria, Tunisia, and Yemen. Subsequent disturbances have rocked Algeria, Iraq, Lebanon, and Sudan. Like all these countries, Saudi Arabia suffers from high unemployment, low economic growth, and very limited popular participation in government. The West cannot solve these nations' problems for them. Finding solutions will require time, local leadership, adequate resources, and, most importantly, political stability.

As we have seen, the current Saudi king is an autocrat who has dismantled many of the established institutional constraints on his prerogatives. His crown prince is an authoritarian who has used force against non-violent opponents. A more repressive kingdom seems to be in the offing. Yet the King and Crown Prince are, above all else, dedicated to preserving the Al Saud dynasty. Both understand that some things will need to change if a monarchy based on God, tradition, and OPEC is to survive in the twenty-first century. After decades of cultural isolation they are moving Saudi Arabia towards global norms on many fronts, and this may eventually include broader political participation.

Several factors offer a possible foundation from which Saudi Arabia might evolve into a more liberal rather than a more repressive country. Unlike every other Arab nation, the kingdom's indigenous institutions were not uprooted and modernized by colonial administrators. Saudi institutions have evolved organically, producing a long-established government with deep local roots and widespread popular legitimacy. It remains the only nation with a recognizable version of the classic Islamic constitutional order in which religious scholars counterbalanced executive authority. Compared with other Arab countries, its judiciary is relatively independent.[14] Saudi Arabia's military is firmly under civilian control; its fledgling, appointed parliament is an established, if weak, part of the political system; it has a large number of Western-educated English-speaking technocrats, the world's most profitable oil company, and the funds to pay some of the world's best consultants to help devise coherent development plans.

Ultimately, the kingdom is passing through a disruptive and potentially destabilizing period of transition, the outcome of which remains very uncertain. What is clear is that in responding to Saudi reforms, the West faces a dilemma. It cannot ignore the assassination of Jamal Khashoggi, the detention of political activists or the war in Yemen—yet it should not let these events overshadow the genuine improvements taking place. The success of these reform efforts is very much in our interest, since Saudi Arabia's political stability, economic development and social liberalization are intertwined; each supportive of and dependent on the others.

We in Western democracies cannot influence how Saudi Arabia evolves unless we are involved in the process. It remains important to set boundaries of behavior that must not be crossed again and encourage compromise solutions to Saudi disputes with Yemen and Qatar. It is equally important to recognize the limitations of outside efforts to accelerate reform. Intimidating or humiliating Saudi Arabia with public shaming and economic sanctions is unlikely to modify the behavior of a people who believe themselves the founders of a great religion and know themselves to be vital to global prosperity. In fact, a shunned or frightened monarchy is one that is more likely to discontinue reform and seek autocratic allies. Ostracizing Saudi Arabia only makes reform more difficult and instability more likely. Encouraging Saudi Arabia's political evolution towards a more accountable, less coercive government will require positive, practical engagement with its monarchy; including firm, sustained support for courts dedicated to the rule of law, universities promoting academic freedom, uncensored media outlets, empowered representative bodies, and independent civic organizations. These efforts will certainly be more successful if they push with the grain of Saudi culture and history rather than against it.

Notes

Chapter 1

1 Robert Lacey, *The Kingdom* (New York: Harcourt, Brace, Jovanovich, 1981), 561. King Abdulaziz is often known as Ibn Saud in the West. The date of his birth is not known with certainty. November 1880 is the commonly accepted date. Both Robert Lacey and Dr. Abdullah al-Uthaymin, the former Dean of the History Faculty at Riyadh's King Saud University, have pointed out inconsistencies with the 1880 date. I have accepted their suggestion that 1876 is the more realistic date.

2 Alexei Vassiliev, *The History of Saudi Arabia* (New York: New York University Press, 2000), 34.

3 Ibid., 73.

4 Abdullah Al Uthaymin, *Muhammed 'Abd al Wahhab, The Man and His Works* (London: I.B. Tauris, 2009). "The Shaykh condemned all forms of innovation, and rejected the views of those who maintained that *bid'a* could be praiseworthy, basing his rejection on the words of the Prophet: 'Every *bid'a* leads people astray.'"

5 Christine Moss Helms, *The Cohesion of Saudi Arabia* (London: Croom Helm, 1981), 103–5.

6 Noah Feldman, *The Fall and Rise of the Islamic State* (Princeton: Princeton University Press, 2008), 2.

7 David Commins, *The Wahhabi Mission and Saudi Arabia* (London: I.B. Tauris, 2006), 5.

8 George Rentz, "Wahhabism and Saudi Arabia," in *The Arabian Peninsula: Society and Politics*, Edited by Derek Hopwood (London: Allen and Unwin, 1972), 56.

9 Vassiliev, *The History of Saudi Arabia*, 78.

10 The Al Saud political leadership has used various titles, including Amir of Riyadh, Sultan of the Nejd, King of Hejaz and King of Nejd, King of Saudi Arabia, and Custodian of the Two Holy Mosques. At least in Riyadh, they have also always been known as the *imam* or the person who leads the community in prayer. In 1921, Abdulaziz took the secular title of sultan in place of the religious title *imam*, which all his predecessors had used as the political leaders of the Wahhabi religious movement. Because the Wahhabi movement was unpopular with the conquered people of the Hejaz, after 1925 Abdulaziz took the even more secular title of king.

11 The most prominent Wahhabi scholar was originally only the chief judge of Riyadh. That position, which is usually held by a member of the Al al-Sheikh family, evolved to become that of the grand mufti of Saudi Arabia.

12 Rentz, "Wahhabism," 59.

13 Vassiliev, *The History of Saudi Arabia*, 80.

14 John Sabini, *Armies in the Sand: The Struggle for Mecca and Medina* (London: Thames and Hudson, 1981), 166. Sabini's account is drawn from H. St. John Philby's *Saudi Arabia* (London: Ernest Benn, 1955) reprinted by Librairie du Liban 1968, and George Forster Sadleir's *Diary of a Journey Across Arabia*, first published in Bombay in 1866 and reprinted by Falcon Oleander in 1977.

15 David Commins, *The Wahhabi Mission and Saudi Arabia* (London: I.B. Tauris, 2006) 37.
16 Vassiliev, *The History of Saudi Arabia*, 155.
17 Sabini, *Armies in the Sand*, 168.
18 Vassiliev, *The History of Saudi Arabia*, 203.
19 Ibid., 202.
20 Madawi Al Rasheed, *A History of Saudi Arabia* (Cambridge: Cambridge University Press, 2002), 25.
21 Jacob Goldberg, *The Foreign Policy of Saudi Arabia: The Formative Years, 1902–1918* (Cambridge MA: Harvard University Press, 1986), 33.
22 David Holden and Richard Johns, *The House of Saud: The Rise and Fall of the Most Powerful Dynasty in the Arab World* (New York: Holt, Rinehart and Winston, 1981), 5.
23 Lacey, *The Kingdom*, 44.
24 Mohammed Al Mana, *Arabia Unified: A Portrait of Ibn Saud* (London: Hutchinson Benham, 1980), 37.
25 Holden and Johns, *The House of Saud*, 7.
26 Ameen Rihani, *Maker of Modern Arabia* (New York: Houghton Mifflin, 1928), 66.
27 There are many accounts of Abdulaziz's capture of Riyadh, all of which vary slightly in detail, but not in basic facts or tone. Other accounts may be found in H. C. Armstrong, *Lord of Arabia* (London: Penguin, 1938); H. St. John Philby, *Arabian Jubilee* (London: Robert Hale, 1952); H. R. P. Dickson, *Kuwait and Her Neighbours* (London: George Allen & Unwin, 1956); David Howarth, *The Desert King: A Life of Ibn Saud* (London: Collins, 1964); Hafiz Wahba, *Arabian Days* (London: Constable, 1928); and Michael Darlow and Barbara Bray, *Ibn Saud* (London: Quartet, 2012).
28 Armstrong, *Lord of Arabia*, 64.
29 Ibid., 65.
30 Vassiliev, *The History of Saudi Arabia*, 222.
31 Al Mana, *Arabia Unified*, 55.
32 Al Uthaymin, *The Man and His Works*, 14.
33 Al Mana, *Arabia Unified*, 48.
34 Ibid., 49.
35 Vassiliev, *The History of Saudi Arabia*, 228.

Chapter 2

1 The Saudi Ikhwan should not be confused with the Egyptian *Ikhwan al-Muslimeen*, or Muslim Brotherhood, movement founded in 1928 by Hasan al-Banna (1906–1946). Although the Saudi Ikhwan no longer exist, the Egyptian Ikhwan created a political party that is now found in many Arab countries. It is a conservative Sunni religious movement that engages in both political activism and social work while calling for the Islamization of both the state and the economy. The contemporary Muslim Brotherhood movement briefly held elected power in Egypt after the 2011 Arab Spring Revolution.
2 H. R. P. Dickson, *Kuwait and Her Neighbours* (London: George Allen & Unwin, 1956), 149.
3 Robert Lacey, *The Kingdom* (New York: Harcourt, Brace, Jovanovich, 1981), 146.
4 Alexei Vassiliev, *The History of Saudi Arabia* (New York: New York University Press, 2000), 228. Sultan ibn Bijad is sometimes referred to as Ibn Humaid, his more complete name being Sultan ibn Bijad ibn Humaid.

5   Abdullah Al Uthaymin, *Muhammed 'Abd al Wahhab, The Man and His Works* (London: I.B. Tauris, 2009), 144.

6   Madawi Al Rasheed, *A History of Saudi Arabia* (Cambridge: Cambridge University Press, 2002), 58–60.

7   Lacey, *The Kingdom*, 108 and Mohammed Al Mana, *Arabia Unified: Portrait of Ibn Saud* (London: Hutchinson Benham, 1980), 56.

8   Vassiliev, *The History of Saudi Arabia*, 232.

9   Donald P. Cole, *Nomads of the Nomads: The Al Murrah Bedouin of the Empty Quarter* (Illinois: AHM Publishing Corporation, 1975), 101.

10 Moojan Momen, *An Introduction to Shi'i Islam* (New Haven: Yale University Press, 1985), 165.

11 Natana DeLong-Bas, *Wahhabi Islam: From Revival and Reform to Global Jihad* (London: I.B. Tauris, 2004), 84.

12 G. Steinberg, "The Shiites in the Eastern Province of Saudi Arabia, 1913–-1953," in *The Twelver Shia in Modern Times: Religious Culture and Political History,* R. Brunner and W. Ende (eds) (Leiden, Brill), 243.

13 Toby Matthiesen, *The Other Saudis: Shiism, Dissent and Sectarianism* (New York: Cambridge University Press, 2015), 8.

14 Vassiliev, *The History of Saudi Arabia*, 237.

15 Gary Troeller, *The Birth of Saudi Arabia: Britain and the Rise of the House of Saud* (London: Frank Cass, 1976), 92–3; Al Mana, *Arabia Unified*, 56; and Vassiliev, *The History of Saudi Arabia*, 245.

16 Lacey, *The Kingdom,* 124.

17 H. St. Philby, *Saudi Arabia* (London: Ernest Benn, 1955), 274.

18 Troeller, *Birth of Saudi Arabia*, 183; and Joseph Kostiner, *The Making of Saudi Arabia 1916-1936* (New York: Oxford University Press, 1993), 55, 57, 61. During the First World War, Britain sought allies among many of the Ottoman Sultan's nominal subjects on the Arabian Peninsula. His Majesty's Government paid subsidies to Al Idris in the Asir, Fahd Beg of the 'Anizah tribe in the north, Sharif Hussein in Mecca, and Abdulaziz in the Nejd. Although Abdulaziz's payment of £5,000 a month was small compared with Hussein's £125,000 a month, he did not actually do much fighting for the British.

19 Elizabeth Monroe, *Philby of Arabia* (New York: Pitman Publishing, 1973), 69–70.

20 John Habib, *Ibn Saud's Warriors of Islam: The Ikhwan of the Najd and their role in the creation of the Saudi Kingdom, 1910–1930* (Leiden: E. J. Brill, 1978), 19.

21 Lacey, *The Kingdom,* 147–8.

22 David Holden and Richard Johns, *The House of Saud: The Rise and Fall of the Most Powerful Dynasty in the Arab World* (London: Sidgwick & Jackson, 1981), 71.

23 Habib, *Ibn Saud's Warriors of Islam,* 20–1.

24 Vassiliev, *The History of Saudi Arabia*, 249.

25 Madawi Al Rasheed, *Politics in an Arabian Oasis: The Rashidis of Saudi Arabia* (London: I.B. Tauris, 1991), 230.

26 Vassiliev, *The History of Saudi Arabia*, 255 and Al Mana, *Arabia Unified*, 58.

27 Al Rasheed, *Politics*, 253.

28 Ibid., 231.

29 Lacey, *The Kingdom,* 171.

30 Troeller, *Birth of Saudi Arabia*, 69; and Vassiliev, *The History of Saudi Arabia*, 255.

31 Dickson, *Kuwait*, 276.

32 Randall Baker, *King Husain and the Kingdom of Hejaz* (Cambridge: Oleander, 1979), 21.

33 Mai Yamani, *Cradle of Islam: The Hejaz and the Quest for and Arabian Identity* (London: I.B. Tauris, 2009), 40.

34 Christine Moss Helms, *The Cohesion of Saudi Arabia* (London: Croom Helm, 1981), 216.

35 Gilbert F. Clayton, *An Arabian Diary*, edited by Robert Collins (Los Angeles: University of California Press, 1969), 40.

36 Al Mana, *Arabia Unified*, 70.

37 Ibid., 71; and Lacey, *The Kingdom*, 188. Mohammed Al Mana, who served as King Abdulaziz's translator for nine years, believes the accounts of Ikhwan brutality were "greatly exaggerated" Hashemite propaganda designed to influence the foreign media. Robert Lacey concluded that whatever the provocation, the slaughter was merciless: "houses were destroyed, shops looted and throats cut." He places the death toll at "more than 300." The author has discussed these events with Saudis from Taif on numerous occasions. Often citing firsthand accounts from their parents or grandparents, they all insist that far more than fifty people were killed in an Ikhwan rampage that additionally destroyed a great deal of private property.

38 Ibid., 8; and Al Rasheed, *A History of Saudi Arabia*, 46.

39 Troeller, *Birth of Saudi Arabia*, 220.

40 Reader Bullard, *Two Kings in Arabia: Letters From Jeddah 1923–5 and 1936–9*, edited by E. C. Hodkins (Reading: Ithaca Press, 1993), 76.

41 H. C. Armstrong, *Lord of Arabia* (London: Penguin, 1938), 243.

42 Reader Bullard, *The Camels Must Go: An Autobiography* (London: Faber and Faber, 1961), 132.

43 Clayton, *An Arabian Diary*, 272–9.

44 Vassiliev, *The History of Saudi Arabia*, 264; and Lacey, *The Kingdom*, 196.

45 Lacey, *The Kingdom*, 199.

46 Vassiliev, *The History of Saudi Arabia*, 200.

47 K. Chaudhry, *The Price of Wealth: Economics and Institutions in the Middle East* (Ithaca: Cornell University Press, 1997), 92, 94.

48 Yamani, *Cradle of Islam*, 14.

49 Michael Darlow and Barbara Bray, *Ibn Saud: The Desert Warrior Who Created the Kingdom of Saudi Arabia* (New York: Skyhorse, 2012), 316.

50 Vassiliev, *The History of Saudi Arabia*, 271; Hafiz Wahba, *Arabian Days* (London: Constable, 1928), 95–6.

51 Chaudhry, *The Price of Wealth*, 75.

52 Bullard, *Two Kings*, 196.

53 David Van der Meulen, *The Wells of Ibn Saud* (John Murray, 1957), 114, 126.

Chapter 3

1 John Bagot Glubb, *War in the Desert* (New York: W. W. Norton, 1961), 126.

2 H. C. Armstrong, *Lord of Arabia Ibn Saud: The Intimate Study of a King* (London: Penguin, 1938), 254.

3 Bray quoted in John S. Habib, *Ibn Saud's Warriors of Islam* (Leiden: E. J. Brill, 1978), 80. Letter from Captain Bray to the India Office, July 28, 1919, Public Records Office, MSS Foreign Office, Vo. 4147, Document No E 129678.

4   Hafiz Wahba, *Arabian Days* (London: Constable, 1928), 134–6.
5   Joseph Kostiner, *The Making of Saudi Arabia 1916–1936* (New York: Oxford University Press, 1993), 115.
6   Habib, *Warriors of Islam*, 70–6.
7   Glubb, *War in the Desert*, 126.
8   Gary Troeller, *The Birth of Saudi Arabia: Britain and the Rise of the House of Saud* (London: Frank Cass, 1976), 130.
9   Glubb, *War in the Desert*, 127.
10 Ameen Rihani, *Maker of Modern Arabia* (New York: Houghton Mifflin, 1928), 55.
11 Alexei Vassiliev, *The History of Saudi Arabia* (New York: New York University Press, 2000), 275.
12 H. R. P. Dickson, *Kuwait and Her Neighbours* (London: George Allen & Unwin, 1956), 285–6.
13 Vassiliev, *The History of Saudi Arabia*, 277.
14 Busaiya was a small fort located roughly seventy-five miles north of the Saudi Iraqi border. It had been built primarily to control the Iraqi branch of the Shammar tribe from raiding into the Nejd. On the night of November 5, 1927, Faisal Daweesh and his Mutair tribesmen massacred the small construction crew building the fort in what could be deemed an Ikhwan freedom-of-navigation exercise, but the fort was soon rebuilt—and others like it added.
15 Armstrong, *Lord of Arabia*, 209.
16 Robert Lacey, *The Kingdom* (New York: Harcourt, Brace, Jovanovich, 1981), 275.
17 Habib, *Warriors of Islam*, 132.
18 Madawi Al Rasheed, *A History of Saudi Arabia* (Cambridge: Cambridge University Press, 2002), 68; and Kostiner, *The Making of Saudi Arabia*, 131.
19 Habib, *Warriors of Islam*, 134–5.
20 Kostiner, *The Making of Saudi Arabia*, 132.
21 Muhammed Asad, *The Road to Mecca* (New York: Simon and Shuster, 1953), 225.
22 Kostiner, *The Making of Saudi Arabia*, 133, extracted from Glubb's March report to the Political and Secret Department of the India Office.
23 Habib, *Warriors of Islam*, 140–1.
24 Mohammed Al Mana, *Arabia Unified: A Portrait of Ibn Saud* (London: Hutchinson Benham, 1980), 107 and 114. Al Mana's account is the only one published in English by an author who was actually present at the battle, though Habib interviewed several survivors in the 1960s.
25 Habib, *Warriors of Islam*, 145.
26 Vassiliev, *The History of Saudi Arabia*, 139.
27 Al Mana, *Arabia Unified*, 127–8.
28 H. St. John Philby's *Saudi Arabia* (London: Ernest Benn, 1955) reprinted by Librairie du Liban 1968, 309.
29 David Howarth, *The Desert King: A Life of Ibn Saud* (London: Collins, 1964), 183.
30 Clive Leatherdale, *Britain and Saudi Arabia 1925–1939: The Imperial Oasis* (London: Frank Cass, 1983), 11.
31 Vassiliev, *The History of Saudi Arabia*, 282.
32 Kostiner, *The Making of Saudi Arabia*, 154–5.
33 Vassiliev, *The History of Saudi Arabia*, 283.
34 Leatherdale, *Britain and Saudi Arabia*, 136; and Vassiliev, *The History of Saudi Arabia*, 283–4.
35 Habib, *Warriors of Islam*, 4.

36 Leslie McLoughlin, *Ibn Saud: Founder of a Kingdom* (New York: St. Martin's Press, 1993), 123.
37 Al Mana, *Arabia Unified*, 203.
38 Vassiliev, *The History of Saudi Arabia*, 285.
39 Kostiner, *The Making of Saudi Arabia*, 171, extracted from the Aden Political Intelligence Summary for the Week ending April 24, 1934 to the Political and Secret Department of the India Office.
40 Al Mana, *Arabia Unified*, 209.
41 McLoughlin, *Founder of a Kingdom*, 131.
42 Kostiner, *The Making of Saudi Arabia*, 171.
43 Thus, Abdulaziz bin Abd al-Rahman al-Saud (Ibn Saud) became Amir of Riyadh in 1902, Sultan of the Nejd in 1921, King of the Hejaz in 1926, and King of Saudi Arabia in 1932.

Part II

1 Dickson quoted in Ali al-Naimi, *Out of the Desert* (New York: Penguin Random House, 2016), 14.

Chapter 4

1 Brian Lees, *A Handbook of the Al Saud Ruling Family of Saudi Arabia* (London: Royal Genealogies, 1980)—still the most accurate and comprehensive study available of the Al Saud.
2 Madawi Al Rasheed, *A History of Saudi Arabia* (Cambridge: Cambridge University Press, 2002), 73.
3 Gary Troeller, *The Birth of Saudi Arabia: Britain and the Rise of the House of Saud* (London: Frank Cass, 1976), Appendix III, 249–56.
4 Alexei Vassiliev, *King Faisal of Saudi Arabia: Personality, Faith and Times* (London: Saqi Books, 2012), 185. Some authors have suggested that Khalid bin Mohammad bin Abdulrahman was murdered during a hunting trip because he was a rival to his cousin, Crown Prince Saud bin Abdulaziz. The author has discussed this point with members of the royal family who dismiss the story as out of hand. They agree that Khalid was hunting gazelle from a speeding car in open desert, that the car flipped, and he was thrown out and killed. Hunting gazelle from a speeding car was an Al Saud pastime until these animals were nearly driven to extinction. King Abdulaziz was a frequent participant in this "sport." Gazelle are now protected from hunting in Saudi Arabia though the Al Saud still speed across the desert following their hunting falcons. This too is a harrowing experience that can easily end unhappily.
5 Leslie McLoughlin, *Ibn Saud: Founder of a Kingdom* (New York: St. Martin's Press, 1993), 192.
6 Mohammed Al Mana, *Arabia Unified: A Portrait of Ibn Saud* (London: Hutchinson Benham, 1980), 213.
7 Parker Hart, *Saudi Arabia and the United States: Birth of a Security Partnership* (Bloomington, IN: Indiana University Press, 1998), 98.
8 Alexander Bligh, *From Prince to King: Royal Succession in the House of Saud in the Twentieth Century* (New York: New York University Press, 1984), 44.

9   Sarah Yizraeli, *The Remaking of Saudi Arabia: The Struggle Between King Saud and Crown Prince Faysal, 1953–1962* (Tel Aviv: Tel Aviv University Press, 1997), 186.

10 Gerald De Gaury, *Faisal: King of Saudi Arabia* (New York: Praeger, 1967), 79.

11 David Holden and Richard Johns, *The House of Saud: The Rise and Fall of the Most Powerful Dynasty in the Arab World* (London: Sidgwick & Jackson, 1981), 174.

12 De Gaury, *Faisal: King of Saudi Arabia*, 79.

13 Elizabeth Monroe, *Philby of Arabia* (New York: Pitman Publishing, 1973), 274.

14 Arthur N. Young, *Saudi Arabia: The Making of a Financial Giant* (New York: New York University Press, 1980), 125, 130.

15 Ibid., 23

16 Robert Lacey, *The Kingdom* (New York: Harcourt, Brace, Jovanovich, 1981), 300; and Mark Weston, *Prophets and Princes: Saudi Arabia from Muhammad to the Present* (New York: Wiley, 2008), 174.

17 José Arnold, *Golden Swords and Pots and Pans* (London: V. Gollancz, 1964), 142 and 222–6.

18 Vassiliev, *King Faisal of Saudi Arabia*, 194.

19 Weston, *Prophets and Princes*, 177; and De Gaury, *Faisal: King of Saudi Arabia*, 95.

20 Reported by Alistair Cooke, *Manchester Guardian*, January 30, 1957.

21 David Holden and Richard Johns, *The House of Saud*, 194.

22 De Gaury, *Faisal: King of Saudi Arabia*, 90–1.

23 Weston, *Prophets and Princes*, 180.

24 David Holden and Richard Johns, *The House of Saud*, 208.

25 Alexei Vassiliev, *The History of Saudi Arabia* (New York: New York University Press, 2000), 358.

26 Ibid., 360.

27 Ibid., 363.

28 De Gaury, *Faisal: King of Saudi Arabia*, 119; Nadav Safran, *Saudi Arabia: The Ceaseless Quest for Security* (New York: Cornell University Press, 1988), 121; and Bruce Riedel, *Kings and Presidents: Saudi Arabia and the United States Since FDR* (Washington, DC: Brookings Institution Press, 2018), 38.

29 Lacey, *The Kingdom*, 346.

30 Vassiliev, *King Faisal of Saudi Arabia*, 254–5; and Weston, *Prophets and Princes*, 190.

31 De Gaury, *Faisal: King of Saudi Arabia*, 65; and Lacey, *The Kingdom*, 356.

32 Lacey, *The Kingdom*, 356; and De Gaury, *Faisal: King of Saudi Arabia*, 130–3.

33 Vassiliev, *King Faisal of Saud Arabia*, 253. The author has also discussed these events with Saudi religious scholars, who were very clear that not all of the *ulama* were prepared to depose King Saud and that at least two who opposed the move were detained in a hotel until the abdication had been publicly announced.

34 Vassiliev, *King Faisal of Saudi Arabia*, 260.

35 Weston, *Prophets and Princes*, 192.

36 Lacey, *The Kingdom*, 351.

37 Vassiliev, *King Faisal of Saudi Arabia*, 251.

38 Rachel Bronson, *Thicker Than Oil: America's Uneasy Partnership with Saudi Arabia* (Oxford: Oxford University Press, 2006), 92.

39 Parker T. Hart, *Saudi Arabia and the United States: Birth of a Security Partnership* (Bloomington, IN: Indiana University Press, 1998), 98.

Chapter 5

1   Joseph A. Kechichian, *Succession in Saudi Arabia* (New York: Palgrave, 2010), 43.

2   Gerald De Gaury, *Faisal: King of Saudi Arabia* (New York: Praeger, 1967), 132.

3   Summer Scott Huyette, *Political Adaption in Saudi Arabia: A study of the Council of Ministers* (Boulder, CO: Westview Press, 1985), 55–7.

4   Alexander Bligh, *From Prince to King: Royal Succession in the House of Saud in the Twentieth Century* (New York: New York University Press, 1984), 89.

5   Their mother, Hussa bint Ahmed al-Sudairi, had twelve children: seven sons and four daughters with King Abdulaziz and one son with his brother, Mohammed; King Abdulaziz married Hussa—the daughter of his ally, Ahmed al-Sudairi—but divorced her when she did not seem able to produce a son. She then married his brother, Mohammed, and bore a son who eventually became commander of the Saudi Navy. King Abdulaziz then decided he wanted to marry Hussa again, and asked his brother to divorce her so that he could do so. Hussa subsequently bore the king the Sudairi Seven, including the current King Salman.

6   Joseph A. Kechichian, *Succession in Saudi Arabia* (New York: Palgrave, 2010), 46.

7   De Gaury, *Faisal: King of Saudi Arabia*, 135.

8   Kechichian, *Succession in Saudi Arabia*, 45.

9   David Holden and Richard Johns, *The House of Saud: The Rise and Fall of the Most Powerful Dynasty in the Arab World* (New York: Holt, Rinehart and Winston, 1981), 201.

10 Mark Weston, *Prophets and Princes: Saudi Arabia from Muhammad to the Present* (New York: Wiley, 2008), 204.

11 Ibid., 180.

12 Bligh, *From Prince to King*, 90.

13 Ibid., 91, citing a lecture given in 1976 at Columbia University by the US Ambassador to Saudi Arabia at the time, James Akins.

14 Robert Lacey, *The Kingdom* (New York: Harcourt, Brace, Jovanovich, 1981), 26.

15   Kechichian, *Succession in Saudi Arabia*, 55.

16 Hasan bin Abdullah Al al-sheikh was minister of higher education, Ibrahim bin Mohammad Al al-sheikh was minister of justice, and Abdulaziz Al al-sheikh was minister of agriculture.

17 Bligh, *From Prince to King*, 92

18 Yaroslav Trofimov, *The Siege of Mecca* (New York: Doubleday, 2007), 79.

19 Quran, Sura 2, Verse 191.

20 Trofimov, *The Siege of Mecca*, 85.

21 Robert Lacey, *Inside the Kingdom: Kings, Clerics, Modernists, Terrorists and the Struggle for Saudi Arabia*, (London: Hutchinson, 2009), 33.

22 Trofimov, *The Siege of Mecca*, 254.

23 Alexei Vassiliev, *The History of Saudi Arabia* (New York: New York University Press, 2000), 465; and Trofimov, *The Siege of Mecca*, 240–1.

24 Lacey, *Inside the Kingdom*, 95.

25 Andrew Scott Cooper, *The Fall of Heaven, The Pahlavis and the Final Days of Imperial Iran* (New York: Henry Holt, 2016), 107, 110.

26 Kechichian, *Succession in Saudi Arabia*, 57.

27 Weston, *Prophets and Princes*, 154.

28 Bureau of Experts at the Council of Ministers Official Translation Department, "Translations of Saudi Laws Series II," Riyadh, 2003, The Basic Law of Governance, Part II, Articles 5 and 6.

29 Madawi Al Rasheed, "Mystique of Monarchy; the Magic of Royal Succession in Saudi Arabia," in *Salman's Legacy*, (ed.) Madawi Al Rasheed (Oxford: Oxford University Press, 2018), 56.
30 Saudi Law of the Allegiance Institution, Article 7.

Chapter 6

This chapter is based on the author's personal experience as well as dozens of conversations over the past five years with a wide range of informed sources, including consultants, bankers, businessmen, diplomats, and government officials—some of whom have worked with Mohammed bin Salman for more than a decade. Only a few specific cases are cited.

1 Author's conversation with a former minister and deputy minister, February 2018.
2 Author's conversation with management consultants in November 2018.
3 Thomas L. Friedman, "Saudi Arabia's Arab Spring, at Last," *The New York Times*, November 23, 2017.
4 Author's conversation with a Saudi central banker, December 2017.
5 Author's conversation with a royal court advisor, June 2016.
6 Author's conversation with management consultants in February 2018.
7 Ibid.
8 NEOM is a combination of the first three letters form the Ancient Greek prefix neo for new; and the first letter of Arabic word *mustaqbal* for future.

Part III

1 The large and diverse religious establishment that exists in Saudi Arabia is led by the *ulama* but, unlike in the Christian Church, there is no ordained clergy in Sunni Islam, nor is there a formal ecclesiastic hierarchy such as exists in Shia Islam.

Chapter 7

1 Christine Moss Helms, *The Cohesion of Saudi Arabia* (Baltimore: The Johns Hopkins University Press, 1981), 60; and author's conversation with a Ministry of the Interior official, February 2017, in Riyadh.
2 Wilfred Thesiger, *Arabian Sands* (New York: Penguin Books, 1978), 17.
3 Carl Raswan, *The Black Tents of Arabia: My Life Amongst the Bedouins*, (New York: Creative Age Press, 1947), 24.
4 Louise E. Sweet, "Camel Raiding of North Arabian Bedouin: A Mechanism of Ecological Adaptation," *American Anthropologist* 67 (1965): 1132–50.
5 William Lancaster, *The Rwala Bedouin Today* (Long Grove: Waveland Press, 1981), 59.
6 Helms, *The Cohesion of Saudi Arabia*, 54.
7 Lancaster, *The Rwala Bedouin*, 95.
8 H. R. P. Dickson, *The Arab of the Desert* (London: George Allen & Unwin Ltd, 1959), 114–16.

9   Helms, *The Cohesion of Saudi Arabia*, 52.
10 Author's conversations with numerous tribal sheiks including those of the 'Anizah in Riyadh, May 2017.
11 Abdulaziz Fahd, "The 'Imama vs the 'Iqal: Hadari-Bedouin Conflict and the Formation of the Saudi State in Counter Narratives History and Contemporary Society and Politics," in *Saudi Arabia and Yemen*, Madawi Al-Rasheed and Robert Vitalis (eds) (London: Palgrave, 2004), 74.
12 Donald Powell Cole, *Nomads of the Nomads, The Al Murrah Bedouin of the Empty Quarter* (Illinois: AHM Publishing Corporation, 1975), 96; ibid. Lancaster, *The Rwala Bedouin*, 87.
13 Author discussions with numerous *sheikhs* of different Saudi tribes or tribal sections.
14 Anthony Cordesman, *Saudi Arabia Enters the Twenty-First Century: The Political, Foreign Policy, Economic, and Energy Dimensions* (Westport CT: Praeger, 2003), 262; and author's discussions.
15 Fahd, "The 'Imama vs the 'Iqal," 55.
16 Lancaster, *The Rwala Bedouin*, 112.
17 Hasan Hamza Hajrah, Public Land Distribution in Saudi Arabia (London: Longman, 1982).
18 Ameen Rihani, *Maker of Modern Arabia* (New York: Houghton Mifflin, 1928), 188.
19 Edward E. Evans-Pritchard, *The Sanusi of Cyrenaica* (Oxford: Clarendon Press, 1949), 62.
20 Thomas Edward Lawrence, *Seven Pillars of Wisdom* (New York: Doubleday, Doran and Company, 1926), 31.
21 Lancaster, *The Rwala Bedouin*, 109.
22 Ibid., 78–81; and conversations with tribal leaders during the author's dozens of trips to rural tribal areas between 1981 and 2019.
23 Lancaster, *The Rwala Bedouin*, 131.
24 Pascal Ménoret, *The Saudi Enigma: A History* (New York: Zed Books Ltd, 2005), 90.
25 Cordesman, *Twenty-First Century*, 170; and the author's many conversations with SANG officers, former officers, and foreign advisors to the SANG.
26 Mordechai Abir, *Saudi Arabia: Government, Society and the Gulf Crisis* (London: Routledge, 1993), xvi and 8, Ahmed A. Shamekh, "Bedouin Settlements," *Ekistics* 43, no. 258 (1977): 249. Accessed February 1, 2020. www.jstor.org/stable/43618835, Mark Weston, *Prophets and Princes: Saudi Arabia from Muhammad to the Present* (New Jersey: Hohn Wiley & Sons, 2008), 235, and World Bank Development statistics.
27 Ibn Khaldûn, *The Muqaddimah: An Introduction to History*, translated from the Arabic by Franz Rosenthal (New York: Pantheon Books, 1958), 136–8.

Chapter 8

1   Alexei Vassiliev, *King Faisal of Saudi Arabia: Personality, Faith and Times* (London: Saqi Books, 2012), 273.
2   David Holden and Richard Johns, *The House of Saud: The Rise and Fall of the Most Powerful Dynasty in the Arab World* (New York: Holt, Rinehart and Winston, 1981), 229. The importation of slaves into Saudi Arabia had been illegal since the 1930s but owning, buying, and selling slaves was still legal until 1962, when they comprised roughly 1 percent of the local population.

3 David Commins, *The Wahhabi Mission and Saudi Arabia* (London: I.B. Tauris, 2006), 5; and www.BBC.com/news/world-middle-east-26487092.

4 www.BBC.com/news/world-middle-east-41260543.

5 A. K. S. Lambton, "Islamic Political Thought," in Joseph Schacht and C. E. Bosworth, *The Legacy of Islam* (Oxford: Oxford University Press, 1974), 404.

6 Saudi Basic Law of Governance, Article 1 Kingdom of Saudi Arabia, Bureau of Experts of the Council of Ministers Official Translation Department, Riyadh, 2003.

7 Natana DeLong-Bas, *Wahhabi Islam: From Revival and Reform to Global Jihad* (New York: Oxford University Press, 2004), 283. Abd al-Wahhab defined monotheism (*tawhid*) as recognizing God alone as the creator and sustainer of the universe. God is unique and cannot be compared in any way to anything else. He is certainly no anthropomorphic, long-bearded elder sitting on a cloud. Worshiping any other beings or objects, and even comparing them to God, is a form of polytheism (*shirk*) and a mortal sin.

8 Saudi Basic Law of Governance, Article 23.

9 Frank Vogel, *Islamic Law and Legal System: Studies of Saudi Arabia* (Leiden: Brill, 2000), 76. Abd al-Wahhab produced no unprecedented opinions, and Saudi authorities today regard him not as an independent thinker in Islamic jurisprudence (*fiqh*), but rather as a preacher of religious reawakening.

10 Noah Feldman, *The Fall and Rise of the Islamic State* (Princeton: Princeton University Press, 2008), 33.

11 DeLong-Bas, *Wahhabi Islam*, 247.

12 Commins, *The Wahhabi Mission*, 25 and 83.

13 Ameen Rihani, *Maker of Modern Arabia* (New York: Houghton Mifflin, 1928), 202.

14 Quran, Sura 4, Verse 59.

15 Lambton, "Islamic Political Thought," 416; Guido Steinberg, "The Wahhabi Ulama and the Saudi State: 1745 to the Present," in Paul Aarts and Gerd Nonneman (eds), *Saudi Arabia in the Balance* (New York: New York University Press, 2005), 17.

16 Saudi Basic Law of Governance, Article 6.

17 Madawi Al Rasheed, *Contesting the Saudi State: Islamic Voices from a New Generation* (Cambridge: Cambridge University Press, 2007), 256.

18 Mohammed Al Sulaimi, "Saudi Gets 23 Years for Takfirism," *Arab News*, May 19, 2016, 2018 Religious Freedom Report, US State Department, 10.

19 Aabdullah Al Uthaymin, *Muhammad ibn Abd al-Wahhab: The Man and his Works* (London: I.B. Tauris, 2009), 131, 144.

20 Hamid Algar, *Wahhabism: A Critical Essay* (Oneonta New York: Islamic Press International, 2002), 31–6, 51, 69.

21 Commins, *The Wahhabi Mission*, 33–4.

22 *Arab News*; "Pope Francis Arrives in UAE for Historic Visit," February 3, 2019, and *Saudi Gazette*, "New Chapter in History of Fraternity and Tolerance," February 3, 2019.

23 Abdulaziz H. Al Fahad, "*From Exclusivism to Accommodation: Doctrinal and Legal Evolution of Wahhabism*," New York University Law Review (May 2004), 496–500.

24 R. Bayly Winder, *Saudi Arabia in the Nineteenth Century* (London: Macmillan, 1965), 20.

25 Robert Lacey, *The Kingdom* (New York: Harcourt, Brace, Jovanovich, 1981), 62.

26 Winder, *Nineteenth Century*, 65–6.

27 Ibid., 87.

28 Commins, *The Wahhabi Mission*, 73. The religious sheikh, Abdullah ibn Amr of Buraydah, was executed in 1908 for strident opposition to the Wahhabi mission.

29 Madawi Al Rasheed, *A History of Saudi Arabia* (Cambridge: Cambridge University Press, 2002), 49.

30 Christine Moss Helms, *The Cohesion of Saudi Arabia* (London: Croom Helm, 1981), 131.

31 Ameen Rihani, *Maker of Modern Arabia* (New York: Houghton Mifflin, 1928), 202–3.

32 Commins, *The Wahhabi Mission*, 76.

33 H. St. John Philby's *Saudi Arabia* (London: Ernest Benn, 1955) reprinted by Librairie du Liban 1968, 268.

34 Guido Steinberg, "The Shiites in the Eastern Province of Saudi Arabia 1913-1953", in Rainer Brunner and Werner Ende (eds), *The Twelver Shia in Modern Times: Religious Culture and Political History* (Leiden: Brill, 2001), 248–54.

35 H. St. John Philby, *Arabia of the Wahhabis* (London: Frank Cass, 1928), 66–7.

36 Mai Yamani, *Cradle of Islam: The Hijaz and the Quest for Identity* (London: I.B. Tauris, 2009), 10.

37 Alexei Vassiliev, *The History of Saudi Arabia* (New York: New York University Press, 2000), 269.

38 David Howarth, *The Desert King: A Life of Ibn Saud* (London: Collins, 1964), 154.

39 Rihani, *Maker of Modern Arabia*, 201.

40 Lacey, *The Kingdom*, 243.

41 Rihani, *Maker of Modern Arabia*, 200; Al-Rasheed, *History of Saudi Arabia*, 68.

42 F. Gregory Gause III, *Oil Monarchies: Domestic and Security Challenges in the Arab Gulf States* (New York: Council on Foreign Relations Press, 1994), 11.

43 Nabil Mouline, "Enforcing and Reinforcing the State's Islam: The Functioning of the Committee of Senior Scholars", in Bernard Haykel, Thomas Hegghammer and Stéphane Lacroix (eds), *Saudi Arabia in Transition* (Cambridge: Cambridge University Press, 2015), 59–63.

44 The Presidency for Proselytization seeks to "promote, defend and propagate Islam." It is headed by the grand mufti and has a large administrative staff that provides research for its senior scholars. The Presidency publishes and distributes its own journal, *The Review of Religious Research*, as well as religious books—most notably, those by Ibn Taymiyyah and Mohammed Abd al-Wahhab—supervises Quran printing in Saudi Arabia, and provides religious instruction for pilgrims during the Hajj. The Commission for Religious Rulings is a smaller body composed of between four and seven scholars, who are easy to reach in person or by mail, phone, and internet. In response to religious questions from the general public, they promptly issue rulings dealing with personal matters.

45 Mouline, "Enforcing and Reinforcing," 63.

46 Gause, *Oil Monarchies*, 17. See also Alexander Bligh, "The Saudi Religious Elite (Ulama) as a Participant in the Political System of the Kingdom," *International Journal of Middle East Studies* 17, no. 1 (February 1985); Joseph Kechichian, "The Role of the Ulama in the Politics of an Islamic State: The Case of Saudi Arabia", *International Journal of Middle East Studies* 18, no. 1 (February 1986).

47 Commins, *The Wahhabi Mission*, 180.

48 Mouline, "Enforcing and Reinforcing," 66.

49 Ali Al Shihabi, *The Saudi Kingdom: Between the Jihadi Hammer and the Iranian Anvil* (Princeton: Marcus Wiener, 2016), 39, 50.

50 Christopher Boucek, *Saudi Fatwa Restrictions and the State-Clerical Relationship*, Carnegie Endowment for International Peace, October 27, 2010, https://carnegieendowment.org/sada/41824.

51 Joseph Schacht, "Islamic Law," in Joseph Schacht and C. E. Bosworth, *The Legacy of Islam* (Oxford: Oxford University Press, 1974), 392. *Sharia* law designates all human activity as

obligatory, recommended, indifferent, reprehensible, or forbidden. For a *sharia* judge (*qadi*), it is more often the letter of the divinely inspired law rather than a human interpretation of its spirit that matters most. *Sharia* law tends to emphasize compensation for victims more than punishment for defendants, but a judge may impose discretionary punishment or *ta'zir* for any action that, in his view, calls for punishment.

52 Noah Feldman, *The Fall and Rise of the Islamic State* (Princeton: Princeton University Press, 2008), 92.

53 Mouline, "Enforcing and Reinforcing," 52–3.

54 Soraya Altorki and Donald Cole, *Arabian Oasis City: The Transformation of Unayzah* (Austin: University of Texas Press, 1989), 91; and Lacey, *The Kingdom*, 176–7.

55 Author's conversation with Ghaith Pharaon, Jeddah, 1992.

56 Mohammed bin Salman interviews, *TIME*, April 2018; Bloomberg, October 2018.

57 John Sabini, *Armies in the Sand: The Struggle for Mecca and Medina* (London: Thames and Hudson, 1981), 79.

58 Commins, *The Wahhabi Mission*, 62.

59 Steinberg, "The Wahhabi Ulama and the Saudi State," 13, 22.

60 Mamoun Fandy, *Saudi Arabia and the Politics of Dissent* (New York: Palgrave, 2001), 61, 89 for detailed accounts of Safar al-Hawali and Salman al-Auda.

61 Steinberg, "The Wahhabi Ulama and the Saudi State," 32.

Chapter 9

1   Kiren Aziz Chaudhry, *The Price of Wealth*, (Ithaca N.Y.: Cornell University Press, 1997), 54.

2   Soraya Altorki and Donald Cole, *Arabian Oasis City: The Transformation of Unayzah* (Austin: University of Texas Press, 1989), 13.

3   Gary Troeller, *The Birth of Saudi Arabia: Britain and the Rise of the House of Saud* (London: Frank Cass, 1976), 183.

4   Madawi Al Rasheed, *Politics in an Arabian Oasis: The Rashidis of Saudi Arabia* (London: I.B. Tauris, 1991), 95, 103, 111, 127, 230.

5   Johann Ludwig Burckhardt, *Notes on the Bedouin and Wahabys* (London: Forgotten Books, 2015), 190–1.

6   Sharaf Sabri, *The House of Saud in Commerce* (New Delhi: I.S. Publications, 2001), 10.

7   Chaudhry, *The Price of Wealth*, 50.

8   Michael Field, *The Merchants* (London: John Murray, 1984), 99.

9   Arthur Young, *Saudi Arabia: The Making of a Financial Giant* (New York: New York University Press, 1983), 126, 130.

10 Michael Field, *From Unayzah to Wall Street: The Story of Suliman S. Olayan* (London: John Murray, 2000), 24.

11 Huda Al Subeaei and Gene Heck, *Mohammed Al Subeaei: A Journey of Poverty and Wealth* (Newport, Isle of Wight: Medina Publishing, 2010), 79.

12 Steven Coll, *The Bin Ladens* (New York: Penguin, 2008), 29.

13 Ragaei El Mallakh, *Saudi Arabia: Rush to Development* (London: Croom Helm, 1982), 112.

14 Mordechai Abir, *Saudi Arabia: Government, Society and the Gulf Crisis* (London: Routledge, 1993), 215.

15 Field, *The Merchants*, 341.

16 Field, *From Unayzah to Wall Street*, 30.

17 Sabri, *The House of Saud in Commerce*, 103; and Fourth Saudi Development Plan, 233.
18 Chaudry, *The Price of Wealth*, 94.
19 Altorki and Cole, *Arabian Oasis City*, 140.
20 Field, *The Merchants*, 112.
21 See Sabri, *The House of Saud in Commerce*, for the most detailed published listing.
22 Steffen Hertog, *Princes, Brokers, and Bureaucrats: Oil and the State in Saudi Arabia*
    (Ithaca: Cornell University Press, 2010), 251.
23 Abir, *Government, Society and the Gulf Crisis*, 115.
24 World Bank, Saudi Arabia Economic Update.

Chapter 10

1   Frank Tachau (ed.), *Political Elites and Political Development in the Middle East* (New
    York: John Wiley, 1975), 16.
2   Mohammed Almana, *Arabia Unified* (London: Hutchinson Benham, 1980), 153.
3   Manfred Wenner, "Saudi Arabia: Survival of Traditional Elites," in Tachau (ed.),
    Political Elites and Political Development, 175.
4   Al Mana, *Arabia Unified*, 174.
5   Steffen Hertog, *Princes, Brokers, and Bureaucrats: Oil and the State in Saudi Arabia*
    (Ithaca: Cornell University Press, 2010), 83.
6   Ibid., 38 and 69.
7   The Law of the Council of Ministers, Article 19, Kingdom of Saudi Arabia, Bureau of
    Experts of the Council of Ministers, Official Translation Department, Translation of
    Saudi Laws Series II, 2003.
8   Norman Walpole, *Area Handbook for Saudi Arabia* (Washington D.C.: US
    Government Printing Office, 1965), 136.
9   Mordechai Abir, *Saudi Arabia: Government, Society and the Gulf Crisis* (London:
    Routledge, 1993), 118.
10 In early 2017, Rotana was owned by Prince Waleed bin Talal. The Middle East
    Broadcasting Center was owned by King Fahd's brother-in-law, Waleed bin Ibrahim Al
    Ibrahim and his son, Abdulaziz bin Fahd. The Arab Radio and Television Network was
    owned by Saleh Kamel. Orbit Showtime was owned in part by the Mawarid Group,
    controlled by Prince Khalid Abdullah bin Abd al-Rahman. Prince Waleed, Saleh
    Kamel, and Waleed al Ibrahim were all involved in the November 2017 anti-
    corruption round-up, and may no longer control these networks; in which case they
    are now even more firmly under government control.
11 Saudi Basic Law of Governance, Article 39.
12 Author's conversation in Riyadh Ministry of the Interior offices, interview, March 2012.
13 Saudi journalist jailed for accusing royal court over corruption, *The Financial Times*,
    February 8, 2018.
14 Saudi economist who criticized Aramco IPO charged with terrorism—Reuters,
    October 1, 2018.
15 There had been several earlier versions of the *Majlis al-Shura* established by King
    Abdulaziz when he conquered the Hejaz. Initially, some members had been elected.
    Although never formally abolished, the once significant powers of the *Majlis al-Shura*
    had been eroded—first by the Viceroy to the Hejaz, Prince Faisal bin Abdulaziz, and
    later by the growing importance of the Council of Ministers. The *Majlis al-Shura* was
    essentially moribund by 1991, when King Fahd announced that a new Consultative

Council would be created. In typical Saudi fashion, it took a year for the members to be appointed and another year before the Council actually met.

16 Author's conversation in Riyadh with *Majlis Al-Shura* member, February 2015.
17 Author's conversation in Riyadh with female former *Majlis Al-Shura* member, April 2016.
18 Author's conversation in Riyadh with Saudi bankers, March 2018.

Chapter 11

1 Madawi Al Rasheed, *A History of Saudi Arabia* (Cambridge: Cambridge University Press, 2002), 15; and J. G. Larimer, *Gazetteer of the Persian Gulf, Oman and Central Arabia*, 2 vol., Government Printing House, Calcutta, 1908.
2 Alexei Vassiliev, *The History of Saudi Arabia* (New York: New York University Press, 2000), 84.
3 The remaining sons of King Abdulaziz are Abd al-Ellah, Ahmed, Muqrin, Mamduh, and King Salman.
4 Michael Herb, *All in the Family: Absolutism, Revolution, and Democracy in the Middle Eastern Monarchies* (Albany: State University of New York Press, 1999), 39.
5 Brian Lees, *A Handbook of the Al Saud Ruling Family of Saudi Arabia* (London: Royal Genealogies, 1980), 39.
6 Leslie McLoughlin, *Ibn Saud: Founder of a Kingdom* (New York: St. Martin's Press, 1993), 56. According to H. St. John Philby, King Abdulaziz had fifteen sisters.
7 Madawi Al Rasheed, "Circles of Power: Royals and Saudi Society," in Paul Aarts and Gerd Nonneman (eds), *Saud Arabia, the Balance, Political Economy, Society and Foreign Affairs* (New York: New York University Press, 2005), 192.
8 Ibn Khaldun, *The Muqaddimah*, translated by Franz Rosenthal, edited by N. J. Dawood (Princeton: Princeton University Press, 2015), 136–8.
9 Mordechai Abir, *Saudi Arabia: Government, Society and the Gulf Crisis* (London: Routledge, 1993), 8.
10 Khalid bin Sultan Al Saud, *Desert Warrior* (New York: Harper Collins, 1995), 47.
11 Joseph A. Kechichian, *Succession in Saudi Arabia* (New York: Palgrave, 2010), 98.
12 Michael Field, *The Merchants* (London: John Murray, 1984), 99–101.
13 Ibid., 113.
14 Mamoun Fandy, *Saudi Arabia and the Politics of Dissent* (New York: Palgrave, 2001), 35.
15 Thomas L. Friedman, "America's Dilemma in Censuring MBS and Not Halting Saudi Reforms," *The New York Times*, October 16, 2018.
16 Field, *The Merchants*, 102.
17 Al Rasheed, "Circles of Power," 188.
18 Bradley Hope and Tom Wright, *Billion Dollar Whale: The Man Who Fooled Wall Street, Hollywood and the World* (New York: Hachette Books, 2018), 61–6.

Chapter 12

1 Mohammed Al Mana, *Arabia Unified: A Portrait of Ibn Saud* (London: Hutchinson Benham, 1980), 91.
2 H. St. John Philby, *Arabian Jubilee* (London: Robert Hale, 1952), 77.

3   Robert Vitals, *America's Kingdom: Mythmaking on the Saudi Oil Frontier* (Stanford: Stanford University Press, 2007), 150.

4   David Holden and Richard Johns, *The House of Saud: The Rise and Fall of the Most Powerful Dynasty in the Arab World* (New York: Holt, Rinehart and Winston, 1981), 278.

5   Thomas Hegghammer, *Jihad in Saudi Arabia* (Cambridge: Cambridge University Press, 2010), 237.

6   "Address to the Nation by Crown Prince Abdullah," May 13, 2003, in "Public Statements by Senior Saudi Officials Condemning Extremism and Promoting Moderation" (Washington, DC: Saudi Embassy, September 2004), 19–20.

7   Thomas Small and Jonathan Hacker, *Path of Blood: The Story of Al Qaeda's War on the House of Saud* (New York: Overlook Press, 2015), 158; and Anthony Cordesman, "Saudi Arabia: Opposition, Islamic Extremism and Terrorism," in *Saudi Arabia Enters the Twenty-First Century: The Political, Foreign Policy, Economic, and Energy Dimensions,* Anthony Cordesman (Westport CT: Praeger, 2003), 4.

8   Robert Lacey, *Inside the Kingdom: Kings, Clerics, Modernists, Terrorists and the Struggle for Saudi Arabia* (London: Hutchinson, 2009), 245.

9   Hegghammer, *Jihad in Saudi Arabia*, 165, 182.

10 Ibid., 175.

11 Ibid., 217.

12 "Saudi US Cooperation in War on Terror up Sharply," *Arab News*, October 25, 2003.

13 Small and Hacker, *Path of Blood*, 379.

14 Christopher Boucek, "Saudi Arabia's 'Soft' Counterterrorism Strategy," paper for the Carnegie Endowment for Peace, Washington, DC, 2008; and Hegghammer, *Jihad in Saudi Arabia*, 218.

15 "The Final Report of the National Commission on the Terrorist Attacks Upon the United States," informally known as the "The 9/11 Commission Report," 171.

16 Jon Alterman, *Understanding Islamic Charities* (Washington, DC: Center for Strategic and International Studies), 2007, 76.

17 Zarate quoted in Small and Hacker, *Path of Blood*, 232.

18 Ibid., 128.

19 Mark Weston, *Prophets and Princes: Saudi Arabia from Muhammad to the Present* (New York: Wiley, 2008), 410.

20 Ibid., 410.

21 Lacey, *Inside the Kingdom*, 257.

22 Angel Rabasa, Stacie L. Pettyjohn, Jeremy J. Ghez, and Christopher Boucek, *Deradicalizing Islamic Extremists* (Santa Monica: Rand Corporation, 2010), 75; "Saudis Say 12% of Rehabilitated Terrorists Have Returned to Terror," *The Washington Post*, November 28, 2014; and John Habib, *Striking Back: The Saudi War Against Terrorism* (CreateSpace Independent Publishing Platform, 2014), 137.

23 Hegghammer, *Jihad in Saudi Arabia*, 219.

24 "How One OFW Survived The Alkhobar Attacks," *Arab News*, June 4, 2004.

25 John Habib, *Ibn Saud's Warriors of Islam: The Ikhwan of the Najd and their role in the creation of the Saudi Kingdom, 1910–1930* (Leiden, E. J. Brill, 1978), 145.

26 Ibid., 159.

Chapter 13

1   Paul Rivlin, *Arab Economies in the Twenty-First Century* (Cambridge: Cambridge University Press, 2009), 220.

2   David Holden and Richard Johns, *The House of Saud: The Rise and Fall of the Most Powerful Dynasty in the Arab World* (New York: Holt, Rinehart and Winston, 1981), 257.

3   www.worldpopulationreview.com/countries/life-expectancy.

4   Roger Owen and Sevket Pamuk, *A History of Middle East Economies in the Twentieth Century* (Cambridge, MA: Harvard University Press, 1999), 214.

5   Ann M. Harrison. *A Tool in His Hand*, (New York: Friendship Press, 1958), 78–9.

6   Toby Craig Jones, *Desert Kingdom: How Oil and Water Forged Modern Saudi Arabia* (Cambridge, MA: Harvard University Press, 2010), 9; and Andrea H. Pampanini, *Desalinated Water in the Kingdom of Saudi Arabia* (New York: Turnaround Associates, 2010), 3.

7   H. St. John Philby, *Arabian Jubilee* (London: Robert Hale, 1952), 78.

8   Author's conversation with Dabbagh family members in Riyadh, October 2014.

9   Roger Owen, *State, Power and Politics in the Making of the Modern Middle East* (London: Routledge, 1993), 81.

10   "Economics of Araby," March 8, 1933, in Philby Papers, Private Papers, Box XXXII/4, Middle East Centre, St. Antony's College, Oxford.

11   H. St. John Philby, *Arabian Days: An Autobiography* (London: Robert Hale, 1948), 291.

12   Robert Lacey, *The Kingdom* (New York: Harcourt, Brace, Jovanovich, 1981), 236.

13   Owen, *State, Power and Politics,* 81.

14   Steffen Hertog, *Princes, Brokers, and Bureaucrats: Oil and the State in Saudi Arabia* (Ithaca: Cornell University Press, 2010), 128.

15   Robert Vitalis, *America's Kingdom: Mythmaking on the Saudi Oil Frontier* (Stanford, CA: Stanford University Press, 2007), 274.

16   Ellen R. Wald, *Saudi, Inc. The Arabian Kingdom's Pursuit of Profit and Power* (New York: Pegasus, 2018), 207.

17   Valérie Marcel, *Oil Titans: National Oil Companies in the Middle East* (London: Chatham House, 2006), 52.

18   Rodney Wilson, "Economic Governance and Reform in Saudi Arabia", in Anoushirvan Ehteshami and Steven Wright (eds), *Reform in the Middle East Oil Monarchies* (Reading: Ithaca Press, 2011), 133.

19   Steffen Hertog, "Petromin: the slow death of statist oil development in Saudi Arabia", http://eprints.lse.ac.uk/29865, 2008, 1–10.

20   Middle East Economic Survey, MEES 42/1974, 4–5; MEES June 27, 1975, 3; MEES May 23, 1975, 7; MEES 49/1974, 7.

21   Holden and Johns, *The House of Saud*, 321.

22   Hertog, *Princes, Brokers, and Bureaucrats*, 128.

23   Ali al-Naimi, *Out of the Desert* (New York: Penguin Random House, 2016), 139.

24   Wald, *Saudi, Inc.*, 221.

25   Marcel, *Oil Titans*, 133.

26   Owen, *State, Power and Politics,* 210.

27   Jones, *Desert Kingdom*, 4.

28   Pampanini, *Desalinated Water*, 11.

29   Hertog, *Princes, Brokers, and Bureaucrats*, 102; and Arnold Thackray and Richard Ulrych, *Building a Petrochemical Industry in Saudi Arabia* (Riyadh: Obeikan Press, 2017), 166.

30   Frank Jungers, *The Caravan Goes On: How Aramco and Saudi Arabia Grew Together* (Surbiton U.K.: Medina Publishing, 2013), 191.

31   Holden and Johns, The House of Saud, 396.

32   Andrea H. Pampanini, *Cities From the Arabian Desert: The Building of Jubail and Yanbu in Saudi Arabia* (Westport, CT: Praeger, 1997), 20.

33 Ibid., 14.

34 "Jubail and Yanbu produce 12% of Saudi GDP," *Construction Week*, September 24, 2013, constructionweekonline.com—article 24387.

35 Thackray and Ulrych, *Building a Petrochemical Industry*, 82.

36 Hertog, *Princes, Brokers, and Bureaucrats*, 128; and Middle East Economic Digest (MEED) Special Report, March 1989, 4.

37 Alexander Tullo, "Global Top 50 Chemical Companies," *Chemical and Engineering News*, July 29, 2019.

38 Thackray and Ulrych, *Building a Petrochemical Industry*, 186–7.

39 Hertog, *Princes, Brokers, and Bureaucrats*, 129, 139; and Andrea H. Pampanini, *Saudi Arabia: Moving Towards a Privatized Economy* (New York: Turnaround Associates, 2005), 124.

40 Wilson, "Economic Governance and Reform," 144.

41 Hertog, *Princes, Brokers, and Bureaucrats*, 241.

42 Wilson, "Economic Governance and Reform," 133 and 137.

43 Summer Scott Huyette, *Political Adaptation in Saudi Arabia: A Study of the Council of Ministers* (London: Westview, 1985), 106.

Chapter 14

1   Abdulaziz bin Abd al-Rahman bin Faisal bin Turki bin Abdullah bin Mohammad Al Saud, the great-great-grandson of the dynasty's founder, became Amir of Riyadh in 1902. He became Sultan of the Nejd in 1921 after the fall of Ha'il, King of the Hejaz in 1925 after the capture of Jeddah, and King of Saudi Arabia in 1932 when he unified the Hejaz and Nejd into one nation. H. V. F. Winstone, *Captain Shakespear* (London, Jonathan Cape, 1976), 42.

2   Reports by Shakespear to Trevor, March 9, 1910, FO 424/223, 88–9. Reports by Shakespear to Cox, April 8, 1911, FO 424/227, 52–4; May 15, 1913, FO 424/2387, 210–11. Cited in Jacob Goldberg, *The Foreign Policy of Saudi Arabia: The Formative Years, 1902–1918* (Cambridge MA: Harvard University Press, 1986), 176, 177.

3   Robert Lacey, *The Kingdom* (New York: Harcourt, Brace, Jovanovich, 1981), 116, 117.

4   Confidential Letter No. S.-13 Shakespear to Cox, January 3, 1915, India Office Records, R/15/5/25. Cited in James P. Piscatori, "Islamic Values and National Interest: The Foreign Policy of Saudi Arabia," in *Islam in Foreign Policy*, Adeed Dawisha (ed.) (Cambridge: Cambridge University Press, 1983), 33.

5   Nadav Safran, *Saudi Arabia: The Ceaseless Quest for Security* (New York: Cornell University Press, 1988), 452.

6   Andrew Ryan, *The Last of the Dragomans* (London: Geoffrey Bles, 1951), 83.

7   Goldberg, *The Foreign Policy of Saudi Arabia*, 82.

8   Ibid., Appendix A. Technically, Abdulaziz became *Wali* of the *vilayet* of the Nejd. A *vilayet* was the largest of four Ottoman administrative regions; the others being *sanjak, qaza,* and *nahiye*. A *vilayet* corresponded to a state or province—modern Iraq, for example, comprised three *vilayets*: Mosul, Baghdad, and Basra. Each was governed by a *wali*.

9   Lacey, *The Kingdom*, 104.

10 Goldberg, *The Foreign Policy of Saudi Arabia*, 184.

11 Ibin Saud to Shakespear, November 28, 1914, FO 371/2479, 292; Safran, *The Ceaseless Quest for Security*, 183.

12 Lacey, *The Kingdom*, 257.

13 Alexei Vassiliev, *The History of Saudi Arabia* (New York: New York University Press, 2000), 323, citing N. I. Proshin, Saudovskaya Arviya, (Saudi Arabia: A Historical and Economic Sketch) (Moscow: Nauka, 1964) 103.

14 David Holden and Richard Johns, *The House of Saud: The Rise and Fall of the Most Powerful Dynasty in the Arab World* (New York: Holt, Rinehart and Winston, 1981), 126.

15 Safran, *The Ceaseless Quest for Security*, 61.

16 "Investigation of the National Defense Program: Additional Report of the Special Committee Investigating the National Defense Program, Navy Purchase of the Navy Middle East Oil," April 28, 1948, Folder: "Israel-Oil Situation in the Middle East", Papers of Henry Morgenthau Jr., Franklin Delano Roosevelt Library. The figure includes money provided directly as well as indirectly through the British. Cited in Rachel Bronson, *Thicker Than Oil: America's Uneasy Partnership with Saudi Arabia* (Oxford: Oxford University Press, 2006), 40.

17 Bronson, *Thicker Than Oil*, 37.

18 Executive Order 8926, February 18, 1943, President Franklin D. Roosevelt to Under Secretary of State Edward Stettinus, Foreign Relations of the United States records, 1943, vol. IV, 859.

19 Bruce Riedel, *Kings and Presidents: Saudi Arabia and the United States Since FDR* (Washington, DC: Brookings Institution Press, 2018), 24.

20 Arabian American Oil Company, "Aramco Annual Reports 1950-52," 13 Box 3 Folder 45, William E. Mulligan Papers, Georgetown Library Special Collections, Georgetown University.

21 "President Truman to King Abdul Aziz ibn Saud of Saudi Arabia", October 31, 1950, Foreign Relations of the United States 1950, V:1190.

22 Thomas Lippman, *Inside the Mirage: America's Fragile Partnership with Saudi Arabia* (Oxford: Westview Press, 2004), 101; and Arthur N. Young, *Saudi Arabia: The Making of a Financial Giant* (New York: New York University Press, 1983), 103.

23 Nathan J. Citino, *From Arab Nationalism to OPEC: Eisenhower, King Saud, and the Making of U.S.-Saudi Relations* (Bloomington, IN: Indiana University Press, 2002), 94.

24 Bronson, *Thicker Than Oil*, 46.

25 Oliver North, *Under Fire: An American Story* (New York: Harper Collins, 1991), 256, 288.

26 Henry Kissinger, *Years of Upheaval*, (Boston, MA: Little Brown, 1982), 633.

27 Riedel, *Kings and Presidents*, 38.

28 Dana Adams Schmidt, *Yemen: The Unknown War* (New York: Holt Rinehart, 1968), 281–2.

29 Warren Bass, *Support Any Friend* (Oxford: Oxford University Press, 2004), 129–30.

30 Schmidt, *The Unknown War*, 193; and Joseph A. Kechichian, *Faisal: Saudi Arabia's King for All Seasons* (Gainesville, FL: University of Florida Press, 2008), 88.

31 Aviad Asher Orkaby, "International History of the Yemen Civil War 1962–1968," doctoral thesis, Harvard University, 2014, 215; and Clive Jones, *Britain and the Yemen Civil War 1962–1965: Ministers, Mercenaries and Mandarins – Foreign Policy and the Limits of Covert Action* (Eastbourne: Sussex Academic Press, 2004).

32 Bronson, Thicker Than Oil, 95.

33 David B. Ottaway, *The King's Messenger: Prince Bandar bin Sultan and America's Tangled Relationship with Saudi Arabia* (New York: Walker Publishing, 2008), 24–36.

34 Robert Lacey, *Inside the Kingdom: Kings, Clerics, Modernists, Terrorists and the Struggle for Saudi Arabia* (London: Hutchinson, 2009), 106.

35 Ibid., 110.

36 Ottaway, *The King's Messenger*, 74.

37 Ibid., 69.

38 Ibid., 75.

39 Riedel, *Kings and Presidents*, 63.

40 Norman Schwarzkopf, *It Doesn't Take a Hero* (New York: Bantam Books, 1992), 302–6; and Bob Woodward, *The Commanders* (New York: Simon and Schuster, 1991), 260.

41 Lippman, *Inside the Mirage*, 301.

42 Woodward, *The Commanders*, 272.

43 Patrick Tyler, "Saudi Arabia Pledges One Billion Dollars to Soviet Union," *The New York Times*, October 9, 1991.

44 H. St. John Philby's *Saudi Arabia* (London: Ernest Benn, 1955) reprinted by Librairie du Liban 1968, 353.

45 Holden and Johns, *The House of Saud*, 252.

46 Bronson, *Thicker Than Oil*, 117. Both Rachel Bronson and the current author discussed this matter (January 2005 and July 2016, respectively) with Ray Close, the CIA station chief at the time, who confirmed that Prince Fahd bin Abdulaziz had made clear to him at the time that the Saudi troops would not engage in the fighting; Dore Gold, *Hatred's Kingdom* (Washington, DC: Regnery Publishing, 2003), 222.

47 Gold, *Hatred's Kingdom*, 222.

48 William A. Eddy, *FDR Meets Ibn Saud* (Washington: American-Middle East Educational and Training Services, 1954), 34–5.

49 Ronald Reagan, Letter to the Speaker of the House of Representatives and the President of the Senate on the Sale of AWACS Aircraft to Saudi Arabia, June 18, 1986, Ronald Reagan Presidential Library website, available at https://www.reaganlibrary.gov/research/speeches/61886e.

50 Thomas L. Friedman, "An Intriguing Signal from the Crown Prince," *The New York Times*, February 17, 2002.

51 Bronson, *Thicker Than Oil*, 144.

52 Ottaway, *The King's Messenger*, 151; and Marwan Muasher, *The Arab Center: The Promise of Moderation* (New Haven, CT: Yale University Press, 2008), 110.

53 Riedel, *Kings and Presidents*, 134.

54 Steve Coll, *Ghost Wars* (New York: Penguin Press, 2004), 397–8, 401.

55 Jim Sciutto, "Congress Releases Secret 28 Pages on Alleged Saudi 9/11 Ties," CNN, July 16, 2016, 0244 GMT. Chairman of the House Intelligence Committee, Devin Nunes, made clear "it is important to note that this section of the report does not put forward vetted conclusions, but rather unverified leads that were later fully investigated by the intelligence community." Senators Richard Bird and Dianne Feinstein, the Chairman and Ranking Minority Members of the Senate Intelligence Committee, issued another statement noting, "These pages include unconfirmed allegations and raw reporting that have been subject to conspiracy for years." "We need to put an end to conspiracy theories and idle speculation that do nothing to shed light on the 9/11 attack." Bruce Riedel, who served on the White House National Security Council staff for four presidents, concludes in his book, *Presidents and Kings*, 207, "The existing evidence alleging Saudi involvement in the 9/11 plot, thus, has been reviewed carefully by the U.S. government more than once. The FBI keeps the investigation of the attacks open to evaluate any new material, but there is no smoking gun that points to any Saudi official."

56 Gilles Kepel, *Jihad, The Trail of Political Islam* (Cambridge, MA: Belknap Press of Harvard University Press, 2002), 314.

57 See Dore Gold's *Hatred's Kingdom* for a detailed account of how Wahhabi and Muslim Brotherhood teachings were fused into a new hybrid ideology that was widely disseminated with Saudi funding. However, the book, which was published in 2003, predates the many efforts that the Saudi government has subsequently taken to correct that problem—particularly under King Salman.

58 Cofer Black, testimony before the House Committee on International Relations Subcommittee on the Middle East and South Asia, May 24, 2004.

59 Michael R. Gordon, "Saudis Warn Against Attack on Iraq by the United States," *The New York Times*, March 17, 2002.

60 John Solomon, "Saudis Secretly Provided Extensive U.S. Help During Iraq War," *Associated Press*, April 24, 2004; Bronson, *Thicker Than Oil*, 239; and Robert Jordan, *Desert Diplomat* (Lincoln, NE: Potomac Books, an imprint of the University of Nebraska Press, 2015), 138–9.

61 Christopher Blanchard, *Saudi Arabia: Background and U.S. Relations*, (Washington, DC: Congressional Research Service, 2016), 39.

62 Peter Baker, "Off the Cuff Obama Line Puts U.S. in a Bind in Syria," *The New York Times*, May 4, 2013.

63 Jeffrey Goldberg, "The Obama Doctrine," *Atlantic*, April 2016.

Chapter 15

1 U.S. Energy Information Administration website, www.eia.gov, Asharq Al Awsat, "Aramco Oil Production Costs Lowest in the World," November 15, 2019.

2 Ali al-Naimi, remarks to Center for Strategic and International Studies, Washington, DC, April 30, 2013.

3 William B. Quandt, *Saudi Arabia's Oil Policy* (Washington, DC: Brookings Institution, 1982), 133; A. M. Jaffe and J. Elass, "Saudi Aramco: National Flagship with Global Responsibilities," paper prepared for the James A. Baker III Institute for Public Policy, Rice University, Houston, TX, 2007, 76; and *Petroleum Intelligence Weekly*, August 13, 1990.

4 Ibrahim Muhanna, remarks delivered by Ministry of Petroleum spokesman and advisor to the Institute of International Finance, MENA Region Economic Forum, Manama, Bahrain, September 29, 2014.

5 U.S. Department of Energy, Energy Information Administration, Washington, DC data sets www.eia.gov.

6 Ali al-Naimi, Energy Legacy, remarks delivered at Center for Strategic and International Studies (CSIS), Washington, DC, April 30, 2013.

7 *The Financial Times*, 8 May 2018.

8 Daniel Yergin, *The Prize: The Epic Quest for Oil, Money & Power* (New York: Simon and Schuster, 1991), 591–4; and the U.S. Department of Energy, Energy Information Administration, Washington DC data sets, www.eia.gov.

9 William B. Quandt, *Peace Process: American Diplomacy and the Arab-Israeli Conflict Since 1967* (Washington, DC: Brookings Institution Press, 1993), 164.

10 Mohamed Heikal, *The Road to Ramadan* (London: Collins, 1975), 231, 232, 246.

11 Rachel Bronson, *Thicker Than Oil: America's Uneasy Partnership with Saudi Arabia* (Oxford: Oxford University Press, 2006), 118; and Nadav Safran, *Saudi Arabia: The Ceaseless Quest for Security* (New York: Cornell University Press, 1988), 156–7.

12 Robert Lacey, *The Kingdom* (New York: Harcourt, Brace, Jovanovich, 1981), 410.

13 Safran, *The Ceaseless Quest for Security*, 256.

14 Yergin, *The Prize*, 613; and Heikal, *The Road to Ramadan*, 67.

15 Jeffrey Robinson, *Yamani: The Inside Story* (London: Simon and Schuster, 1988), 94; and Anthony Cave Brown, *Oil, God and Gold* (New York: Houghton Mifflin, 1999), 296.

16 "Excerpts from the Kissinger New Conference," *The New York Times*, November 22, 1973.

17 David Holden and Richard Johns, *The House of Saud: The Rise and Fall of the Most Powerful Dynasty in the Arab World* (New York: Holt, Rinehart and Winston, 1981), 328.

18 Robinson, *Yamani: The Inside Story*, 91.

19 Bronson, *Thicker Than Oil*, 117; Safran, The Ceaseless Quest for Security, 158; and Robinson, *Yamani: The Inside Story*, 94.

20 Henry Kissinger, *Years of Upheaval*, (Boston, MA: Little Brown, 1982), Chapter 11.

21 Andrew Scott Cooper, *The Oil Kings: How the U.S., Iran & Saudi Arabia Changed the Balance of Power in the Middle East* (New York: Simon and Schuster, 2011), 118.

22 Cave Brown, *Oil, God and Gold*, 296.

23 Frank Jungers, *The Caravan Goes On: How Aramco and Saudi Arabia Grew Together* (Surbiton U.K.: Medina Publishing, 2013), 158. CALTEX was a joint venture between Standard Oil of California and Texaco. Today it is a brand name used by Chevron in the Asia-Pacific region.

24 Bronson, *Thicker than Oil*, 121.

25 Ibid., 127.

26 U.S. Army Corp of Engineers, www.USACE.army.mil/about/history.

27 Natana J. DeLong-Bas, *Wahhabi Islam: From Revival and Reform to Global Jihad* (New York: Oxford University Press, 2004), 246.

28 Alexei Vassiliev, *The History of Saudi Arabia* (New York: New York University Press, 2000), 266, citing AFP, USSR, 1929, section 127, inventory 1, file 2, dossier 18, 3–5 and 10.

29 Martin Kramer, *Islam Assembled: The Advent of the Muslim Congresses*, (New York: Columbia University Press, 1986), 122.

30 Willard Beling (ed.), *King Faisal and the Modernisation of Saudi Arabia* (London: Croom Helm, 1980), 186, note 6.

31 James P. Piscatori, "Islamic Values and National Interest: The Foreign Policy of Saudi Arabia," in *Islam in Foreign Policy*, Adeed Dawisha (ed.) (Cambridge: Cambridge University Press, 1983), 40.

32 Beling, *Modernisation of Saudi Arabia*, 188, note 13.

33 James Piscatori, "Ideological Politics in Saudi Arabia," *in Islam in the Political Process*, Piscatori (ed) (Cambridge: Cambridge University Press, 1983), 57

34 Bernard Lewis, *The Political Language of Islam*, (Chicago: University of Chicago Press, 1988), 3.

35 Robert Lacey, *Inside the Kingdom: Kings, Clerics, Modernists, Terrorists and the Struggle for Saudi Arabia*, (London: Hutchinson, 2009), 95.

36 The ministry is sometime referred to as the Ministry of Islamic Affairs, Call and Guidance, with the Arabic word *dawah* being translated as "Call." The word's literal meaning is to issue a summons or offer an invitation, which, in the Wahhabi context, means to summon or invite someone to become a Muslim.

37 David B. Ottaway, *The King's Messenger: Prince Bandar bin Sultan and America's Tangled Relationship with Saudi Arabia* (New York: Walker Publishing, 2008), 185–6;

and "U.S. Eyes Money Trails of Saudi-Backed Charities," *The Washington Post*, August 19, 2004.

38 President Jimmy Carter, The State of the Union Address delivered before a Joint Session of Congress, January 23, 1980.

39 "Reflections on Soviet Intervention in Afghanistan", memorandum for the president from Zbigniew Brzezinski, December 26, 1979, released by the Cold War International History Project.

40 Bronson, *Thicker than Oil*, 149.

41 Zbigniew Brzezinski, *Power and Principle: Memoirs of the National Security Adviser 1977–1981* (New York: Farrar Straus Giroux, 1983), 449.

42 William Simpson, *The Prince: The Secret Story of the World's Most Intriguing Royal: Prince Bandar Bin Salman* (New York: Regan Books, 2006), 112.

43 Dana Burde, *Schools for Conflict or for Peace in Afghanistan* (New York: Columbia University Press, 2014), 77.

44 Bruce Riedel, *Kings and Presidents: Saudi Arabia and the United States Since FDR* (Washington, DC: Brookings Institution Press, 2018), 79.

45 Steve Coll, *Ghost Wars, The Secret History of the CIA, Afghanistan and Bin Ladin, from the Soviet Invasion to September 10, 2001* (New York: Penguin Press, 2004), 152.

46 Anthony Cordesman, "Saudi Arabia: Opposition, Islamic Extremism and Terrorism," in *Saudi Arabia Enters the Twenty-First Century: The Political, Foreign Policy, Economic, and Energy Dimensions,* Anthony Cordesman (Westport CT: Praeger, 2003), 4.

47 Coll, *Ghost Wars*, 201.

48 Saifullah Paracha, "The Guantanamo Docket," May 2, 2018, *The New York Times*.

49 Coll, *Ghost Wars*, 152.

50 Peter Tomsen, *The Wars of Afghanistan: Messianic Terrorism, Tribal Conflicts, and the Failures of Great Powers* (New York: Public Affairs, 2011), 248. Tomsen's source is Milton Bearden, CIA Station Chief in Islamabad 1986–9.

51 Mohammad Yousaf and Mark Adkin, *The Bear Trap: Afghanistan's Untold Story* (London: Leo Cooper, 1991), 106.

Chapter 16

1 James D. Lunt, *Glubb Pasha: A Biography* (London: Harvill, 1984), 46.

2 Robert Lacey, *The Kingdom* (New York: Harcourt, Brace, Jovanovich, 1981), 370.

3 Alexei Vassiliev, *The History of Saudi Arabia* (New York: New York University Press, 2000), 335.

4 Nasser Ibrahim Rashid and Esber Ibrahim Shaheen, *King Fahd and Saudi Arabia's Great Revolution* (Joplin, MO: International Institute of Technology, 1987), 78.

5 George Lipsky, *Saudi Arabia: Its People, Its Society, Its Culture* (New Haven: HRAF Press, 1959), 277.

6 Karen Elliot House, *On Saudi Arabia: Its People, Past, Religion, Fault Lines and Future* (New York: Alfred Knopf, 2012), 144.

7 Robert Lacey, *The Kingdom*, 368.

8 Author's conversation with Ministry of Education official during the early 1980s.

9 House, *On Saudi Arabia*, 141.

10 Saudi Ministry of Education, www.MOE.GOV.SA, Goals and Objectives.

11 Author's conversation with Saudi clerics, November 2015.

12 "The Diplomat," Saudi Ministry of Foreign Affairs, March 2007, 30. In 2001, "Regulations on Procedures Before the *Sharia* Courts" were issued to outline a process for entering pleas, giving evidence, and calling witnesses. Defendants were, for the first time, granted the right to legal representation, something that *sharia* judges and the *ulama* had long considered unnecessary. The following year, "Regulations on the Practice of Law" specified the training and licensing requirements for attorneys and, for the first time, defined the concept of attorney client privilege. Later in 2002, "Regulations for Criminal Procedures" were issued that outlined, among other things, a defendant's rights during investigation, interrogation, and incarceration

13 Joseph A. Kéchichian, *Legal and Political Reforms in Saudi Arabia* (New York: Routledge, 2013), 27.

14 David Commins, *The Wahabi Mission and Saudi Arabia* (New York: I.B. Tauris & Co., 2006), 33.

15 Robert Lacey, *Inside the Kingdom: Kings, Clerics, Modernists, Terrorists and the Struggle for Saudi Arabia*, (London: Hutchinson, 2009), 271.

16 Kéchichian, *Legal and Political Reforms*, 87.

17 Elisabeth Rosenthal, "Pope Meets King of Saudi Arabia," *The New York Times*, November 6, 2007.

18 Caryle Murphy, "Saudi King Set to Lead Interfaith Talks," *Christian Science Monitor*, July 8, 2008; and Lisa Abend, "Saudis Host Interfaith Conference," *Christian Science Monitor*, July 18, 2008.

19 Sohail Karam, "Saudi Cleric Backs Gender Segregation with Fatwah," Reuters, February 24, 2011 .

20 Katherine Zoepf, "Talk of Women's Rights Divides Saudi Arabia," *The New York Times*, May 31, 2010.

21 Ibid.

22 Robert L. Wagner, "Saudi Female Journalist Defies Stereotypes," *News Tilt*, April 13, 2010. Wagner was managing editor of both of Saudi Arabia's English language newspapers, *Arab News* and *Saudi Gazette*. Robert L Wagner was the managing editor of Saudi Arabia's two leading English language papers, *The Arab News* and *Saudi Gazette*. This article can be found online at roblwagner archives under "Saudi Female Journalist Defies Stereotypes."

23 John L. Esposito with Natana DeLong-Bas, *Women in Muslim Family Law* (Syracuse, NY: Syracuse University Press, 2001).

24 Natana DeLong-Bas, *Wahhabi Islam: From Revival and Reform to Global Jihad* (Oxford: Oxford University Press, 2004), 125.

25 Ibid., 131.

26 Kéchichian, *Legal and Political Reforms*, 60, citing "Saudi Women Gain Court Access," *Al Jazeera*, February 21, 2010.

27 Ben Hubbard, "Saudi Crown Prince in His Own Words," *The New York Times*, May 18, 2018; and "Interview with Saudi Crown Prince," *TIME* magazine, April 5, 2018.

28 To view an example of drifting, simply google search "How to Change a Tire in Saudi Arabia."

29 World Bank 2020, *Women, Business and the Law 2020,* Women, Business and the Law. Washington, DC: World Bank. doi:10.1596/978-1-4648-1532-4.

30 "Saudi Arabia Arrests Key Activists in Human Rights Crack Down," *The Guardian*, May 25, 2018; and "Detained Women's Rights Activists to be Put on Trial," *The Guardian*, March 2, 2019.

Chapter 17

1   U.S. Government, Energy Information Administration, https://www.eia.gov/dnav/pet/ pet_pri_spt_s1_a.htm.
2   Steffen Hertog, "Challenges to the Saudi Distributional State in the Age of Austerity," in *Salman's Legacy,* Madawi Al Rasheed (ed) (Oxford: Oxford University Press, 2018), 73.
3   U.S. Government, Energy Information Administration, https://www.eia.gov/dnav/pet/ hist/LeafHandler.ashx?n=PET&s=MCRFPUS1&f=M.
4   Saudi Ministry of Environment, Water and Agriculture, cited on https://www.ceicdata. com/en/saudi-arabia/electricity-statistics/electricity-consumption
5   Author's conversation with Saudi Saline Water Conversion Corporation official, October 2015.
6   Author's conversation with Saudi Petroleum Ministry officials, March 2014.
7   "Back to Work in A New Economy, Background to Saudi Labor," Market Harvard Kennedy Schools, April 2015, 19 as well as International Labor Organization and Saudi Ministry of Labor Statistics.
8   The World Bank, https://data.worldbank.org/indicator/SP.RUR.TOTL. ZS?locations=SA.
9   Simon Kerr, "Saudi Arabia Pledges to 'End Addiction to Oil'," *The Financial Times,* April 26, 2016.
10 Vivian Nereim, "Saudi Arabia Budget Deficit is $98 in 2015," Bloomberg, December 28, 2015.
11 OECD, https://data.oecd.org/gga/general-government-deficit.htm.
12 ibid. Hertog, "Challenges to the Saudi Distribution State," 80; and "Energy Subsidy Reform: Lessons and Implications," paper prepared for the International Monetary Fund, Washington, D.C, 2013.
13 Elaine Moore, "First Saudi Bond Sale Raises $17.5 bn in Emerging Market Record," *The Financial Times,* October 26, 2016.
14 International Monetary Fund, https://www.imf.org/external/datamapper/CG_DEBT_ GDP@GDD/CHN/FRA/DEU/ITA/JPN/GBR/USA/ATG/UZB.
15 Author's conversation with Ma'adeen executive, February 2018.
16 *Saudi Gazette,* 13 May 2017, citing Saudi government statistics.
17 Jadwa Investments, "Saudi Labor Market Update For First Quarter 2019," citing General Authority for Statistics figures and OECD (2020), Youth unemployment rate (indicator). doi: 10.1787/c3634df7-en.
18 Andrew England, "A Long Road to Labour Reform," *The Financial Times,* March 1, 2019, 9.
19 Nicholas Parasie, "Fitch Cuts Saudi Credit Rating," *The Wall Street Journal,* March 22, 2017 and Arif Sharif "Saudi Arabia's Credit Rating Cut by Fitch," Bloomberg, March 22, 2017.
20 Author's conversation with Saudi Aramco executive, October 2018.
21 The World Bank Data Sets, data.worldbank.org.
22 Adel al-Turaifi, "Saudi Arabia Takes on Radical Islam," *The Wall Street Journal,* March 20, 2018.

Chapter 18

1   Hezbollah, whose name means "The Party of God," is a Shia political and military organization based in Lebanon that seeks to create an Islamic state in Lebanon

modeled on that in Iran. The civilian arm of Hezbollah is deeply involved in Lebanese politics, and has a record of providing social services to underprivileged social groups. However, Hezbollah also has well-established military units and relations with both Tehran and Damascus. It has been designated a terrorist organization by the United States, Britain, and Saudi Arabia.

2 Press Conference with Secretary of State John Kerry, March 5, 2015.

3 Arab News, "Saudi Arabia Tell Iran to Stop Meddling in Iraqi Affairs," March 30, 2016.

4 There are roughly 25 million Saudis and 30 million Yemenis. Yemeni GDP per capita was roughly $1,500 in 2015, but as a result of the civil war has fallen sharply to barely $500. Mexico's GDP per capital was roughly $10,000 in 2019.

5 "Remarks of Foreign Minister Adel al-Jubair at the Munich Security Conference February 19, 2017," https://www.saudiembassy.net/statements and Arab News, February 20, 2017.

6 Remarks of President Donald Trump delivered to the Arab, Islamic, American Summit in Riyadh Saudi Arabia, May 21, 2017, https://www.whitehouse.gov/briefings-statements/.

7 "Donald Trump and Saudi Arabia Sign Agreements to Counter Iran," *The Wall Street Journal*, May 20, 2017; "Aramco Signs Fifty Billion With US Companies," *Oil and Gas Journal*, May 22, 2017; and "Blackstone Unveils Forty Billion Dollar Infrastructure Fund In Saudi Arabia," Forbes, May 20, 2017.

8 "Sixty Minutes interview with Mohammed bin Salman," CBS News, September 29, 2019.

9 International Energy Agency, "Oil Market Report," March 15, 2017.

10 Maayan Groisman, "Palestinians Celebrate Terror Attack in Tel Aviv, Saudis Strongly Condemn," *The Jerusalem Post*, June 9, 2016.

11 Kristian Coates Ulrichsen, "Israel and the Arab Gulf States: Drivers and Direction of Change," paper prepared for the James A. Baker III Baker Institute for Public Policy, Rice University, Houston, TX, September 2016, 2, 7–8.

12 David Sanger, "Saudi Arabia and Israel Share Common Opposition," *The New York Times*, June 4, 2015.

13 Jonathan Ferziger and Peter Waldman, "How Do Israeli Tech Firms Do Business in Saudi Arabia? Very Quietly," *Business Week*, February 2, 2017; and Amos Harel, "Cyber Firm Negotiated with Saudis," Haaretz, November 25, 2018.

14 The father of Qatar's current amir deposed his own father in 1995. The Saudi government opposed this coup. Amir Hamad bin Khalifa al-Thani (ruled 1995 to 2013) then sought to assert Qatar's independence from its much larger neighbor and preserve good relations with Iran, with which Qatar shares the North Pars gas field. Resulting tensions led Saudi Arabia to withdraw its ambassador from Doha between 2002 and 2008.

    During the 2011 Arab Spring, Qatar and its media outlet, *Al Jazeera*, supported the protestors, especially in Egypt, while Saudi Arabia largely backed the existing governments. In addition, Qatar gave refuge to the pro-Muslim Brotherhood preacher, Yousef al-Qaradawi, who vehemently encouraged the Arab Spring uprising against Saudi ally, Egyptian President Hosni Mubarak. A 2014 GCC agreement was meant to end these tensions, but Riyadh believes that Qatar never met its 2014 commitments.

15 The author was present at the 2018 Future Investment conference in Riyadh.

16 Marieke Brandt, *Tribes and Politics in Yemen: A History of the Houthi Conflict* (Oxford: Oxford University Press, 2017), 61, 67.

17 Robert Ward, "Yemen Rebels and Saudis Clash on Border," *The New York Times*, November 6, 2009; and Riedel Bruce, "Who are the Houthis and Why Are We Fighting Them"? Brookings, December 18, 2017.

18 Through the unanimous adoption of resolution 2201 on February 15, 2015, the UN Security Council strongly deplored actions by the Houthis—who had gained control of the capital, Sanaa, the previous September—and demanded that they refrain from further unilateral actions that could undermine the political transition and security in the country. The resolution further demanded that Houthi rebels in Yemen "immediately and unconditionally" withdraw from government institutions, safely release President Abdu Rabbu Mansour Hadi and all others from house arrest, and engage in good faith in UN-brokered negotiations designed to keep the country on a steady path towards democratic transition.

   UN Security Council resolution 2216 of April 14, 2015 called upon the Houthis to refrain from any provocations or threats to neighboring states; release the country's Minister for Defense, all political prisoners and individuals under house arrest or arbitrarily detained; and end the recruitment of children. It also imposed sanctions— including a general assets freeze, travel ban, and arms embargo—on Abdul Malik al Houthi and Ahmed Ali Abdullah Saleh, son of the president who stepped down in 2011. The resolution called upon all Yemeni parties to abide by the Gulf Cooperation Council initiative and to resume the UN-brokered political transition.

19 A comparison between the Houthis and Hezbollah can be misleading— Lebanese and Iranian Shia are both from the Ithnashari (Twelver) sect; they share deep historic and family relationships. The Houthis, on the other hand, belong to the Shia Zaydi (Fiver) sect whom the Iranians long regarded as heretics. This deep, historical antagonism between the Ithnasharis and Zaydis makes Iranian support for the Houthis more pragmatic than visceral.

20 Leon Panetta, *Worthy Fights: A Memoir of Leadership in War and Peace* (New York: Penguin Press, 2014), 433.

21 Yonah Jeremy Bob, "Israel Needs to Attack Iran to Stop Ring of Fire," *The Jerusalem Post*, November 11, 2019.

Chapter 19

1 David Commins, *The Wahabi Mission and Saudi Arabia* (New York: I.B. Tauris & Co., 2006), 171

2 Toby Matthiesen, *The Other Saudis: Shiism, Dissent and Sectarianism* (New York: Cambridge University Press, 2015), 117

3 Yaroslav Trofimov, *The Siege of Mecca* (New York: Doubleday, 2007), 199–200.

4 Jacob Goldberg, The Shi'i Minority in Saudi Arabia, in *Shi'ism and Social Protest,* Juan Cole and Nikki Keddie, eds. (New Haven: Yale University Press, 1986), 244.

5 The author was in Riyadh on March 11, 2011 staying two blocks from Cairo Square. Although he did not go to the planned Day of Rage, he spoke with several people who had driven through the square and reported that nothing was happening. Although there were more traffic police on hand than usual, there were no riot police or other security forces present.

6 The National Dialogue is a series of public meetings initiated by King Abdullah to discuss controversial subjects such as unemployment, healthcare, women's rights, and

the Shia. See Mark C. Thompson, *Saudi Arabia and the Path to Political Change: National Dialogue and Civil Society* (London: I. B. Tauris, 2014).

7   *Gulf News*, Saudi Arrests Princes, Minster and Business Figures, November 5, 2017.
8   *Saudi Gazette*, Princes and Ministers Detained, November 5, 2017.
9   Ahmed Al Omran, Saudis Conclude Anti-Corruption Drive, *The Financial Times*, January 31, 2019.
10 Ibid.
11 *Saudi Gazette*, New Era of Transparency, November 5, 2017.
12 Mohammed Bin Salman, Bloomberg Interview 5, 2018.
13 Ignatius, David "How a Chilling Saudi Cyber War Ensnared Jamal Khashoggi," *The Washington Post,* December 7, 2018.
14 Author's conversations with Saudi citizens in Riyadh October 2019.

Chapter 20

1   See Jon Alterman. "Ties that Bind: Family, Tribe and Nation and the Rise of Individualism, Center for Strategic and International Studies 2019, 16–19. See also discussion in Chapter 7.
2   Steven Cunningham, "US posts First Month in 70 Years as a net Petroleum Exporter," November 29, 2019. This report is somewhat misleading as the United States still imports more than 7 mbd of crude oil, but exports even more crude oil and refined petroleum products—making it a net petroleum exporter.
3   *Gulf News*, "King Salman Inaugurates Shura Council," December 13, 2017.
4   Interview with Mohammed bin Salman, *TIME* magazine, April 5, 2018.
5   See John Hannah, "Saudi Arabia Can Win Islam's War of Ideas," *Foreign Policy*, March 15, 2018.
6   US State Department 2018 Country Reports of Terrorisms, 151.
7   Letter from Dr Abd al-Karim al-Issa to Dr Robert Satloff, Executive Director, Washington Institute for Near East Policy, www.washingtoninstitute.org/press-room/view/al-issa-on-pittsburgh-synagogue-attack.
8   Dore Gold, *Hatred's Kingdom: How Saudi Arabia Supports the New Global Terrorism* (Washington, DC: Regnery Publishing, 2003).
9   US State Department 2018 Country Reports on Terrorism, 149.
10 Daniel Benjamin, "Not His Father's Saudi Arabia," Foreign Affairs, October 18, 2018.
11 Jeffrey Goldberg, Saudi Crown Prince, "Iran's Supreme Leader Makes Hilter Look Good," *The Atlantic*, April 2, 2018 and Ben Hubbard , Saudi Crown Prince Says Israelis Have Right to "Their Own Land," *The New York Times*, April 3, 2018
12 Ibid., Goldberg.
13 Stephane Lacroix, "Understanding Stability and Dissent in the Kingdom: The Role of the Jama'at in Saudi Politics", in *Saudi Arabia in Transition: Insights on Social, Political, Economic and Religious Change,* Bernard Haykel, Thomas Hegghammer and Stéphane Lacroix (eds) (Cambridge: Cambridge University Press, 2015), 178–9.
14 The Davos World Economic Forum's 2019 Global Competitiveness Report ranked Saudi Arabia number 16 out of 141 countries surveyed for judicial independence.

Bibliography

Aarts, Paul and Gerd Nonneman. *Saudi Arabia in the Balance: Political Economy, Society, Foreign Affairs*. New York: New York University Press, 2005.

Aarts, Paul and Roelants. Carolien. *Saudi Arabia: A Kingdom in Peril*. London: Hurst, 2015.

Abir, Mordechai. *Saudi Arabia in the Oil Era: Regime and Elites; Conflict and Collaboration*. Kent: Croom Helm, 1988.

Abir, Mordechai. *Saudi Arabia: Government, Society and the Gulf Crisis*. London: Routledge, 1993.

Adair, John. *The Leadership of Muhammad*. London: Kogan Page, 2010.

Al-Awaji, Ibrahim Mohamed. *Bureaucracy and Society in Saudi Arabia*. Unpublished PhD Thesis, University of Virginia, 1971.

Al-Bashir, Faisal S. *A Structural Econometric Model of the Saudi Arabian Economy: 1960–1970*. New York: John Wiley & Sons, 1977.

Al-Haj Ali, Mohammad Said. *Saudi Arabian Monetary Agency: A Review of its Accomplishments*. Saudi Arabia: Saudi Arabian Monetary Agency, 2001.

Al-Hazimi, Mansour, Salma Khadra Jayyusi and Ezzat Khattab. *Beyond the Dunes: An Anthology of Modern Saudi Literature*. London: I.B. Tauris & Co, 2006.

Al-Munajjed, Mona. *Saudi Women Speak*. Beirut: Arab Institute for Research and Publishing, 2006.

Al-Naimi, Ali. *Out of the Desert*. Penguin, Random House. 2016.

Al-Rasheed, Madawi. *Contesting the Saudi State: Islamic Voices from a New Generation*. Cambridge: Cambridge University Press, 2007.

Al-Rasheed, Madawi and Robert Vitalis. *Counter-Narratives: History, Contemporary Society, and Politics in Saudi Arabia and Yemen*. New York: Palgrave, 2004.

Al-Rasheed, Madawi. *A History of Saudi Arabia*. Cambridge: Cambridge University Press, 2002.

Al-Rasheed, Madawi. *Politics in an Arabian Oasis: The Rashidis of Saudi Arabia*. London: I.B. Tauris & Co., 1997.

Al-Rasheed, Madawi. *Muted Modernists: The Struggle Over Divine Politics in Saudi Arabia*. Oxford: Oxford University Press, 2015.

Al-Rasheed, Madawi ed. *Salman's Legacy: The Dilemmas of a New Era in Saudi Arabia*. Oxford: Oxford University Press, 2018.

Al-Saud, Faisal bin Salman. *Iran, Saudi Arabia and the Gulf: Power Politics in Transition*. London: I.B. Tauris & Co., 2003.

Al-Saud, Khaled bin Sultan. *Desert Warrior: A Personal View of the Gulf War by the Joint Forces Commander*. New York: HarperCollins, 1995.

Al-Semmari, Fahd ed. *A History of the Arabian Peninsula*. London: I.B Tauris & Co., 2010.

Al-Shihabi Ali, *The Saudi Kingdom: Between the Jihadi Hammer and the Iranian Anvil*. Princeton: Markus Wiener, 2016.

Al-Sudairi, Abd al-Rahman bin Ahmed. *The Desert Frontier of Arabia, Al-Jawf Through the Ages*. London: Stacey International, 1995.

Al-Uthaymin, Abd Allah Salih. *Muhammad ibn Abd al-Wahhab: The Man and his Works.* New York: I.B. Tauris & Co., 2009.

Algar, Hamid. *Wahhabism: A Critical Essay.* New York: Islamic Publications International, 2002.

Algosaibi, Ghazi A. *Yes, (Saudi) Minister! A Life in Administration.* London: The London Centre of Arab Studies, 1999.

Allen, Charles. *God's Terrorists: The Wahhabi Cult and the Hidden Roots of Modern Jihad.* United Kingdom: Da Capo Press, 2006.

Almana, Mohammed. *Arabia Unified: Portrait of Ibn Saud.* London: Hutchinson Benham Limited, 1980.

Alsubeaei, Huda and Gene W. Heck. *Mohammed Alsubeaei: A Journey of Poverty and Wealth.* Isle of Wight: Medina Publishing Ltd, 2010.

Antonius, George. *The Arab Awakening.* New York: Putnam, 1946.

Altorki, Soraya and Donald P. Cole. *Arabian Oasis City: The Transformation of Unayzah.* Austin: University of Texas Press, 1989.

Armerding, Paul. *Doctors for the Kingdom: The Work of the American Mission Hospitals in the Kingdom of Saudi Arabia.* Michigan: Wm. B. Eerdmans, 2003.

Armstrong, H. C. *Lord of Arabia.* London: Penguin, 1938.

Asad, Muhammad. *The Road to Mecca.* Gibraltar: Dar al-Andalus Ltd, 1980.

Asad, Muhammad. *This Law of Ours and Other Essays.* Gibraltar: Dar al-Andalus, 1987.

Asher, Michael. *Lawrence: The Uncrowned King of Arabia.* London: Penguin Books Ltd, 1998.

Asher, Michael. *Thesiger.* Dubai: Motivate Publishing, 1994.

Azzam, Henry T. *Saudi Arabia: Economic Trends, Business Environment and Investment Opportunities.* London: Euromoney Books, 1993.

Badeau, John S. *The American Approach to the Arab World.* New York: Harper Collins, 1968.

Bakr, Mohammed A. *A Model in Privatization: Successful Change Management in the Ports of Saudi Arabia.* London: London Centre of Arab Studies, 2001.

Banafe, Ahmed and Rory Macleod. *The Saudi Arabian Monetary Agency, 1952–2016.* London: Palgrave Macmillian, 2017.

Barger, Thomas C. *Out in the Blue: Letters from Arabia 1937 to 1940.* Vista California: Selwa Press, 2000.

Barr, James. *Lords of the Desert: Britain's Struggle with American to Dominate the Middle East.* London: Simon and Schuster, 2018.

Beling, Willard A. *King Faisal and the Modernisation of Saudi Arabia.* London: Croom Helm, 1980.

Bligh, Alexander. *From Prince to King: Royal Succession in the House of Saud in the Twentieth Century.* New York: New York University Press, 1984.

Brandt, Marieke. *Tribes and Politics in Yemen: A History of the Houthi Conflict.* Oxford: Oxford University Press, 2017.

Bronson, Rachel. *Thicker than Oil: America's Uneasy Partnership with Saudi Arabia.* New York: Oxford University Press, 2006.

Broug, Eric. *Islamic Geometric Design.* London: Thames & Hudson, 2008.

Brown, Anthony Cave. *Oil, God and Gold: The story of Aramco and the Saudi Kings.* New York: Houghton Mifflin Company, 1999.

Brown Nathan J. *Constitutions in a Nonconstitutional World: Arab Basic Laws and the Prospects for Accountable Government.* Albany: State University of New York Press, 2002.

Bullard, Reader. *Two Kings in Arabia: Letters from Jeddah 1923–5 and 1936–9*. Reading: Ithaca Press, 1993.

Bullard, Reader. *The Camels Must Go: An Autobiography*. London: Faber and Faber, 1961.

Burde, Dana. *Schools for Conflict or for Peace in Afghanistan*. New York: Columbia University Press, 2014.

Burton, Edward. *Business and Entrepreneurship in Saudi Arabia: Opportunities for Partnering and Investing in Emerging Businesses*. Hoboken, N.J.: Wiley & Sons, 2016.

Carter, J.R.L. *Leading Merchant Families of Saudi Arabia*. England: Scorpion Publications in association with The D R Llewellyn Group, 1979.

Champion, Daryl. *The Paradoxical Kingdom: Saudi Arabia and the Momentum of Reform*. New York: Columbia University Press, 2003.

Chaudhry, Kiren Aziz. *The Price of Wealth: Economies and Institutions in the Middle East*. Ithaca: Cornell University Press, 1997.

Cintino, Nathan J. *From Arab Nationalism to OPEC: Eisenhower, King Saud and the Making of the U.S. Saudi Relations*. Indiana University Press: Bloomington, 2002.

Clayton, Gilbert Falkingham. *An Arabian Diary*. Los Angeles: University of California Press, 1969.

Cobbold, Lady Evelyn. *Pilgrimage to Mecca*. Riyadh: King Abdulaziz Public Library, 2008.

Cockburn Patrick. *The Rise of Islamic State, ISIS and the New Sunni Revolution*. New York: Verso, 2014.

Cole, Donald Powell. *Nomads of the Nomads: The Al Murrah Bedouin of the Empty Quarter*. Illinois: AHM Publishing Corporation, 1975.

Coll, Steve. *Ghost Wars: The Secret History of the CIA, Afghanistan and Bin Laden, from the Soviet Invasion to September 10, 2001*. New York: The Penguin Press, 2004.

Coll, Steve. *The Bin Ladens: An Arabian Family in the American Century*. New York: The Penguin Press, 2008.

Commins, David. *The Wahabi Mission and Saudi Arabia*. London: I. B. Tauris & Co., 2006.

Cooper, Ander Scot, *The Fall of Heaven: The Pahlavis and the Final Days of Imperial Iran*. New York: Henry Hold and Co. 2016.

Cooper, Andrew Scott. *The Oil Kings: How the U.S., Iran, and Saudi Arabia Changed the Balance of Power in the Middle East*. New York: Simon and Schuster, 2011.

Cordesman, Anthony H. *Saudi Arabia Enters the Twenty-First Century: The Political, Foreign Policy, Economic, and Energy Dimensions*. Connecticut: Praeger Publishers, 2003.

Cuddihy, Kathy. *Saudi Customs and Etiquette*. Riyadh: King Fahd National Library, 1999.

Darlow, Michael and Barbara Bray. *Ibn Saud: The Desert Warrior Who Created the Kingdom of Saudi Arabia*. New York: Skyhorse Publishing, 2012.

Davidson, Christopher M. *After the Sheikhs: The Coming Collapse of the Gulf Monarchies*. New York: Oxford University Press, 2013.

Dawisha, Adeed. *Islam in Foreign Policy*. Cambridge: The Press Syndicate of the University of Cambridge, 1983.

de Gaury, Gerald. *Arabia Phoenix*. London: George G. Harrap & Co, 1946.

de Gaury, Gerald. *Faisal: King of Saudi Arabia*. New York: Frederick A. Praeger, 1967.

de Gaury, Gerald. *Review of the Anizah Tribe*. Beirut: Kutub Press, 2005.

de Gaury, Gerald. *Rulers of Mecca*. New York: Roy Publisher, 1954.

DeLong-Bas, Natana J. *Wahhabi Islam: From Revival to Reform to Global Jihad*. New York: Oxford University Press, 2004.

Dickson, H. R. P. *The Arab of the Desert*. London: George Allen & Unwin Ltd, 1959.

Djerejian, Edward P. *Danger and Opportunity: An American Ambassador's Journey through the Middle East*. New York: Threshold Editions, 2008.

Doughty, Charles M. *Travels in Arabia Deserta, Volume 1.* New York: Dover Publications, 1979.

Eddy, William A. *F.D.R Meets Ibn Saud.* California: Selwa Press, 2005; originally published in Washington DC by American-Middle East Educational and Training Services, 1954.

Ehteshami, Anoushirvan and Steven Wright (eds). *Reform in the Middle East Oil Monarchies.* Reading: Ithaca Press, 2011.

Esposito, John. *Islam: The Straight Path.* New York: Oxford University Press, 1988.

Fandy, Mamoun. *Saudi Arabia and the Politics of Dissent.* New York: Palgrave, 2001.

Farsy, Fouad. *Modernity and Tradition: The Saudi equation.* London: Kegan Paul International Ltd, 1990.

Fatany, Samar H. *Modernizing Saudi Arabia.* South Carolina: CreateSpace Independent Publishing Platform, 2013.

Feldman, Noah, *The Rise and Fall of the Islamic State.* Princeton University Press, 2008.

Field, Michael. *From Unayzah to Wall Street: The Story of Suliman Saleh Olayan.* London: John Murray, 2000.

Field, Michael. *Inside the Arab World.* London: John Murray, 1994.

Field, Michael. *The Merchants: The Big Business Families of Arabia.* London: John Murray, 1984.

Frankopan, Peter. *The Silk Roads: A New History of the World.* New York: Alfred Knopf, 2016.

Freedom House Center for Religious Freedom. *Saudi Publications on Hate Ideology Invade American Mosques.* Washington D.C.: Freedom House, January 2005.

Freeman, Chas W. Jr. *America's Misadventures in the Middle East.* Charlottesville, Virginia: Just World Books, 2012.

Gause, F. Gregory. *Oil Monarchies: Domestic and Security Challenges in the Arab Gulf States.* New York: Council on Foreign Relations Press, 1994.

Gibb, Hamilton A. R. *Mohammedanism.* Oxford: Oxford University Press, 1969.

Gibb, Hamilton A.R. *Studies on the Civilization of Islam,* edited by Stanford J. Shaw and William R. Polk. Princeton: Princeton University Press, 1982.

Glubb, John Bagot. *War in the Desert: An R.A.F Frontier Campaign.* New York: W.W. Norton & Company, 1961.

Gold, Dore. *Hatred's Kingdom: How Saudi Arabia Supports the New Global Terrorism.* Washington, DC: Regnery Publishing, 2003.

Goldberg, Jacob. *The Foreign Policy of Saudi Arabia: The Formative Years, 1902–1918.* Massachusetts: Harvard University Press, 1986.

Graves, Robert. *Lawrence and the Arabs.* London: Jonathan Cape, 1927.

Guazzone, Laura and Daniela Pioppi. *The Arab State and Neo-Liberal Globalization: The Restructuring of the State Power in the Middle East.* Reading: Ithaca Press, 2012.

Habib, Ghazi. *Strategic Management of Services in the Arab Gulf States.* Berlin: Walter de Gruyter, 1989.

Habib, John. *Ibn Saud's Warriors of Islam: The Ikhwan of the Najd and Their Role in the Creation of the Saudi Kingdom, 1910–1930.* Leiden: E.J. Brill, 1978.

Habib, John. *Striking Back: The Saudi War Against Terrorism.* South Carolina: CreateSpace Independent Publishing Platform, 2013.

Halliday, Fred. *Arabia Without Sultans.* Baltimore: Penguin Books, 1975.

Harrison Ann M. *A Tool in His Hand.* New York: Friendship Press, 1958.

Hart, Parker T. *Saudi Arabia and the United States: Birth of a Security Partnership.* Indiana: Indiana University Press, 1998.

Haykel, Bernard ed. *Saudi Arabia in Transition: Insights on Social, Political, Economic and Religious Change.* Cambridge: Cambridge University Press, 2015.

Heck, Gene W. and Omar Bahlaiwa. *Saudi Arabia: An Evolving Modern Economy*. Riyadh: Alpha Graphics, 2006.

Hegghammer, Thomas. *Jihad in Saudi Arabia: Violence and Pan-Islamism since 1979*. New York: Cambridge University Press, 2010.

Helms, Christine Moss. *The Cohesion of Saudi Arabia*. Maryland: The Johns Hopkins University Press, 1981.

Herb, Michael. *All in the Family: Absolutism, Revolution, and Democracy in the Middle Eastern Monarchies*. Albany: State University of New York Press, 1999.

Hertog, Steffen. *Princes, Brokers, and Bureaucrats: Oil and the State in Saudi Arabia*. New York: Cornell University Press, 2010.

Holden, David and Richard Johns. *The House of Saud: The Rise and Rule of the Most Powerful Dynasty in the Arab World*. New York: Holt, Rinehart and Winston, 1981.

Homoud, Sami Hasan. *Islamic Banking: The Adaption of Banking Practice to Conform with Islamic Law*. London: Arabian Information, 1985.

Hope, Bradley and Tom Wright. *Billion Dollar Whale: The Man Who Fooled Wall Street, Hollywood, and the World*. New York: Hachette Books, 2018.

Hopwood, Derek. *The Arabian Peninsula: Society and Politics*. London: George Allen and Unwin, 1972.

Hourani, Albert. *Arabic Thought in the Liberal Age 1798–1939*. Oxford: Oxford University Press, 1970.

Hourani, Albert. A History of the Arab Peoples. Cambridge: Harvard University Press, 1991.

House, Karen Elliott. *On Saudi Arabia: Its People, Past, Religion, Fault Lines and Future*. New York: Alfred A. Knopf, 2012.

Howarth, David. The Desert King Ibn Saud and His Arabia. New York: McGraw-Hill, 1964.

Hubbard, Ben. *MBS The Rise to Power of Mohammed bin Salman*. New York: Tim Dugan Books, 2020.

Huyette, Summer Scott. *Political Adaptation in Saudi Arabia: A Study of the Council of Ministers (Westview Special Studies on the Middle East)*. London: Westview Press, 1985.

Ibn Khaldun, translated by Franz Rosenthal, edited by N. J. Dawood, *The Muqaddimah: An Introduction to History*. Princeton: Princeton University Press, 2015.

Izzard, Molly. *The Gulf: Arabia's Western Approaches*. London: John Murray, 1979.

Jones, Toby Craig. *Desert Kingdom: How Oil and Water Forged Modern Saudi Arabia*. Massachusetts: Harvard University Press, 2010.

Jordan, Robert W. *Desert Diplomat: Inside Saudi Arabia Following 9/11*. Nebraska: Potomac Books, 2015.

Jungers, Frank. *The Caravan Goes On: How Aramco and Saudi Arabia Grew Together*. Surbiton U.K: Medina Publishing, 2013.

Kanoo, Khalid M. *The House of Kanoo: Century of an Arabian Family Business*. London: The London Center for Arab Studies, 1997.

Kechichian, Joseph A. *Succession in Saudi Arabia*. New York: Palgrave, 2010.

Kechichian, Joseph. *Legal and Political Reforms in Saudi Arabia*. London: Routledge, 2013.

Kelly, J. B. *Arabia, The Gulf and the West: A Critical View of the Arabs and their Oil Policy*. New York: Basic Books, 1980.

Kepel, Gilles. *Jihad: The Trail of Political Islam*. Massachusetts: The Belknap Press of Harvard University Press, 2002.

Khan, Riz. *Alwaleed: Businessman Billionaire Prince*. New York: HarperCollins, 2005.

Kissinger, Henry. *World Order: Reflections on the Character of Nations and the Course of History*. New York: Penguin 2014.

Kostiner, Joseph. *The Making of Saudi Arabia 1916–1936: From Chieftaincy to Monarchical State*. Oxford: Oxford University Press, 1993.

Lacey, Robert. *Inside the Kingdom: Kings, Clerics, Modernists, Terrorists and the Struggle for Saudi Arabia*. London: Hutchinson, 2009.

Lacey, Robert. *The Kingdom: Arabia and the House of Saud*. New York: Harcourt Brace Jovanovich, Publishers, 1981.

Lacroix, Stephene, *Awakening Islam: The Politics of Religious Dissent in Contemporary Saudi Arabia*. Cambridge MA: Harvard University Press, 2011.

Lancaster, William. *The Rwala Bedouin Today*. Illinois: Waveland Press, 1981.

Lawrence, T. E. *Seven Pillars of Islam*. New York: Doubleday, Doran and Company, 1926.

Leatherdale, Clive. *Britain and Saudi Arabia 1925–1939: The Imperial Oasis*. London: Frank Cass, 1983.

Lee, Rev. Samuel. *The Travels of Ibn Battuta in the Near East, Asia and Africa 1325–1354*. New York: Dover Publications, Inc., 2004.

Lees, Brian. *A Handbook of the Al Sa'ud Ruling Family of Saudi Arabia*. London: Royal Genealogies, 1980.

Levorsen, A. I. *Geology of Petroleum*. London: W. H. Freeman and Co, 1954.

Lewis, Bernard. *The Middle East and the West*. New York: Harper Row, 1963.

Lewis, Bernard. *The Arabs in History*. London: Hutchinson, 1975.

Lewis, Bernard. *The Crisis of Islam Holy War and Unholy Terror*. London: Weidenfeld & Nicolson, 2003.

Lewis, Bernard. *The Political Language of Islam*. Chicago: University of Chicago Press, 1988.

Lippman, Thomas W. *Arabian Knight: Colonel Bill Eddy USMC and the Rise of American Power in the Middle East*. California: Selwa Press, 2008.

Lippman, Thomas W. *Inside the Mirage: America's Fragile Partnership with Saudi Arabia*. Oxford: Westview Press, 2004.

Lippman, Thoman W. *Saudi Arabia on the Edge: The Uncertain Future of an American Ally*. Washington, D.C.: Potomac Books, 2012.

Long, David E. *Culture and Customs of Saudi Arabia*. Westport CT: Greenwood Publishing, 2005.

Long, David E., *The United States and Saudi Arabia: Ambivalent Allies*. Boulder Colorado: Westview Press, 1985.

Long David E. *The Washington Papers: Saudi Arabia*. Washington D.C: Center for Strategic and International Studies Georgetown University, 1975.

Mallakh, Ragaei el. *Saudi Arabia: Rush to Development*. London: Croom Helm Ltd., 1982.

Mallakh, Ragaei el and Dorothea H. El Mallakh. *Saudi Arabia: Energy, Developmental Planning, and Industrialization*. Lexington: D.C. Health and Company, 1982.

Mansfield, Peter. *A History of the Middle East*. London: Penguin Books, 1991.

Marcel, Valérie. *Oil Titans: National Oil Companies in the Middle East*. London: Chatham House, 2006.

Matthiesen, Toby. *The Other Saudis: Shiism, Dissent and Sectarianism*. New York: Cambridge University Press, 2015.

McLoughlin, Leslie. *Ibn Saud: Founder of a Kingdom*. London: St Martins Press, 1993.

Ménoret, Pascal. *The Saudi Enigma: A History*. New York: Zedd Books Ltd, 2005.

Momen, Moojan. *An Introduction to Shi'i Islam*. New Haven: Yale University Press, 1985.

Monroe, Elizabeth. *Philby of Arabia*. New York: Pitman Publishing, 1973.

Morgon, Clyde. *Personal Pilot to King Ibn Saud*. booklocker.com, 2010.

Morris, James. *The Hashemite Kings*. London: Faber and Faber, 1959.

Mouline, Nabil. *The Clerics of Islam Religious Authority and Political Power in Saudi Arabia*. New Haven: Yale University Press, 2014.

Nasr, Vali. *The Shia Revival: How Conflicts within Islam Will Shape the Future*. New York: W. W. Norton & Co., 2006.

Nydell, Margaret K. *Understanding Arabs: A Guide for Modern Times*. Massachusetts: Intercultural Press, 2006.

O'Sullivan, Edmund. *The New Gulf: How Modern Arabia is Changing the World for Good*. Dubai: Motivate Publishing, 2008.

Obaid, Nawaf. *The Oil Kingdom at 100: Petroleum Policymaking in Saudi Arabia*. Washington D.C.: Washington Institute for Near East Policy, 2000.

Ottaway, David B. *The King's Messenger: Prince Bandar bin Sultan and America's Tangled Relationship with Saudi Arabia*. New York: Walker & Co., 2008.

Owen, Roger. *State, Power and Politics in the Making of the Modern Middle East*. London: Routledge, 1993.

Owen, Roger and Şevket Pamuk. *A History of Middle East Economies in the Twentieth Century*. Massachusetts: Harvard University Press, 1999.

Pampanini, Andrea H. *Cities from the Arabian Desert: The Building of Jubail and Yanbu in Saudi Arabia*. Connecticut: Praeger Publishers, 1997.

Pampanini, Andrea H. *Desalinated Water in the Kingdom of Saudi Arabia*. New York: Turnaround Associates, 2010.

Pampanini, Andrea H. *Saudi Arabia: Moving Towards a Privatized Economy*. New York: Turnaround Associates, 2005.

Patrick, Neil ed. *Saudi Arabian Foreign Policy: Conflict and Cooperation*. London: I.B. Tauris, 2017.

Peters, F.E. *Mecca: A Literary History of the Muslim Holy Land*. New Jersey: Princeton University Press, 1994.

Philby, H. St. John. *Arabia of the Wahhabis*. London: Frank Cass, 1928.

Philby, H. St. John. *Arabian Days: An Autobiography*. London: Robert Hale, 1948.

Philby, H. St. John. *Arabian Jubilee*. New York: John Day, 1953.

Philby, H. St. John. *Saudi Arabia*. London: Ernest Benn, 1955, reprinted by Librairie du Liban 1968.

Piscatori, James P. *Islam in the Political Process*. Cambridge: Cambridge University Press, 1983.

Polk, William R. and William J. Mares. *Passing Brave*. New York: Alfred A. Knopf, 1973.

Quandt, William B. *Peace Process: American Diplomacy and the Arab–Israeli Conflict since 1967*. Washington, D.C.: Brookings Institution Press, 1993.

Quandt, William B. *Saudi Arabia in the 1980s: Foreign Policy, Security, and Oil*. Washington, D.C.: Brookings Institution Press, 1981.

Qutb, Sayyid. *Milestones*. Damascus: Dar al Ilm, 1995.

Ramady, Mohamed A. *The Saudi Arabian Economy: Policies, Achievements and Challenges*. New York: Springer, 2005.

Rashid, Dr. Nasser Ibrahim and Dr. Esber Ibrahim Shaheen. *King Fahd and Saudi Arabia's Great Evolution*. Missouri: International Institute of Technology, 1987.

Raswan, Carl R. *Black Tents of Arabia: My Life Among the Bedouins*. New York: Creative Age Press, 1947.

Rentz, George S. *The Birth of the Islamic Reform Movement in Saudi Arabia*. Riyadh: The King Abdulaziz Public Library, 2004.

Riedel, Bruce. *Kings and Presidents: Saudi Arabia and the United States Since FDR*. Brookings Institution Press, 2018.

Rihani, Ameen. *Maker of Modern Arabia*. New York: Houghton Mifflin, 1928.

Rivlin, Paul. *Arab Economies in the Twenty-First Century*. Cambridge: Cambridge University Press, 2009.

Robinson, Jeffrey. *Yamani: The Inside Story*. London: Simon and Schuster, 1988.

Rogan, Eugene. *The Arabs: A History*. London: Penguin Books, 2009.

Rogan, Eugene. *The Fall of the Ottomans: The Great War in the Middle East, 1914–1920*. London: Allen Lane, 2015.

Rogerson, Barnaby. *The Prophet Muhammad: A Biography*. New Jersey: HiddenSpring, 2003.

Ruthven, Malise. *Islam in the World*. New York: Penguin Books, 1984.

Sabini, John. *Armies in the Sand: The Struggle for Mecca and Medina*. London: Thames and Hudson, 1981.

Sabri, Sharaf. *The House of Saud in Commerce: A Study of Royal Entrepreneurship in Saudi Arabia*. New Delhi: I. S. Publications, 2001.

Sadleir, George Forster, Diary of a Journey Across Arabia. Bombay: Educational Society Press, 1866.

Safran, Nadav. *Saudi Arabia: The Ceaseless Quest for Security*. New York: Cornell University Press, 1988.

Sanger, Richard H. *The Arabian Peninsula*. New York: Cornell University Press, 1954.

Schacht, Joseph and C.E. Bosworth. The Legacy of Islam. Oxford: Clarendon Press, 1974.

Sheehan, Vincent. *Faisal: The King and his Kingdom*. Kentucky: Fons Vitae, 2007.

Sieff, Martin. *The Politically Incorrect Guide to the Middle East*. Washington, D.C.: Regnery Publishing, Inc., 2008.

Steffof, Rebecca. *Faisal*. Kentucky: Fons Vitae, 2007.

Stegner, Wallace. *Discovery! The Search for Arabian Oil*. Beirut: Middle East Export Press 1971.

Tachau, Frank. Political Elites and Political Development in the Middle East. New York: John Wiley, 1975.

Teitelbaum, Joshua. *Holier than Thou Saudi Arabia's Islamic Opposition*. Washington DC: Washington Institute for Near East Policy, 2000.

Tenet, George. *At the Center of the Storm: My Years at the CIA*. New York: HarperCollins, 2007.

Thackray, Arnold. *Building a Petrochemical Industry in Saudi Arabia*. Riyadh: Obeikan Press, 2017.

Thesiger, Wilfred. *Arabian Sands*. New York: Penguin Books, 1978.

Thesiger, Wilfred. *The Life of My Choice*. New York: W. W. Norton & Co., 1987.

Thomas, Lowell. *With Lawrence in Arabia*. New York: P. F. Collier & Son Corporation, 1924.

Trench, Richard. *Arabian Travellers: The European Discovery of Arabia*. Massachusetts: Salem House Publishers, 1986.

Troeller, Gary. *The Birth of Saudi Arabia: Britain and the Rise of the House of Sa'ud*. London: Frank Cass & Co., 1976.

Trofimov, Yaroslav. *The Siege of Mecca: The Forgotten Uprising in Islam's Holiest Shrine*. New York: Doubleday, 2007.

Van der Meulen, David. *The Wells of Ibn Sa'ud*. London: Kegan Paul International, 2000.

Vassiliev, Alexei. *King Faisal of Saudi Arabia: Personality, Faith and Times*. London: Saqi Books, 2012

Vassilliev, Alexei. *The History of Saudi Arabia*. New York: New York University Press, 2000.

Victor, David G. ed. *Oil and Governance: State-Owned Enterprises and the World Energy Supply*. Cambridge: Cambridge University Press, 2014.

Vitalis, Robert. *America's Kingdom: Mythmaking on the Saudi Oil Frontier*. Stanford: Stanford University Press, 2007.

Vogel, Frank E. and Samuel L. Hayes, III. *Islamic Law and Finance: Religion, Risk, and Return*. The Hague: Kluwer Law International, 1998.

Wald, Ellen R. *Saudi, Inc.: The Arabian Kingdom's Pursuit of Profit and Power*. New York: Pegasus Books, 2018.

Wallach, Janet. *Desert Queen: The Extraordinary Life of Gertrude Bell: Adventurer, Adviser to Kings, Ally of Lawrence of Arabia*. New York: Anchor Books, 1996.

Wells, Colin. *The Complete Idiot's Guide To Saudi Arabia*. Indiana: Alpha Books, 2003.

Weston, Mark. *Prophets and Princes: Saudi Arabia from Muhammad to the Present*. New Jersey: Hohn Wiley & Sons, 2008.

Wilberding, Stevie. *Guidebook to the Ruins of Dir'aiyah*. Riyadh: S. Van C. Wilberding Publishing, 1987.

Wilson, Jeremy. *Lawrence of Arabia: The Authorized Biography of T.E. Lawrence*. New York: Atheneum, 1990.

Wilson, Peter and Douglas F. Graham. *Saudi Arabia: The Coming Storm*. Armonk, New York: M. E. Sharpe Inc., 1994.

Winder, R. Bayly. *Saudi Arabia in the Nineteenth Century*. London: Macmillan, 1965.

Winstone, H. V. F. *Captain Shakespear*. London: Jonathan Cape Limited, 1976.

Woodward, Bob. *The Commanders*. New York: Simon & Schuster, 1991.

Wright, Lawrence. *The Looming Tower: Al-Qaeda's Road to 9/11*. London: Penguin Books, 2006.

Yamani, Mai. *Cradle of Islam: The Hijaz and the Quest for an Arabian Identity*. London: I.B. Tauris & Co., 2009.

Yergin, Daniel. *The Prize: The Epic Quest for Oil, Money & Power*. New York: Simon and Schuster, 1991.

Yergin, Daniel. *The Quest: Energy, Security, and the Remaking of the Modern World*. London: Penguin, 2012.

Yizraeli, Sarah. *Politics and Society in Saudi Arabia: The Crucial Years of Development, 1960–1982*. London: C. Hurst & Company, 2012.

Yizraeli, Sarah. *The Remaking of Saudi Arabia*. Tel Avivi: The Moshe Dayan Center for Middle Eastern and African Studies, Tel Aviv University, 1997.

Young, Arthur N. *Saudi Arabia: The Making of a Financial Giant*. New York: New York University Press, 1983.

Zogby James. *Arab Voices: What they are Saying to Us, and Why it Matters*. New York, Palgrave MacMillan, 2010.

Index

The letter *c* after an entry indicates a page that includes a chart.
The letter *f* after an entry indicates a page that includes a figure.
The letter *m* after an entry indicates a page that includes a map.
Diacritics and the articles al-, al and Al are disregarded in the alphabetical order of main entries.
All royal titles are Saudi Arabian unless detailed otherwise.